Student Solutions Manual

for

Hungerford's

Contemporary Precalculus
A Graphing Approach

Fourth Edition

D0139088

Student Solutions Manual

for

Hungerford's

Contemporary Precalculus
A Graphing Approach

Fourth Edition

Fred Safier
City College of San Francisco

THOMSON
™
BROOKS/COLE

Australia • Canada • Mexico • Singapore • Spain • United Kingdom • United States

COPYRIGHT © 2004 Brooks/Cole, a division of
Thomson Learning, Inc. Thomson Learning™ is a
trademark used herein under license.

ALL RIGHTS RESERVED. No part of this work covered
by the copyright hereon may be reproduced
or used in any form or by any means—graphic, electronic,
or mechanical, including but not limited to photocopying,
recording, taping, Web distribution, information
networks, or information storage and retrieval systems—
without the written permission of the publisher.

Printed in the United States of America
2 3 4 5 6 7 07 06 05 04 03

ISBN: 0-534-40344-1

Printer: Victor Graphics

For more information about our products,
contact us at:
Thomson Learning Academic Resource Center
1-800-423-0563

For permission to use material from this text,
contact us by:
Phone: 1-800-730-2214
Fax: 1-800-730-2215
Web: http://www.thomsonrights.com

For more information contact:
Brooks/Cole-Thomson Learning, Inc.
10 Davis Drive
Belmont, CA 94002-3098
USA

Asia
Thomson Learning
5 Shenton Way #01-01
UIC Building
Singapore 068808

Australia/ New Zealand
Thomson Learning
102 Dodds Street
Southbank, Victoria 3006
Australia

Canada
Nelson
1120 Birchmount Road
Toronto, Ontario M1K 5G4
Canada

Europe/Middle East/South Africa
Thomson Learning
High Holborn House
50/51 Bedford Row
London WC1R 4LR
United Kingdom

Latin America
Thomson Learning
Seneca, 53
Colonia Polanco
11560 Mexico D.F.
Mexico

Spain/ Portugal
Paraninfo
Calle/Magallanes, 25
28015 Madrid, Spain

CONTENTS

Chapter 1
Basics

1.1 The Real Number System

1.

```
   -10 -9 -8 -7 -6 -5 -4 -3 -2 -1  0  1  2  3  4  5  6  7  8  9  10
   ├──┼──┼──┼──┼──┼──┼──┼──┼──┼──┼──┼──┼──┼──┼──┼──┼──┼──┼──┼──┤►
           ↑    ↑↑          ↑   ↑↑     ↑↑                    ↑
          -7  -5 -4.75     -1  0 1/2  2.25 8/3               10
```

3. $6,785,000,000 = 6.785 \times 1,000,000,000 = 6.785 \times 10^9$ [Decimal point is moved 9 places to the left and 10 is raised to the power 9.]

5. $5,910,000,000,000 = 5.91 \times 1,000,000,000,000 = 5.91 \times 10^{12}$ [Decimal point is moved 12 places to the left and 10 is raised to the power 12.]

7. $.000000002 = 2 \times \dfrac{1}{1,000,000,000} = 2 \times 10^{-9}\,\text{m}$ [Decimal point is moved 9 places to the right and 10 is raised to -9.]

9. $1.50 \times 10^{11} = 1.50 \times 100,000,000,000 = 150,000,000,000\,\text{m}$

11. $1.6726 \times 10^{-19} = 1.6726 \times \dfrac{1}{10,000,000,000,000,000,000}$
$= .00000000000000000016726\,\text{kg}$

13. a. Debt:
5626 billion $= 5626 \times 10^9 = 5.626 \times 10^3 \times 10^9 = 5.626 \times 10^{12}$ dollars
Population:
281.3 million $= 281.3 \times 10^6 = 2.813 \times 10^2 \times 10^6 = 2.813 \times 10^8$ persons

b. $\dfrac{\text{Debt}}{\text{Population}} = \dfrac{5.626 \times 10^{12}}{2.813 \times 10^8} = 2 \times 10^4 = \$20,000$ per person.

15. $-17 < 14$ **17.** $x \geq 0$ **19.** $t > 0$ **21.** $c \leq 3$

23. $-6 < -2$ **25.** $3/4 = .75$ **27.** $1/3 > .33$ **29.** $b + c = a$

31. a lies to the right of b. **33.** $a < b$

35.

```
   -10 -9 -8 -7 -6 -5 -4 -3 -2 -1  0  1  2  3  4  5  6  7  8  9  10
   ├──┼──┼──┼──┼──┼──┼──┼──┼──┼──┼──(──┼──┼──┼──┼──┼──┼──┼──]──┼──┼►
                                    0                       8
```

37.

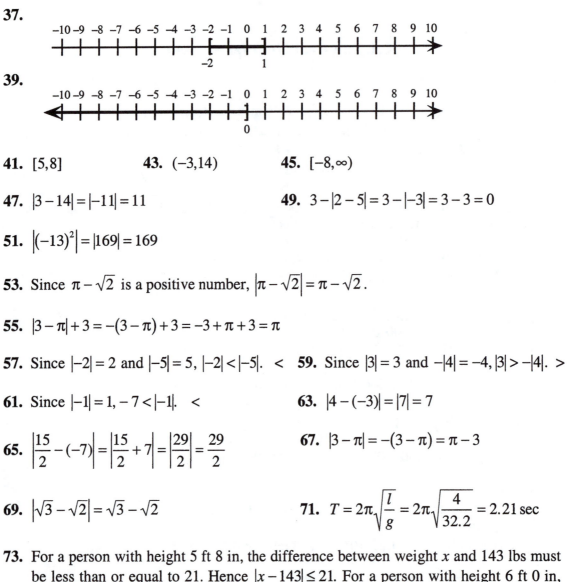

39.

41. $[5,8]$ **43.** $(-3,14)$ **45.** $[-8,\infty)$

47. $|3-14|=|-11|=11$ **49.** $3-|2-5|=3-|-3|=3-3=0$

51. $\left|(-13)^2\right|=|169|=169$

53. Since $\pi-\sqrt{2}$ is a positive number, $\left|\pi-\sqrt{2}\right|=\pi-\sqrt{2}$.

55. $|3-\pi|+3=-(3-\pi)+3=-3+\pi+3=\pi$

57. Since $|-2|=2$ and $|-5|=5$, $|-2|<|-5|$. $<$ **59.** Since $|3|=3$ and $-|4|=-4$, $|3|>-|4|$. $>$

61. Since $|-1|=1$, $-7<|-1|$. $<$ **63.** $|4-(-3)|=|7|=7$

65. $\left|\dfrac{15}{2}-(-7)\right|=\left|\dfrac{15}{2}+7\right|=\left|\dfrac{29}{2}\right|=\dfrac{29}{2}$ **67.** $|3-\pi|=-(3-\pi)=\pi-3$

69. $\left|\sqrt{3}-\sqrt{2}\right|=\sqrt{3}-\sqrt{2}$ **71.** $T=2\pi\sqrt{\dfrac{l}{g}}=2\pi\sqrt{\dfrac{4}{32.2}}=2.21\,\text{sec}$

73. For a person with height 5 ft 8 in, the difference between weight x and 143 lbs must be less than or equal to 21. Hence $|x-143|\le 21$. For a person with height 6 ft 0 in, the difference between the weight x and 163 lbs must be less than or equal to 26. Hence $|x-163|\le 26$.

75. Since $t^2\ge 0$, $\left|t^2\right|=t^2$.   **77.** If $b\ge 3$, $b-3\ge 0$, hence $|b-3|=b-3$.

79. If $c<d$, $c-d<0$, hence $|c-d|=-(c-d)=d-c$.

81. Since $|u-v|=|v-u|$, $|u-v|-|v-u|=|u-v|-|u-v|=0$.

83. Since $(c-d)^2\ge 0$, $\left|(c-d)^2\right|=(c-d)^2=c^2-2cd+d^2$.

85. $|x-5|<4$ **87.** $|x+4|\le 17$ **89.** $0<|c|<|b|$

91. The distance from x to 3 is less than 2.   **93.** x is at most 3 units from -7.

37.

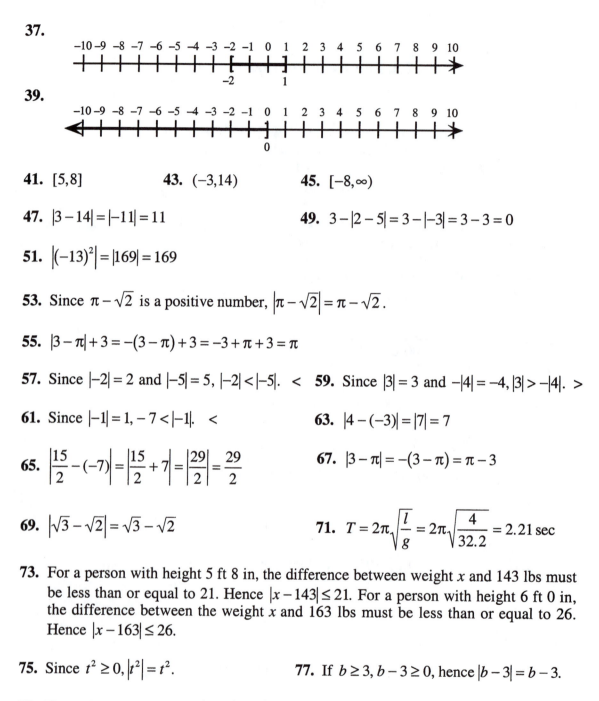

39.

41. $[5,8]$ **43.** $(-3,14)$ **45.** $[-8,\infty)$

47. $|3-14|=|-11|=11$ **49.** $3-|2-5|=3-|-3|=3-3=0$

51. $\left|(-13)^2\right|=|169|=169$

53. Since $\pi-\sqrt{2}$ is a positive number, $\left|\pi-\sqrt{2}\right|=\pi-\sqrt{2}$.

55. $|3-\pi|+3=-(3-\pi)+3=-3+\pi+3=\pi$

57. Since $|-2|=2$ and $|-5|=5$, $|-2|<|-5|$. $<$ **59.** Since $|3|=3$ and $-|4|=-4$, $|3|>-|4|$. $>$

61. Since $|-1|=1$, $-7<|-1|$. $<$ **63.** $|4-(-3)|=|7|=7$

65. $\left|\dfrac{15}{2}-(-7)\right|=\left|\dfrac{15}{2}+7\right|=\left|\dfrac{29}{2}\right|=\dfrac{29}{2}$ **67.** $|3-\pi|=-(3-\pi)=\pi-3$

69. $\left|\sqrt{3}-\sqrt{2}\right|=\sqrt{3}-\sqrt{2}$ **71.** $T=2\pi\sqrt{\dfrac{l}{g}}=2\pi\sqrt{\dfrac{4}{32.2}}=2.21\,\text{sec}$

73. For a person with height 5 ft 8 in, the difference between weight x and 143 lbs must be less than or equal to 21. Hence $|x-143|\le 21$. For a person with height 6 ft 0 in, the difference between the weight x and 163 lbs must be less than or equal to 26. Hence $|x-163|\le 26$.

75. Since $t^2\ge 0$, $\left|t^2\right|=t^2$.   **77.** If $b\ge 3$, $b-3\ge 0$, hence $|b-3|=b-3$.

79. If $c<d$, $c-d<0$, hence $|c-d|=-(c-d)=d-c$.

81. Since $|u-v|=|v-u|$, $|u-v|-|v-u|=|u-v|-|u-v|=0$.

83. Since $(c-d)^2\ge 0$, $\left|(c-d)^2\right|=(c-d)^2=c^2-2cd+d^2$.

85. $|x-5|<4$ **87.** $|x+4|\le 17$ **89.** $0<|c|<|b|$

91. The distance from x to 3 is less than 2. **93.** x is at most 3 units from -7.

95. (a) — iii
    (b) — i
    (c) — ii
    (d) — v
    (e) — iv

97. The distance of x from 0 is equal to 1. Thus $x = 1$ or $x = -1$.

99. The distance of x from 2 is equal to 1. Thus $x = 2 + 1 = 3$ or $x = 2 - 1 = 1$.

101. The distance of x from $-\pi$ is equal to 4. Thus $x = -\pi + 4$ or $x = -\pi - 4$.

103. x is between -7 and 7.

105. The distance of x from 5 is less than 2. Thus x is between $5 - 2$ and $5 + 2$. $3 < x < 7$.

107. The distance of x from -2 is greater than or equal to 3. Thus $x \geq -2 + 3$ or $x \leq -2 - 3$. $x \geq 1$ or $x \leq -5$.

109. If, and only if, none of the three numbers a,b,c is different from 0, then $|a| = 0, |b| = 0, |c| = 0,$ and $|a| + |b| + |c| = 0$. Otherwise $|a| + |b| + |c| > 0$.

1.1A Decimal Representation of Real Numbers

1. .777...

3. 1.642857142857...

5. .052631578947368...

7.
$$100d = 37.3737...$$
$$d = \quad .3737...$$
$$\overline{99d = 37}$$
$$d = \frac{37}{99}$$

9.
$$10,000d = 766342.4242...$$
$$100d = \quad 7663.4242...$$
$$\overline{9900d = 758,679}$$
$$d = \frac{758,679}{9900} = \frac{252,893}{3300}$$

11.
$$1000d = 135.135135...$$
$$d = \quad .135135...$$
$$\overline{999d = 135}$$
$$d = \frac{135}{999} = \frac{5}{37}$$

13.
$$10,000d = 523127.2727...$$
$$100d = \quad 5231.2727...$$
$$\overline{9900d = 517,896}$$
$$d = \frac{517,896}{9900} = \frac{14,386}{275}$$

15. Not possible (repeating decimal)

17. Exactly (terminating decimal)

19. Not possible (irrational)

21. Exactly (integer)

23. Let $d = .7499\ldots$
$$1000d = 749.99\ldots$$
$$100d = 74.99\ldots$$
$$\overline{900d = 675}$$
$$d = \frac{675}{900} = \frac{3}{4}$$

Let $d = .7500\ldots$
$$1000d = 750.00\ldots$$
$$100d = 75.00\ldots$$
$$\overline{900d = 675}$$
$$d = \frac{675}{900} = \frac{3}{4}$$

25. .0588235294117647 0588235294117647…

27. .03448275862068965517241379310344827586206896551724137931 0344827586206896551724137931…

29. 6.0212765957446808510638297872340425531914893617 02127659574 46808510638297872340425531914893617…

31. a. With each terminating decimal $.d_1d_2\ldots d_n$ associate the non-terminating, non-repeating decimal $.d_1d_2\ldots d_n1234567891011\ldots$ Thus there are at least as many irrational numbers as terminating decimals.

b. With each repeating decimal $.(d_1d_2\ldots d_n)$ repeated, associate the non-terminating, non-repeating decimal $.d_1d_2\ldots d_n0d_1d_2\ldots d_n00\ldots$ Thus there are at least as many irrational numbers as repeating decimals.

1.2 Solving Equations Algebraically

1. $3x + 2 = 26$
$$3x = 24$$
$$x = 8$$
$$8$$

3. $3x + 2 = 9x + 7$
$$-6x + 2 = 7$$
$$-6x = 5$$
$$x = -\frac{5}{6}$$
$$-\frac{5}{6}$$

5. $\dfrac{3y}{4} - 6 = y + 2$
$$3y - 24 = 4y + 8$$
$$-y - 24 = 8$$
$$-y = 32$$
$$y = -32$$
$$-32$$

7.
$$x = 3y - 5$$
$$x + 5 = 3y$$
$$\frac{x + 5}{3} = y$$
$$y = \frac{x + 5}{3}$$

9.
$$A = \frac{h}{2}(b + c)$$
$$\frac{2A}{h} = b + c$$
$$\frac{2A}{h} - c = b$$
$$b = \frac{2A}{h} - c$$

11.
$$V = \frac{\pi d^2 h}{4}$$
$$4V = \pi d^2 h$$
$$\frac{4V}{\pi d^2} = h$$
$$h = \frac{4V}{\pi d^2}$$

13.
$$x^2 - 8x + 15 = 0$$
$$(x - 3)(x - 5) = 0$$
$$x - 3 = 0 \text{ or } x - 5 = 0$$
$$x = 3 \qquad x = 5$$
$$3, 5$$

15.
$$x^2 - 5x = 14$$
$$x^2 - 5x - 14 = 0$$
$$(x + 2)(x - 7) = 0$$
$$x + 2 = 0 \text{ or } x - 7 = 0$$
$$x = -2 \qquad x = 7$$
$$-2, 7$$

17.
$$2y^2 + 5y - 3 = 0$$
$$(2y - 1)(y + 3) = 0$$
$$2y - 1 = 0 \text{ or } y + 3 = 0$$
$$y = \frac{1}{2} \qquad y = -3$$
$$\frac{1}{2}, -3$$

19.
$$4t^2 + 9t + 2 = 0$$
$$(4t + 1)(t + 2) = 0$$
$$4t + 1 = 0 \text{ or } t + 2 = 0$$
$$t = -\frac{1}{4} \qquad t = -2$$
$$-\frac{1}{4}, -2$$

21.
$$3u^2 + u = 4$$
$$3u^2 + u - 4 = 0$$
$$(3u + 4)(u - 1) = 0$$
$$3u + 4 = 0 \text{ or } u - 1 = 0$$
$$u = -\frac{4}{3} \qquad u = 1$$
$$-\frac{4}{3}, 1$$

23.
$$x^2 - 2x = 12$$
$$x^2 - 2x + 1 = 13$$
$$(x - 1)^2 = 13$$
$$x - 1 = \pm\sqrt{13}$$
$$x = 1 \pm \sqrt{13}$$
$$1 - \sqrt{13}, 1 + \sqrt{13}$$

25. $x^2 - x - 1 = 0$

$x^2 - x = 1$

$x^2 - x + \dfrac{1}{4} = \dfrac{5}{4}$

$\left(x - \dfrac{1}{2}\right)^2 = \dfrac{5}{4}$

$x - \dfrac{1}{2} = \pm\sqrt{\dfrac{5}{4}}$

$x = \dfrac{1}{2} \pm \dfrac{\sqrt{5}}{2}$

$\dfrac{1 - \sqrt{5}}{2}, \dfrac{1 + \sqrt{5}}{2}$

27. $a = 1, b = 4, c = 1$

$b^2 - 4ac = 4^2 - 4(1)(1) = 12.$

Two real solutions.

29. $9x^2 = 12x + 1$

$9x^2 - 12x - 1 = 0$

$a = 9, b = -12, c = -1$

$b^2 - 4ac = (-12)^2 - 4(9)(-1) = 180$

Two real solutions.

31. $25t^2 + 49 = 70t$

$25t^2 - 70t + 49 = 0$

$a = 25, b = -70, c = 49$

$b^2 - 4ac = (-70)^2 - 4(25)(49) = 0$

One real solution.

33. $x^2 - 4x + 1 = 0$

$a = 1, b = -4, c = 1$

$x = \dfrac{-(-4) \pm \sqrt{(-4)^2 - 4(1)(1)}}{2(1)}$

$x = \dfrac{4 \pm \sqrt{12}}{2}$

$x = \dfrac{4 \pm 2\sqrt{3}}{2}$

$x = 2 \pm \sqrt{3}$

$2 - \sqrt{3}, 2 + \sqrt{3}$

35. $x^2 + 6x + 7 = 0$

$a = 1, b = 6, c = 7$

$x = \dfrac{-6 \pm \sqrt{6^2 - 4(1)(7)}}{2(1)}$

$x = \dfrac{-6 \pm \sqrt{8}}{2}$

$x = \dfrac{-6 \pm 2\sqrt{2}}{2}$

$x = -3 \pm \sqrt{2}$

$-3 - \sqrt{2}, -3 + \sqrt{2}$

37. $x^2 + 6 = 2x$

$x^2 - 2x + 6 = 0$

$a = 1, b = -2, c = 6$

$b^2 - 4ac = (-2)^2 - 4(1)(6) = -20$

No real solutions.

39. $4x^2 - 4x = 7$

$4x^2 - 4x - 7 = 0$

$a = 4, b = -4, c = -7$

$x = \dfrac{-(-4) \pm \sqrt{(-4)^2 - 4(4)(-7)}}{2(4)}$

$x = \dfrac{4 \pm \sqrt{128}}{8}$

$x = \dfrac{4 \pm 8\sqrt{2}}{8}$

$x = \dfrac{1 \pm 2\sqrt{2}}{2}$

$\dfrac{1 - 2\sqrt{2}}{2}, \dfrac{1 + 2\sqrt{2}}{2}$

41. $4x^2 - 8x + 1 = 0$

$a = 4, b = -8, c = 1$

$x = \dfrac{-(-8) \pm \sqrt{(-8)^2 - 4(4)(1)}}{2(4)}$

$x = \dfrac{8 \pm \sqrt{48}}{8}$

$x = \dfrac{8 \pm 4\sqrt{3}}{8}$

$x = \dfrac{2 \pm \sqrt{3}}{2}$

$\dfrac{2 - \sqrt{3}}{2}, \dfrac{2 + \sqrt{3}}{2}$

43. $x^2 + 9x + 18 = 0$

$(x + 3)(x + 6) = 0$

$x + 3 = 0$ or $x + 6 = 0$

$x = -3 \qquad\qquad x = -6$

$-3, -6$

45. $4x(x+1)=1$

$4x^2+4x=1$

$4x^2+4x-1=0$

$a=4, b=4, c=-1$

$x=\dfrac{-4\pm\sqrt{4^2-4(4)(-1)}}{2(4)}$

$x=\dfrac{-4\pm\sqrt{32}}{8}$

$x=\dfrac{-4\pm4\sqrt{2}}{8}$

$x=\dfrac{-1\pm\sqrt{2}}{2}$

$\dfrac{-1-\sqrt{2}}{2},\dfrac{-1+\sqrt{2}}{2}$

47. $2x^2=7x+15$

$2x^2-7x-15=0$

$(2x+3)(x-5)=0$

$\qquad 2x+3=0 \ \text{ or } \ x-5=0$

$\qquad x=-\dfrac{3}{2}\qquad x=5$

$\qquad -\dfrac{3}{2},5$

49. $t^2+4t+13=0$

$a=1, b=4, c=13$

$b^2-4ac=4^2-4(1)(13)=-36$

No real solution.

51. $\dfrac{7x^2}{3}=\dfrac{2x}{3}-1$

$7x^2=2x-3$

$7x^2-2x+3=0$

$a=7, b=-2, c=3$

$b^2-4ac=(-2)^2-4(7)(3)=-80$

No real solution.

53. $4.42x^2-10.14x+3.79=0$

$x=\dfrac{-(-10.14)\pm\sqrt{(-10.14)^2-4(4.42)(3.79)}}{2(4.42)}$

$x=0.47, 1.82$

55. $3x^2-82.74x+570.4923=0$

$x=\dfrac{-(-82.74)\pm\sqrt{(-82.74)^2-4(3)(570.4923)}}{2(3)}$

$x=13.79$

57. To solve $y^4 - 7y^2 + 6 = 0$, let $u = y^2$,

$$y^4 - 7y^2 + 6 = 0$$

$$u^2 - 7u + 6 = 0$$

$$(u-1)(u-6) = 0$$

$$u - 1 = 0 \text{ or } u - 6 = 0$$

$$u = 1 \qquad u = 6$$

Since $u = y^2$, we have the equivalent statements:

$$y^2 = 1 \qquad y^2 = 6$$

$$y = \pm 1 \qquad y = \pm\sqrt{6}$$

$$1, -1, \sqrt{6}, -\sqrt{6}$$

59. To solve $x^4 - 2x^2 - 35 = 0$, let $u = x^2$,

$$x^4 - 2x^2 - 35 = 0$$

$$u^2 - 2u - 35 = 0$$

$$(u-7)(u+5) = 0$$

$$u - 7 = 0 \text{ or } u + 5 = 0$$

$$u = 7 \qquad u = -5$$

Since $u = x^2$, we have the equivalent statements:

$$x^2 = 7 \qquad x^2 = -5$$

$$x = \pm\sqrt{7} \quad \text{no real solution}$$

$$-\sqrt{7}, \sqrt{7}$$

61. To solve $2y^4 - 9y^2 + 4 = 0$, let $u = y^2$.

$$2y^4 - 9y^2 + 4 = 0$$

$$2u^2 - 9u + 4 = 0$$

$$(2u-1)(u-4) = 0$$

$$2u - 1 = 0 \text{ or } u - 4 = 0$$

$$u = \frac{1}{2} \qquad u = 4$$

Since $u = y^2$, we have the equivalent statements:

$$y^2 = \frac{1}{2} \qquad y^2 = 4$$

$$y = \pm\sqrt{\frac{1}{2}} \qquad y = \pm 2$$

$$y = \pm\frac{\sqrt{2}}{2}$$

$$-\frac{\sqrt{2}}{2}, \frac{\sqrt{2}}{2}, -2, 2$$

63. To solve $10x^4 + 3x^2 = 1$, let $u = x^2$.

$$10x^4 + 3x^2 = 1$$

$$10u^2 + 3u = 1$$

$$10u^2 + 3u - 1 = 0$$

$$(5u-1)(2u+1) = 0$$

$$5u - 1 = 0 \text{ or } 2u + 1 = 0$$

$$u = \frac{1}{5} \qquad u = -\frac{1}{2}$$

Since $u = x^2$, we have the equivalent statements:

$$x^2 = \frac{1}{5} \qquad x^2 = -\frac{1}{2}$$

$$x = \pm\sqrt{\frac{1}{5}} \qquad \text{no real solution}$$

$$x = \pm\frac{\sqrt{5}}{5}$$

$$-\frac{\sqrt{5}}{5}, \frac{\sqrt{5}}{5}$$

65.
$$1 - \frac{3}{x} = \frac{40}{x^2}$$

$$x^2 \cdot 1 - x^2 \cdot \frac{3}{x} = x^2 \cdot \frac{40}{x^2} \qquad x \neq 0$$

$$x^2 - 3x = 40$$

$$x^2 - 3x - 40 = 0$$

$$(x - 8)(x + 5) = 0$$

$$x - 8 = 0 \quad \text{or} \quad x + 5 = 0$$

$$x = 8 \qquad\qquad x = -5$$

$$8, -5$$

67.
$$\frac{2}{x^2} - \frac{5}{x} = 4$$

$$x^2 \cdot \frac{2}{x^2} - x^2 \cdot \frac{5}{x} = 4x^2 \qquad x \neq 0$$

$$2 - 5x = 4x^2$$

$$0 = 4x^2 + 5x - 2$$

$$x = \frac{-5 \pm \sqrt{5^2 - 4(4)(-2)}}{2(4)}$$

$$x = \frac{-5 \pm \sqrt{57}}{8}$$

$$\frac{-5 - \sqrt{57}}{8}, \frac{-5 + \sqrt{57}}{8}$$

69.
$$Y = C + I + G + (X - M)$$

$$Y = 120 + .9Y + 140 + 150 + \big[60 - (20 + .2Y)\big]$$

$$Y = 120 + .9Y + 290 + 60 - 20 - .2Y$$

$$Y = 450 + .7Y$$

$$.3Y = 450$$

$$Y = 1500$$

71. In this case v_0 is 0 and h_0 is 640, so that the height equation is
$$h = -16t^2 + 640$$
We must solve
$$0 = -16t^2 + 640$$

$$16t^2 = 640$$

$$t^2 = 40$$

$$t = \pm\sqrt{40} = \pm 2\sqrt{10} = \pm 6.32$$
6.32 seconds

73. In this case v_0 is 800 and h_0 is 0, so that the height equation is $h = -16t^2 + 800t$

   a. Solve $3200 = -16t^2 + 800t$

   $16t^2 - 800t + 3200 = 0$

   $t^2 - 50t + 200 = 0$

   $t = \dfrac{50 \pm \sqrt{(-50)^2 - 4(1)(200)}}{2(1)}$

   $t = \dfrac{50 \pm \sqrt{1700}}{2}$

   $t = 4.38, 45.61$
   4.38 seconds for the rocket to first reach 3200 feet.

   b. Solve $0 = -16t^2 + 800t$

   $0 = t^2 - 50t$

   $0 = t(t - 50)$

   $t = 0$ or $t = 50$
   50 seconds for the rocket to reach the ground again.

75. **a.** At sea level, evaluate a for $h = 0$; $a = 2116.1$ pounds per square foot.
   For the top of Mt. Everest, evaluate

   $a = .8315h^2 - 73.93h + 2116.1$ for $h = 29.035$

   $a = .8315(29.035)^2 - 73.93(29.035) + 2116.1$

   $a = 670.5$ pounds per square foot.

   b. Solve $1223.43 = .8315h^2 - 73.93h + 2116.1$

   $0 = .8315h^2 - 73.93h + 892.67$

   $h = \dfrac{-(-73.93) \pm \sqrt{(-73.93)^2 - 4(.8315)(892.67)}}{2(.8315)}$

   $h = 74.502$ or 14.409
   About 14,400 feet high.

77. Solve $804.2 = .08x^2 - 13.08x + 927$

   $0 = .08x^2 - 13.08x + 122.8$

   $x = \dfrac{-(-13.08) \pm \sqrt{(-13.08)^2 - 4(.08)(122.8)}}{2(.08)}$

   $x = \dfrac{13.08 \pm 11.48}{2(.08)}$

   $x = 153.5$ or 10
   Only the answer 10 represents a past year: 1960.

79. This is a quadratic equation with $a = 1, b = k, c = 25$. It will have exactly one real solution only if $b^2 - 4ac = 0$. Thus

$$k^2 - 4(1)(25) = 0$$
$$k^2 - 100 = 0$$
$$k^2 = 100$$
$$k = \pm 10$$

-10 or 10

81. This is a quadratic equation with $a = k, b = 8, c = 1$. It will have exactly one real solution only if $b^2 - 4ac = 0$. Thus

$$8^2 - 4(k)(1) = 0$$
$$64 - 4k = 0$$
$$k = 16$$

16

83. Since $b^2 - 4ac$ is a perfect square, the solutions of the equation are rational numbers.

85. Since $b^2 - 4ac$ is not a perfect square, the solutions of the equation are not rational numbers.

87. Since 4 is a solution of the equation, $4^2 - 5 \cdot 4 + k = 0$, hence $k - 4 = 0$ and $k = 4$. Also $1^2 - 5 \cdot 1 + k = 0$, which also yields $k = 4$.

1.2A Absolute Value Equations

1. The equation is equivalent to

$$2x + 3 = 9 \quad \text{or} \quad -(2x + 3) = 9$$
$$2x = 6 \qquad\qquad -2x - 3 = 9$$
$$x = 3 \qquad\qquad\quad -2x = 12$$
$$x = -6$$

Checking each possible solution, we see
$$|2 \cdot 3 + 3| = 9 \text{ and } |2 \cdot (-6) + 3| = |-9| = 9$$
Thus both 3 and –6 are solutions.

3. The equation is equivalent to

$$6x - 9 = 0$$
$$x = \frac{9}{6} = \frac{3}{2}$$

Checking, we see
$$\left| 6\left(\frac{3}{2}\right) - 9 \right| = 0$$

$\frac{3}{2}$ is the solution.

5. The equation is equivalent to

$$2x + 3 = 4x - 1 \quad \text{or} \quad -(2x + 3) = 4x - 1$$

$$-2x = -4 \qquad\qquad -2x - 3 = 4x - 1$$

$$x = 2 \qquad\qquad\qquad -6x = 2$$

$$x = -\frac{1}{3}$$

Checking, we see that 2 is a solution
because $|2 \cdot 2 + 3| = 7$ and $4 \cdot 2 - 1 = 7$
However, $-\frac{1}{3}$ is not a solution because

$$\left|2\left(-\frac{1}{3}\right) + 3\right| = \frac{7}{3} \quad \text{but} \quad 4\left(-\frac{1}{3}\right) - 1 = -\frac{7}{3}$$

7. The equation is equivalent to

$$x - 3 = x \quad \text{or} \quad x - 3 = -x$$

$$-3 = 0 \qquad\qquad -3 = -2x$$

$$\text{no solution} \qquad\qquad x = \frac{3}{2}$$

Checking, we see that $\frac{3}{2}$ is a

solution because $\left|\frac{3}{2} - 3\right| = \frac{3}{2}$

9. The equation is equivalent to

$$x^2 + 4x - 1 = 4 \quad \text{or} \quad x^2 + 4x - 1 = -4$$

$$x^2 + 4x - 5 = 0 \qquad\qquad x^2 + 4x + 3 = 0$$

$$(x + 5)(x - 1) = 0 \qquad\qquad (x + 3)(x + 1) = 0$$

$$x = -5 \text{ or } x = 1 \qquad\qquad x = -3 \text{ or } x = -1$$

All four of these numbers are solutions of the original equation.

11. The equation is equivalent to

$$x^2 - 5x + 1 = 3 \qquad\qquad \text{or} \qquad\qquad x^2 - 5x + 1 = -3$$

$$x^2 - 5x - 2 = 0 \qquad\qquad\qquad\qquad x^2 - 5x + 4 = 0$$

$$x = \frac{-(-5) \pm \sqrt{(-5)^2 - 4(1)(-2)}}{2(1)} \qquad (x - 1)(x - 4) = 0$$

$$x = \frac{5 \pm \sqrt{33}}{2} \qquad\qquad\qquad x = 1 \text{ or } x = 4$$

$$x = \frac{5 + \sqrt{33}}{2} \quad \text{or} \quad x = \frac{5 - \sqrt{33}}{2}$$

All four of these numbers are solutions of the original equation.

13. Substituting the given values for $\bar{p}$ and n, we obtain

$$|CL - .02| = 3\sqrt{\frac{.02(1 - .02)}{200}}$$

$$|CL - .02| = .0297$$

This equation is equivalent to

$$CL - .02 = .0297 \quad \text{or} \quad -(CL - .02) = .0297$$

$$CL = .0497 \qquad\qquad\qquad -CL + .02 = .0297$$

$$CL = -.0097$$

The upper control limit is .0497 and the lower control limit is –.0097 (which may be interpreted as 0 depending on the situation).

1.2B Variation

1. Area varies directly as the square of the radius: the constant of variation is π.

3. Area varies jointly as the length and the width; the constant of variation is 1.

5. Volume varies jointly as the square of the radius and the height; the constant of variation is $\pi / 3$.

7. $a = k/b$

9. $z = kxyw$

11. $d = k\sqrt{h}$

13.
$$v = ku$$
$$8 = k(2)$$
$$k = \frac{8}{2} = 4$$

15.
$$v = \frac{k}{u}$$
$$8 = \frac{k}{2}$$
$$k = 16$$

17.
$$t = krs$$
$$24 = k(2)(3)$$
$$k = \frac{24}{6} = 4$$

19.
$$w = kxy^2$$
$$96 = k \cdot 3 \cdot 4^2$$
$$k = \frac{96}{3 \cdot 4^2} = 2$$

21.
$$T = \frac{kpv^3}{u^2}$$
$$24 = \frac{k \cdot 3 \cdot 2^3}{4^2}$$
$$k = \frac{24 \cdot 4^2}{3 \cdot 2^3} = 16$$

23.
$$r = kt$$
$$6 = k(3)$$
$$k = 2$$
Therefore,
the variation equation
is $r = 2t$. When $t = 2$,
$r = 2(2) = 4$.

25.
$$b = \frac{k}{x}$$
$$9 = \frac{k}{3}$$
$$k = 27$$
Therefore,
the variation equation
is $b = \dfrac{27}{x}$.
When $x = 12$,
$$b = \frac{27}{12} = 2.25$$

27.
$$w = k(u + v^2)$$
$$200 = k(1 + 7^2)$$
$$200 = 50k$$
$$k = 4$$
Therefore,
the variation equation
is $w = 4(u + v^2)$.
When $w = 300$ and $v = 5$,
$$300 = 4(u + 5^2)$$
$$300 = 4(u + 25)$$
$$75 = u + 25$$
$$u = 50$$

29.
$$r = \frac{k}{st}$$
$$12 = \frac{k}{3 \cdot 1}$$
$$k = 36$$
Therefore,
the variation equation
is $r = \dfrac{36}{st}$.
When $s = 6$ and $t = 2$,
$$r = \frac{36}{6 \cdot 2}$$
$$r = 3$$

31. Let t = tax, x = AGI, Then,

$$t = kx$$

$$1008 = k(24,000)$$

$$k = \frac{1008}{24,000}$$

Therefore,

the variation equation is $t = \frac{1008}{24,000} x$.

When $x = 39,000$,

$$t = \frac{1008}{24,000}(39,000)$$

$$t = \$1638$$

33.

$$d = kW$$

$$15.75 = k(7)$$

$$k = 2.25$$

Therefore, the variation equation is $d = 2.25W$.

When $d = 27$,

$$27 = 2.25W$$

$$W = \frac{27}{2.25}$$

$$W = 12 \text{ pounds}$$

35. Let P = pressure, V = volume. Then

$$P = \frac{k}{V}$$

$$50 = \frac{k}{200}$$

$$k = 10,000$$

Therefore, the variation equation is

$P = \frac{10,000}{V}$. When $V = 125$,

$$P = \frac{10,000}{125}$$

$$P = 80 \text{ kilograms per sq centimeter}$$

37. Let d = distance, t = time. Then

$$d = kt^2$$

$$100 = k(2.5)^2$$

$$k = 16$$

Therefore, the variation equation is

$d = 16t^2$. When $t = 5$,

$$d = 16(5)^2$$

$$d = 400 \text{ feet}$$

39. Let W = weight, h = height, r = radius. Then

$$W = khr^2$$

$$250 = k(20)(5)^2$$

$$k = \frac{250}{20 \cdot 5^2}$$

$$k = \frac{1}{2}$$

Therefore, the variation equation is

$W = \frac{1}{2}hr^2$. When $W = 960$ and $r = 8$,

$$960 = \frac{1}{2}h \cdot 8^2$$

$$h = \frac{2 \cdot 960}{8^2}$$

$$h = 30 \text{ inches}$$

41. Let F = force, W = weight, v = speed.

$$F = \frac{kWv^2}{r}$$

$$1500 = \frac{k(1000)50^2}{200}$$

$$k = \frac{(1500)(200)}{1000(50)^2}$$

$$k = .12$$

Therefore, the variation equation is

$F = \frac{.12Wv^2}{r}$.

When $W = 1000$, $v = 100$, $r = 320$

$$F = \frac{.12(1000)100^2}{320}$$

$$F = 3750 \text{ kilograms}$$

43. Let L = the maximum safe load.

Then $L = \dfrac{kwh^2}{l}$

$1000 = \dfrac{k \cdot 4 \cdot 2^2}{6}$

$k = \dfrac{6000}{4 \cdot 2^2}$

$k = 375$

Therefore, the variation equation

is $L = \dfrac{375wh^2}{l}$.

a. When $w = 4$, $h = 4$, $l = 10$,

$L = \dfrac{375 \cdot 4 \cdot 4^2}{10}$

$L = 2400$ pounds

b. When $w = 4$, $h = 4$, $L = 6000$

$6000 = \dfrac{375 \cdot 4 \cdot 4^2}{l}$

$6000l = 24,000$

$l = 4$ feet

1.3 The Coordinate Plane

1. $A(-3,3)$; $B(-1.5,3)$; $C(-2.5,0)$; $D(-1.5,-3)$; $E(0,2)$; $F(0,0)$; $G(2,0)$; $H(3,1)$; $I(3,-1)$

3. $(-6,3)$ **5.** $\left(4, \dfrac{1}{2} \times 4\right) = (4,2)$

7.

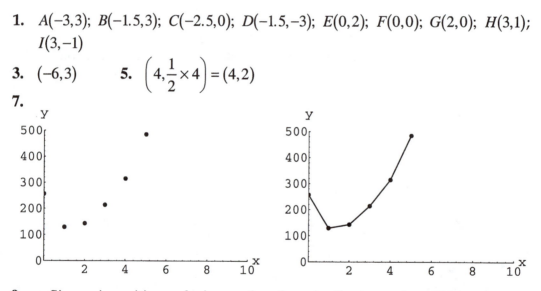

9. a. Since x is positive and y is negative, the point lies in quadrant IV.
b. Since y is negative, the point lies in quadrant III or IV.

11. a.   **b.**

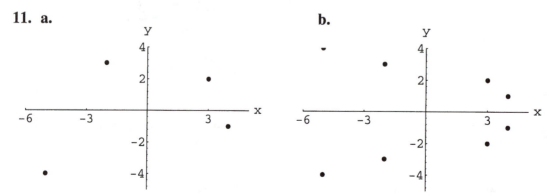

c. The new points are found by reflecting the original points in the x-axis.

13. $\text{distance} = \sqrt{(x_1 - x_2)^2 + (y_1 - y_2)^2}$

$= \sqrt{(-3-2)^2 + (5-(-7))^2}$

$= \sqrt{(-5)^2 + 12^2}$

$= \sqrt{25 + 144} = \sqrt{169} = 13$

The midpoint is

$\left(\dfrac{x_1 + x_2}{2}, \dfrac{y_1 + y_2}{2}\right) = \left(\dfrac{-3+2}{2}, \dfrac{5+(-7)}{2}\right)$

$= \left(-\dfrac{1}{2}, -1\right)$

15. $\text{distance} = \sqrt{(x_1 - x_2)^2 + (y_1 - y_2)^2}$

$= \sqrt{(1-2)^2 + (-5-(-1))^2}$

$= \sqrt{(-1)^2 + (-4)^2}$

$= \sqrt{1 + 16} = \sqrt{17}$

The midpoint is

$\left(\dfrac{x_1 + x_2}{2}, \dfrac{y_1 + y_2}{2}\right) = \left(\dfrac{1+2}{2}, \dfrac{-5+(-1)}{2}\right)$

$= \left(\dfrac{3}{2}, -3\right)$

17. $\text{distance} = \sqrt{(x_1 - x_2)^2 + (y_1 - y_2)^2}$

$= \sqrt{(\sqrt{2} - \sqrt{3})^2 + (1-2)^2}$

$= \sqrt{2 - 2\sqrt{2}\sqrt{3} + 3 + 1}$

$= \sqrt{6 - 2\sqrt{6}}$

The midpoint is

$\left(\dfrac{x_1 + x_2}{2}, \dfrac{y_1 + y_2}{2}\right) = \left(\dfrac{\sqrt{2} + \sqrt{3}}{2}, \dfrac{1+2}{2}\right)$

$= \left(\dfrac{\sqrt{2} + \sqrt{3}}{2}, \dfrac{3}{2}\right)$

19. $\text{distance} = \sqrt{(x_1 - x_2)^2 + (y_1 - y_2)^2}$

$= \sqrt{(a-b)^2 + (b-a)^2}$

$= \sqrt{2(a-b)^2}$

$= |a - b|\sqrt{2}$

The midpoint is

$\left(\dfrac{x_1 + x_2}{2}, \dfrac{y_1 + y_2}{2}\right) = \left(\dfrac{a+b}{2}, \dfrac{b+a}{2}\right)$

21. The perimeter is the sum of the distances:

between $(0,0)$ and $(4,0) = 4$

between $(4,0)$ and $(2,2) = \sqrt{(4-2)^2 + (0-2)^2} = \sqrt{4+4} = \sqrt{8} = 2\sqrt{2}$

between $(2,2)$ and $(4,5) = \sqrt{(2-4)^2 + (2-5)^2} = \sqrt{4+9} = \sqrt{13}$

between $(4,5)$ and $(0,5) = 4$

between $(0,5)$ and $(0,0) = 5$

Thus, the perimeter $= 4 + 2\sqrt{2} + \sqrt{13} + 4 + 5 = 13 + 2\sqrt{2} + \sqrt{13}$.

23. The lengths of the sides of the triangle are the distances:

between $(0,0)$ and $(1,1) = \sqrt{(0-1)^2 + (0-1)^2} = \sqrt{1+1} = \sqrt{2}$

between $(1,1)$ and $(2,-2) = \sqrt{(1-2)^2 + (1-(-2))^2} = \sqrt{1+9} = \sqrt{10}$

between $(2,-2)$ and $(0,0) = \sqrt{(2-0)^2 + (-2-0)^2} = \sqrt{4+4} = \sqrt{8}$

Since $\left(\sqrt{2}\right)^2 + \left(\sqrt{8}\right)^2 = \left(\sqrt{10}\right)^2$, the triangle is a right triangle. The length of the hypotenuse is $\sqrt{10}$ units.

25. a. In this graph, year 2000 is represented by 0.

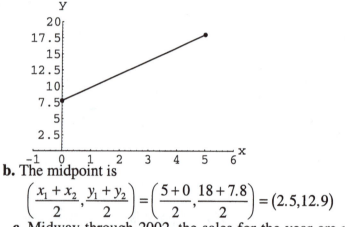

b. The midpoint is
$$\left(\frac{x_1 + x_2}{2}, \frac{y_1 + y_2}{2}\right) = \left(\frac{5+0}{2}, \frac{18+7.8}{2}\right) = (2.5, 12.9)$$

c. Midway through 2002, the sales for the year are projected to be 12.9 billion. The assumptions are that the actual growth is linear, as well as that one can assign a value to a partial year's sales.

27. a. The quarterback has coordinates $(20,10)$. The receiver has coordinates $\left(48\frac{1}{3}, 45\right)$. The length of the pass can be calculated from the distance formula:
$$distance = \sqrt{(x_1 - x_2)^2 + (y_1 - y_2)^2}$$
$$= \sqrt{\left(20 - 48\frac{1}{3}\right)^2 + (10 - 45)^2}$$
$$\approx 45 \text{ yds}$$

(The pass would be recorded as the change in y coordinate only, however, or 35 yds.)

b. The midpoint is
$$\left(\frac{x_1 + x_2}{2}, \frac{y_1 + y_2}{2}\right) = \left(\frac{20 + 48\frac{1}{3}}{2}, \frac{10 + 45}{2}\right)$$
$$= \left(34\frac{1}{6}, 27\frac{1}{2}\right)$$

29. Since $3(1) - (-2) - 5 = 3 + 2 - 5 = 0, (1,-2)$ satisfies the equation $3x - y - 5 = 0$ and the point is on the graph.

31. Since $3 \cdot 2 + 6 = 6 + 6 = 12$, $(6, 2)$ satisfies the equation $3y + x = 12$ and the point is on the graph.

33. Since $(3 - 2)^2 + (4 + 5)^2 = 1^2 + 9^2 = 82 \neq 4$, $(3, 4)$ does not satisfy the equation $(x - 2)^2 + (y + 5)^2 = 4$ and the point is not on the graph.

35. To find the x-intercepts, set $y = 0$ and solve for x.
$$x^2 - 6x + 5 = 0$$
$$(x - 1)(x - 5) = 0$$
$x = 1$ or $x = 5$
To find the y-intercepts, set $x = 0$ and solve for y.
$$y + 5 = 0$$
$$y = -5$$

37. To find the x-intercepts, set $y = 0$ and solve for x.
$$(x - 2)^2 = 9$$
$$x - 2 = \pm 3$$
$$x - 2 = 3 \ \text{ or } \ x - 2 = -3$$
$$x = 5 \qquad\qquad x = -1$$
To find the y-intercepts, set $x = 0$ and solve for y.
$$(-2)^2 + y^2 = 9$$
$$y^2 = 5$$
$$y = \pm\sqrt{5}$$
$$y = \sqrt{5} \ \text{ or } \ y = -\sqrt{5}$$

39. To find the x-intercepts, set $y = 0$ and solve for x.
$$9x^2 + 90x = 0$$
$$9x(x + 10) = 0$$
$$9x = 0 \ \text{ or } \ x + 10 = 0$$
$$x = 0 \qquad\qquad x = -10$$
To find the y-intercepts, set $x = 0$ and solve for y.
$$16y^2 - 128y = 0$$
$$16y(y - 8) = 0$$
$$16y = 0 \ \text{ or } \ y - 8 = 0$$
$$y = 0 \qquad\qquad y = 8$$

41. $(x - (-3))^2 + (y - 4)^2 = 2^2$
$(x + 3)^2 + (y - 4)^2 = 4$

43. $x^2 + y^2 = \sqrt{2}^2$
$x^2 + y^2 = 2$

45.

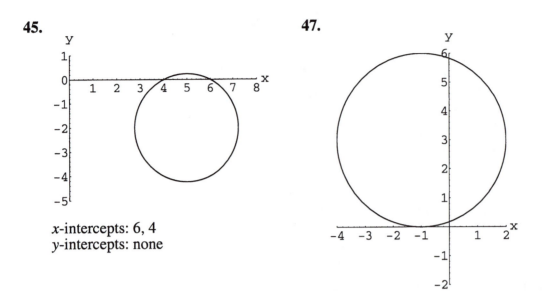

x-intercepts: 6, 4
y-intercepts: none

47.

x-intercept: −1
y-intercepts: $3 \pm 2\sqrt{2}$

49.
$$x^2 + y^2 + 8x - 6y - 15 = 0$$
$$\left(x^2 + 8x\right) + \left(y^2 - 6y\right) = 15$$
$$\left(x^2 + 8x + 16\right) + \left(y^2 - 6y + 9\right) = 15 + 16 + 9$$
$$(x+4)^2 + (y-3)^2 = 40$$
Center: $(-4,3)$, Radius: $\sqrt{40} = 2\sqrt{10}$

51.
$$x^2 + y^2 + 6x - 4y - 15 = 0$$
$$\left(x^2 + 6x\right) + \left(y^2 - 4y\right) = 15$$
$$\left(x^2 + 6x + 9\right) + \left(y^2 - 4y + 4\right) = 15 + 9 + 4$$
$$(x+3)^2 + (y-2)^2 = 28$$
Center: $(-3,2)$, Radius: $\sqrt{28} = 2\sqrt{7}$

53.
$$x^2 + y^2 + 25x + 10y = -12$$
$$\left(x^2 + 25x\right) + \left(y^2 + 10y\right) = -12$$
$$\left(x^2 + 25x + \frac{625}{4}\right) + \left(y^2 + 10y + 25\right) = -12 + \frac{625}{4} + 25$$
$$\left(x + \frac{25}{2}\right)^2 + (y+5)^2 = \frac{677}{4}$$
Center: $\left(-\frac{25}{2}, -5\right)$, Radius: $\frac{\sqrt{677}}{2}$

55. a. Approximately 7.3 million.
    b. Approximately 2007; approximately 7.8 million.
    c. 2000 — 2003, 2012 — .

57. B   **59.** Depending on the time scale, A or C.

61. The radius of the circle is the distance from $(2,2)$ to $(0,0)$.
$$\sqrt{(2-0)^2 + (2-0)^2} = \sqrt{4+4} = \sqrt{8}$$
The equation of the circle with center (2, 2) and radius $\sqrt{8}$ is
$$(x-2)^2 + (y-2)^2 = \left(\sqrt{8}\right)^2$$
$$x^2 - 4x + 4 + y^2 - 4y + 4 = 8$$
$$x^2 + y^2 - 4x - 4y = 0$$

63. The radius of the circle is the distance from $(1,2)$ to $(3,0)$.
The equation of the circle with center $(1,2)$ and radius $\sqrt{8}$ is
$$(x-1)^2 + (y-2)^2 = \left(\sqrt{8}\right)^2$$
$$x^2 - 2x + 1 + y^2 - 4y + 4 = 8$$
$$x^2 + y^2 - 2x - 4y - 3 = 0$$

65. The radius of the circle is the distance from $(-5,4)$ to the x-axis, i.e. 4.
The equation of the circle with center $(-5,4)$ and radius 4 is
$$\left(x-(-5)\right)^2 + (y-4)^2 = 4^2$$
$$(x+5)^2 + (y-4)^2 = 16$$
$$x^2 + 10x + 25 + y^2 - 8y + 16 = 16$$
$$x^2 + y^2 + 10x - 8y + 25 = 0$$

67. The center of the circle is the midpoint of the diameter.
$$\left(\frac{3+1}{2}, \frac{3+(-1)}{2}\right) = (2,1)$$
The radius of the circle is the distance from the center $(2,1)$ to $(3,3)$.
$$\sqrt{(2-3)^2 + (1-3)^2} = \sqrt{1+4} = \sqrt{5}$$
The equation of the circle with center $(2,1)$ and radius $\sqrt{5}$ is
$$(x-2)^2 + (y-1)^2 = \left(\sqrt{5}\right)^2$$
$$x^2 - 4x + 4 + y^2 - 2y + 1 = 5$$
$$x^2 + y^2 - 4x - 2y = 0$$

69. Interchange the x-coordinates of the given points to obtain $(2,1)$ and $(-3,-4)$. These points will be the endpoints of the other diagonal.

71. If (c,d) and (c,k) are adjoining vertices, then the square has length of side $|d - k|$. Then two squares could be formed with two sides of this length perpendicular to the given line segment. One would lie $|d - k|$ units on one side of the given segment, with vertices $(c + |d - k|, d)$ and $(c + |d - k|, k)$. The other would lie $|d - k|$ units on the other side, with vertices $(c - |d - k|, d)$ and $(c - |d - k|, k)$. The third square is formed if (c,d) and (c,k) are opposite vertices. Then the midpoint of this diagonal would be $\left(c, \frac{d+k}{2}\right)$, and a second diagonal of equal length, perpendicular to the given diagonal, would have endpoints $\left(c - \frac{|d-k|}{2}, \frac{d+k}{2}\right)$ and $\left(c + \frac{|d-k|}{2}, \frac{d+k}{2}\right)$.

73. Using the hint, we have the distance from $(x,0)$ to $(3,4)$ is 5. Thus

$$\sqrt{(x-3)^2 + (0-4)^2} = 5$$
$$(x-3)^2 + 16 = 25$$
$$(x-3)^2 = 9$$
$$x - 3 = 3 \quad \text{or} \quad x - 3 = -3$$
$$x = 6 \qquad\qquad x = 0$$

The points are $(6,0)$ and $(0,0)$.

75. The distance from $(3,y)$ to $(-2,-5)$ is 6. Thus

$$\sqrt{(3-(-2))^2 + (y-(-5))^2} = 6$$
$$25 + (y+5)^2 = 36$$
$$(y+5)^2 = 11$$
$$y + 5 = \sqrt{11} \qquad \text{or} \qquad y + 5 = -\sqrt{11}$$
$$y = -5 + \sqrt{11} \qquad\qquad y = -5 - \sqrt{11}$$

The points are $\left(3, -5 + \sqrt{11}\right)$ and $\left(3, -5 - \sqrt{11}\right)$.

77. The distance from $(0,0)$ to $(3,2)$ equals the distance from $(x,0)$ to $(3,2)$. Thus

$$\sqrt{(0-3)^2 + (0-2)^2} = \sqrt{(x-3)^2 + (0-2)^2}$$
$$\sqrt{9+4} = \sqrt{(x-3)^2 + 4}$$
$$13 = (x-3)^2 + 4$$
$$(x-3)^2 = 9$$
$$x - 3 = 3 \quad \text{or} \quad x - 3 = -3$$
$$x = 6 \qquad\qquad x = 0$$

$x = 6$, since $x = 0$ is already a vertex.

79. From the figure, the coordinates of M are $\left(\frac{0+s}{2}, \frac{r+0}{2}\right) = \left(\frac{s}{2}, \frac{r}{2}\right)$. Then the distances from M to the three vertices are given by

$$\sqrt{\left(0 - \frac{s}{2}\right)^2 + \left(0 - \frac{r}{2}\right)^2} = \sqrt{\left(\frac{s}{2}\right)^2 + \left(\frac{r}{2}\right)^2} = \sqrt{\frac{s^2 + r^2}{4}}$$

$$\sqrt{\left(s - \frac{s}{2}\right)^2 + \left(0 - \frac{r}{2}\right)^2} = \sqrt{\left(\frac{s}{2}\right)^2 + \left(\frac{r}{2}\right)^2} = \sqrt{\frac{s^2 + r^2}{4}}$$

$$\sqrt{\left(0 - \frac{s}{2}\right)^2 + \left(r - \frac{r}{2}\right)^2} = \sqrt{\left(\frac{s}{2}\right)^2 + \left(\frac{r}{2}\right)^2} = \sqrt{\frac{s^2 + r^2}{4}}$$

Thus the three distances are equal.

81. Sketch a figure.

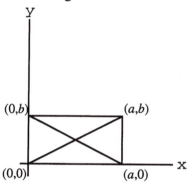

The diagonals have lengths

$$\sqrt{(0 - a)^2 + (b - 0)^2} = \sqrt{a^2 + b^2}$$

$$\sqrt{(0 - a)^2 + (0 - b)^2} = \sqrt{a^2 + b^2}$$

Thus the diagonals have the same length.

83. The circles have centers at $(k, 0)$ on the x-axis and radius k, therefore, they are tangent to the y-axis at the origin.

85. The midpoint of the line segment joining (c, d) and $(-c, -d)$ is given by

$$\left(\frac{x_1 + x_2}{2}, \frac{y_1 + y_2}{2}\right) = \left(\frac{c + (-c)}{2}, \frac{d + (-d)}{2}\right) = (0, 0)$$

that is, the origin. Therefore, (c, d) and $(-c, -d)$ lie on a straight line through the origin and equidistant from the origin. Since the x- and y-coordinates both have opposite signs, the points are on opposite sides of the origin.

1.4 Lines

1. a. C   **b.** B   **c.** B   **d.** D

3. Slope $= \dfrac{y_2 - y_1}{x_2 - x_1} = \dfrac{7-2}{3-1} = \dfrac{5}{2}$

5. Slope $= \dfrac{y_2 - y_1}{x_2 - x_1} = \dfrac{2-0}{\frac{3}{4}-\frac{1}{4}} = \dfrac{2}{\frac{1}{2}} = 4$

7. Slope $= \dfrac{y_2 - y_1}{x_2 - x_1}$

$-2 = \dfrac{4-t}{9-0}$

$-18 = 4 - t$

$t = 22$

9. Slope $= \dfrac{y_2 - y_1}{x_2 - x_1}$

$-2 = \dfrac{(-3t+7)-5}{6-(t+1)}$

$-2 = \dfrac{-3t+2}{5-t}$

$-10 + 2t = -3t + 2$

$t = \dfrac{12}{5}$

11. Sketch a figure:

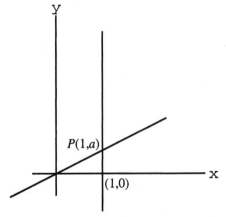

From the figure, the slope m of line L is given by $\dfrac{a-0}{1-0} = m$

Thus $a = m$, and the second coordinate of P is the slope of L.

13. Solve $2x - y + 5 = 0$ for y in terms of x.

$2x - y + 5 = 0$

$2x + 5 = y$

slope: 2, y-intercept: 5

15. Solve for y in terms of x.

$3(x-2) + y = 7 - 6(y+4)$

$3x - 6 + y = 7 - 6y - 24$

$7y = -3x - 11$

$y = -\dfrac{3}{7}x - \dfrac{11}{7}$

slope: $-\frac{3}{7}$, y-intercept: $-\frac{11}{7}$

17. Substitute 1 for m and (3,5) for (x_1, y_1) in the point-slope equation.

$$y - y_1 = m(x - x_1)$$
$$y - 5 = 1(x - 3)$$
$$y - 5 = x - 3$$
$$y = x + 2$$

19. Substitute -1 for m and (6,2) for (x_1, y_1) in the point-slope equation.

$$y - y_1 = m(x - x_1)$$
$$y - 2 = (-1)(x - 6)$$
$$y - 2 = -x + 6$$
$$y = -x + 8$$

21. The slope of the line is given by

$$m = \frac{y_2 - y_1}{x_2 - x_1} = \frac{-2 - (-5)}{-3 - 0} = -1$$

Use the slope and one of the points to find the equation of the line.

$$y - y_1 = m(x - x_1)$$
$$y - (-2) = -1(x - (-3))$$
$$y + 2 = -x - 3$$
$$y = -x - 5$$

23. The slope of the line is given by

$$m = \frac{y_2 - y_1}{x_2 - x_1} = \frac{3 - \frac{2}{3}}{\frac{1}{3} - \frac{4}{3}} = -\frac{7}{3}$$

Use the slope and one of the points to find the equation of the line.

$$y - y_1 = m(x - x_1)$$
$$y - \frac{2}{3} = -\frac{7}{3}\left(x - \frac{4}{3}\right)$$
$$y - \frac{2}{3} = -\frac{7}{3}x + \frac{28}{9}$$
$$y = -\frac{7}{3}x + \frac{34}{9}$$

25.

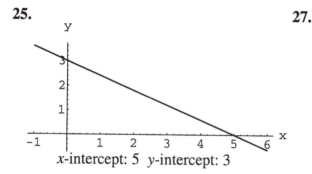

x-intercept: 5 y-intercept: 3

27.

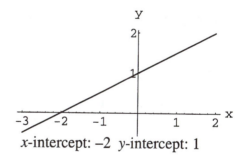

x-intercept: -2 y-intercept: 1

29.

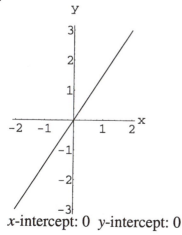

x-intercept: 0 y-intercept: 0

31. Slope of line $PQ = \dfrac{-1-5}{-1-2} = 2$

Slope of line $RS = \dfrac{1-2}{6-4} = -\dfrac{1}{2}$

Since $2\left(-\tfrac{1}{2}\right) = -1$, line PQ is perpendicular to line RS.

33. Slope of line $PQ = \dfrac{-1-\frac{1}{3}}{1-(-3)} = -\dfrac{1}{3}$

Slope of line $RS = \dfrac{-\frac{2}{3}-0}{4-2} = -\dfrac{1}{3}$

Since these slopes are equal, line PQ is parallel to line RS.

35. Write both equations in the slope-intercept form.

$$2x + y - 2 = 0 \qquad\qquad 4x + 2y + 18 = 0$$

$$y = -2x + 2 \qquad\qquad y = -2x - 9$$

Both lines have slope -2, hence they are parallel.

37. Write both equations in the slope-intercept form.

$$.5x + y = -3$$
$$y = 2x + 4$$
$$y = -.5x - 3$$

Since $2(-.5) = -1$, the lines are perpendicular.

39. Slope of the line containing $(9,6)$ and $(-1,2) = \dfrac{2-6}{-1-9} = \dfrac{2}{5}$

Slope of the line containing $(-1,2)$ and $(1,-3) = \dfrac{-3-2}{1-(-1)} = -\dfrac{5}{2}$

Slope of the line containing $(9,6)$ and $(1,-3) = \dfrac{-3-6}{1-9} = \dfrac{9}{8}$

Since $\left(\tfrac{2}{5}\right)\left(-\tfrac{5}{2}\right) = -1$, the first two lines are perpendicular, and the points are vertices of a right triangle.

41. Substitute 3 for m and $(-2,1)$ for (x_1, y_1) in the point-slope form.

$$y - y_1 = m(x - x_1)$$
$$y - 1 = 3(x - (-2))$$
$$y - 1 = 3x + 6$$
$$y = 3x + 7$$

43. First find the slope of the given line.

$$3x - 2y = 5$$
$$y = \dfrac{3}{2}x - \dfrac{5}{2}$$

The slope is $\tfrac{3}{2}$. This is also the slope of the required line.

Now substitute $\tfrac{3}{2}$ for m and $(2,3)$ for (x_1, y_1) in the point-slope equation.

$$y - y_1 = m(x - x_1)$$
$$y - 3 = \dfrac{3}{2}(x - 2)$$
$$y - 3 = \dfrac{3}{2}x - 3$$
$$y = \dfrac{3}{2}x$$

45. The line passes through $(5,0)$ and $(0,-5)$. Hence its slope is given by

$$\frac{y_2 - y_1}{x_2 - x_1} = \frac{-5 - 0}{0 - 5} = 1$$

Substitute 1 for m and -5 for b in the slope-intercept equation.

$$y = mx + b$$

$$y = 1x + (-5)$$

$$y = x - 5$$

47. First find the slope of the line through $(0,1)$ and $(2,3)$.

$$\frac{y_2 - y_1}{x_2 - x_1} = \frac{3 - 1}{2 - 0} = 1$$

Therefore, the slope of the required perpendicular line is -1.
Now substitute -1 for m and $(-1,3)$ for (x_1, y_1) in the point-slope equation.

$$y - y_1 = m(x - x_1)$$

$$y - 3 = -1(x - (-1))$$

$$y - 3 = -x - 1$$

$$y = -x + 2$$

49. $(3,-2)$ is on the line $kx - 2y + 7 = 0$ if its coordinates satisfy the equation. Substituting, we obtain

$$3k - 2(-2) + 7 = 0$$

$$3k + 11 = 0$$

$$k = -\frac{11}{3}$$

51. Since C is $(0,0)$ and P is $(3,4)$, the slope of radius CP is given by

$$\frac{y_2 - y_1}{x_2 - x_1} = \frac{4 - 0}{3 - 0} = \frac{4}{3}$$

Therefore the slope of the tangent line is $-\frac{3}{4}$. The equation of the tangent line is found by substituting $-\frac{3}{4}$ for m and $(3,4)$ for (x_1, y_1) in the point-slope form.

$$y - y_1 = m(x - x_1)$$

$$y - 4 = -\frac{3}{4}(x - 3)$$

$$y - 4 = -\frac{3}{4}x + \frac{9}{4}$$

$$y = -\frac{3}{4}x + \frac{25}{4}$$

53. Since C is $(1,3)$ and P is $(2,5)$, the slope of radius CP is given by

$$\frac{y_2 - y_1}{x_2 - x_1} = \frac{5-3}{2-1} = 2$$

Therefore the slope of the tangent line is $-\frac{1}{2}$. The equation of the tangent line is found by substituting $-\frac{1}{2}$ for m and $(2,5)$ for (x_1, y_1) in the point-slope form.

$$y - y_1 = m(x - x_1)$$

$$y - 5 = -\frac{1}{2}(x - 2)$$

$$y - 5 = -\frac{1}{2}x + 1$$

$$y = -\frac{1}{2}x + 6$$

55. Put the equations of the lines into slope-intercept form.

$$Ax + By + C = 0 \qquad\qquad Ax + By + D = 0$$

$$By = -Ax - C \qquad\qquad By = -Ax - D$$

$$y = -\frac{A}{B}x - \frac{C}{B} \qquad\qquad y = -\frac{A}{B}x - \frac{D}{B}$$

Thus, both lines have the slope $-A/B$. Therefore the lines are parallel.

57. a. The required equation is of the form $y = mx + b$. Since $y = 54$ when $x = 0$, $b = 54$.
Since $y = 63$ when $x = 8$, we have

$$y = mx + 54$$

$$63 = m \cdot 8 + 54$$

$$m = \frac{9}{8}$$

Thus, $y = \frac{9}{8}x + 54$

b. Substitute 3 for x in the equation:

$$y = \frac{9}{8}x + 54$$

$$y = \frac{9}{8}(3) + 54$$

$$y = 57.4 \text{ million cars}$$

c. Substitute 60 for y in the equation:

$$y = \frac{9}{8}x + 54$$

$$60 = \frac{9}{8}x + 54$$

$$x = 5.33$$

If $x = 5$ corresponds to 2005, $x = 5.33$ may be interpreted as 2005 or 2006.

59. a. Use data points $(0, 136.2)$ and $(10, 177.1)$.
The slope is given by

$$\frac{y_2 - y_1}{x_2 - x_1} = \frac{177.1 - 136.2}{10 - 0} = 4.09$$

Substitute 4.09 for m and 136.2 for b in the slope-intercept form.

$$y = mx + b$$

$$y = 4.09x + 136.2$$

b. Substitute 14 for x.

$$y = 4.09x + 136.2$$

$$y = 4.09(14) + 136.2$$

$$y = 193.5$$

c. The yearly rate of increase is given by the slope, 4.09 units per year.

61. The problem is solved using inches: 4 ft 10 in $= 4 \times 12 + 10 = 58$ in, and so on.
a. Use the datapoints $(7, 150)$ and $(17, 200)$.
The slope is given by

$$\frac{y_2 - y_1}{x_2 - x_1} = \frac{200 - 150}{17 - 7} = 5$$

Substitute 5 for m and $(7, 150)$ for (x_1, y_1) in the point-slope form.

$$y - y_1 = m(x - x_1)$$

$$y - 150 = 5(x - 7)$$

$$y - 150 = 5x - 35$$

$$y = 5x + 115$$

b. Substitute 2 for x in this equation to obtain $y = 5 \cdot 2 + 115 = 125$ pounds, the maximum healthy weight for a person of height 5 ft. Substitute 14 for x in this equation to obtain $y = 5 \cdot 14 + 115 = 185$ pounds, the maximum healthy weight for a person of height 6 ft.

c. Substitute 220 for y in the equation $y = 5x + 115$, and solve for x:

$$220 = 5x + 115$$

$$105 = 5x$$

$$x = 21$$

21 inches above 4 ft 10 in is 6 ft 7 in.

63. a. Use the datapoints $(0, 1.1)$ and $(20, 4.9)$.
The slope is given by

$$\frac{y_2 - y_1}{x_2 - x_1} = \frac{4.9 - 1.1}{20} = .19$$

Substitute .19 for m and 1.1 for b in the slope-intercept form.

$$y = mx + b$$

$$y = .19x + 1.1$$

2004: Substitute 27 for x in this equation.

$$y = .19(27) + 1.1$$

$$y = 6.23 \text{ million}$$

b. Substitute 7.2 for y in this equation.

$$7.2 = .19x + 1.1$$

$$6.1 = .19x$$

$$x = 32$$

This corresponds to the year 2009.

65. a. The average cost per set is the total cost $145x + 120,000$ divided by the number of sets x, thus

$$A = \frac{145x + 120,000}{x}$$

b. Substitute 175 for A and solve for x.

$$175 = \frac{145x + 120,000}{x}$$

$$175x = 145x + 120,000$$

$$30x = 120,000$$

$$x = 4000 \text{ sets}$$

67. a. The cost c = variable cost + fixed costs = $50x + 110,000$
b. The revenue r = (price per book) × (number of books sold) = $72x$
c. The profit $p = r - c$. Thus,

$$p = 72x - (50x + 110,000)$$

$$p = 72x - 50x - 110,000$$

$$p = 22x - 110,000$$

d. Set $r = c$ and solve for x.

$$72x = 50x + 110,000$$

$$22x = 110,000$$

$$x = 5000 \text{ books}$$

69. a. The rate of emptying is given by the opposite of the slope.

During the first two minutes, the slope $= \dfrac{y_2 - y_1}{x_2 - x_1} = \dfrac{50 - 75}{2 - 0} = -12.5$

Emptying at 12.5 gallons per minute.

During the next three minutes, the slope $= \dfrac{y_2 - y_1}{x_2 - x_1} = \dfrac{25 - 50}{5 - 2} = -8.33$

Emptying at 8.33 gallons per minute.

During the last minute, the slope $= \dfrac{y_2 - y_1}{x_2 - x_1} = \dfrac{0 - 25}{6 - 5} = -25$

Emptying at 25 gallons per minute.

b.

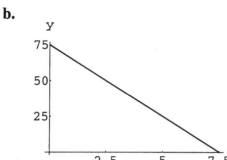

Since the slope is given as –10, substitute –10 for m and 75 for b in the slope-intercept form to obtain $y = -10x + 75$.

71. a. Slope $= \dfrac{0-60}{30-0} = -2$; y-intercept $= 60$

$y = mx + b$

$y = -2x + 60$

b. Slope $= \dfrac{0-80}{40-0} = -2$; y-intercept $= 80$

$y = mx + b$

$y = -2x + 80$

c. Slope $= \dfrac{0-160}{40-0} = -4$; y-intercept $= 160$

$y = mx + b$

$y = -4x + 160$

73. a. Substitute 20 for x in $y = -2x + 60$

$y = -2(20) + 60$

$y = \$20$

b. Substitute 20 for x in $y = -2x + 80$

$y = -2(20) + 80$

$y = \$40$

75. Using the hint, if (x_1, y_1) satisfied both equations, then

$y_1 = mx_1 + b$

$y_1 = mx_1 + c$

$mx_1 + b = mx_1 + c$

$b = c$

Since $b \neq c$, this is a contradiction, and no point can lie on both lines, that is, the lines are parallel.

77. Sketch a figure:

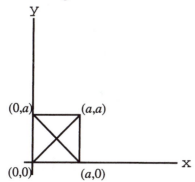

The slopes of the diagonals are given by

$\dfrac{a-0}{a-0} = 1$ and $\dfrac{0-a}{a-0} = -1$.

Since $1(-1) = -1$, we conclude that the diagonals are perpendicular.

Chapter 1 Review Exercises

1. **a.** >   **b.** <   **c.** <   **d.** >   **e.** =

3. **a.** $12,320,000,000,000,000 = 1.232 \times 10,000,000,000,000,000 = 1.232 \times 10^{16}$

    b. $.0000000000789 = 7.89 \times \dfrac{1}{100,000,000,000} = 7.89 \times 10^{-11}$

5. **a.** $-10 < y < 0$ **b.** $0 \le x \le 10$

7. **a.** $|x - (-7)| < 3$ or $|x + 7| < 3$ **b.** $|y| > |x - 3|$

9. The equation is equivalent to
    $$x - 5 = 3 \quad \text{or} \quad x - 5 = -3$$
    $$x = 8 \qquad \qquad x = 2$$
    Checking each possible solution, we see
    $$|8 - 5| = 3 \text{ and } |2 - 5| = |-3| = 3$$
    Thus both 8 and 2 are solutions.

11. The equation is equivalent to
    $$x + 3 = \frac{5}{2} \quad \text{or} \quad x + 3 = -\frac{5}{2}$$
    $$x = -\frac{1}{2} \qquad \qquad x = -\frac{11}{2}$$
    Checking each possible solution, we see
    $$\left| -\frac{1}{2} + 3 \right| = \frac{5}{2} \text{ and } \left| -\frac{11}{2} + 3 \right| = \left| -\frac{5}{2} \right| = \frac{5}{2}$$
    Thus both $-\frac{1}{2}$ and $-\frac{11}{2}$ are solutions.

13. The inequality is equivalent to
    $$-2 \le x + 2 \le 2$$
    $$-4 \le x \le 0$$

15. **a.** $|\pi - 7| = -(\pi - 7) = 7 - \pi$ **b.** $\left| \sqrt{23} - \sqrt{3} \right| = \sqrt{23} - \sqrt{3}$

17. **a.** $x > -8$, thus $(-8, \infty)$ **b.** $x \le 5$, thus $(-\infty, 5]$

19. Let $d = .2828\ldots$
    $$100d = 28.2828\ldots$$
    $$\underline{d = .2828\ldots}$$
    $$99d = 28$$
    $$d = \frac{28}{99}$$

21. $$2\left(\frac{x}{5} + 7 \right) - 3x = \frac{x + 2}{5} - 4$$
    $$\frac{2x}{5} + 14 - 3x = \frac{x + 2}{5} - 4$$
    $$2x + 70 - 15x = x + 2 - 20$$
    $$70 - 13x = x - 18$$
    $$-14x = -88$$
    $$x = \tfrac{44}{7}$$
    $$\tfrac{44}{7}$$

23. Since $b^2 - 4ac = (-2)^2 - 4(3)(5) = -56$, the equation has no real solutions.

25.
$$5z^2 + 6z = 7$$
$$5z^2 + 6z - 7 = 0$$

$$z = \frac{-6 \pm \sqrt{6^2 - 4(5)(-7)}}{2(5)}$$

$$z = \frac{-6 \pm \sqrt{176}}{10} = \frac{-3 \pm 2\sqrt{11}}{5}$$

$$\frac{-3 + 2\sqrt{11}}{5}, \frac{-3 - 2\sqrt{11}}{5}$$

27. Let x = width, $\frac{3}{4}x$ = height, d = diagonal, A = area. Then,

$$A = x\left(\frac{3}{4}x\right) = \frac{3}{4}x^2$$

$$d = \sqrt{x^2 + \left(\frac{3}{4}x\right)^2}$$

$$d = \frac{5}{4}x, \text{ hence } x = \frac{4}{5}d$$

$$A = \frac{3}{4}\left(\frac{4}{5}d\right)^2$$

$$A = \frac{12}{25}d^2$$

If $d_1 = 14, A_1 = \frac{12}{25} \cdot 14^2$. If $d_2 = 21, A_2 = \frac{12}{25} \cdot 21^2$.

$$\frac{A_2}{A_1} = \left(\frac{12}{25} \cdot 21^2\right) \div \left(\frac{12}{25} \cdot 14^2\right)$$

$$\frac{A_2}{A_1} = 21^2 \div 14^2$$

$$A_2 = \frac{9}{4}A_1$$

29. $20x^2 + 12 = 31x$
$20x^2 - 31x + 12 = 0$

$b^2 - 4ac = (-31)^2 - 4(20)(12) = 1$
Since $b^2 - 4ac$ is positive, the equation has two real solutions.

31. $y = kx$
$36 = k(12)$

$k = 3$
Therefore, the variation equation is $y = 3x$. When $x = 2$, $y = 3(2) = 6$.

33. $T = \dfrac{kR^2}{S}$

$.6 = \dfrac{k(3)^2}{15}$

$9 = k(3)^2$

$k = 1$

35. To solve $x^4 - 11x^2 + 18 = 0$, let $u = x^2$

$x^4 - 11x^2 + 18 = 0$

$u^2 - 11u + 18 = 0$

$(u - 2)(u - 9) = 0$

$u - 2 = 0 \quad \text{or} \quad u - 9 = 0$

$u = 2 \qquad\qquad u = 9$

Since $u = x^2$, we have the equivalent statements: $\quad x^2 = 2 \qquad x^2 = 9$

$x = \pm\sqrt{2} \quad x = \pm 3$

$\sqrt{2}, -\sqrt{2}, 3, -3$

37. The equation is equivalent to

$3x - 1 = 4 \quad \text{or} \quad 3x - 1 = -4$

$3x = 5 \qquad\qquad 3x = -3$

$x = \dfrac{5}{3} \qquad\qquad x = -1$

Checking each possible solution, we see $\left|3\left(\frac{5}{3}\right) - 1\right| = |5 - 1| = 4$ and $|3(-1) - 1| = |-4| = 4$.

Thus, both $\frac{5}{3}$ and -1 are solutions.

39. $\text{distance} = \sqrt{(x_1 - x_2)^2 + (y_1 - y_2)^2}$

$= \sqrt{(1 - 4)^2 + (-2 - 5)^2}$

$= \sqrt{(-3)^2 + (-7)^2}$

$= \sqrt{9 + 49} = \sqrt{58}$

41. $\text{distance} = \sqrt{(x_1 - x_2)^2 + (y_1 - y_2)^2}$

$= \sqrt{(c - (c - d))^2 + (d - (c + d))^2}$

$= \sqrt{(c - c + d)^2 + (d - c - d)^2}$

$= \sqrt{d^2 + (-c)^2} = \sqrt{c^2 + d^2}$

43. The midpoint is

$\left(\dfrac{x_1 + x_2}{2}, \dfrac{y_1 + y_2}{2}\right) = \left(\dfrac{c + 2d - c}{2}, \dfrac{d + c + d}{2}\right) = \left(\dfrac{2d}{2}, \dfrac{c + 2d}{2}\right) = \left(d, \dfrac{c}{2} + d\right)$

45. a. The radius of the circle is the distance from $(2, -3)$ to $(1, 1)$.

$\sqrt{(2 - 1)^2 + (-3 - 1)^2} = \sqrt{1^2 + (-4)^2} = \sqrt{17}$

b. The equation of the circle with center $(2, -3)$ and radius $\sqrt{17}$ is

$(x - 2)^2 + (y - (-3))^2 = \left(\sqrt{17}\right)^2$

$(x - 2)^2 + (y + 3)^2 = 17$

$x^2 - 4x + 4 + y^2 + 6y + 9 = 17$

$x^2 + y^2 - 4x + 6y - 4 = 0$

47.

49. (b) and (d)

51. (c)

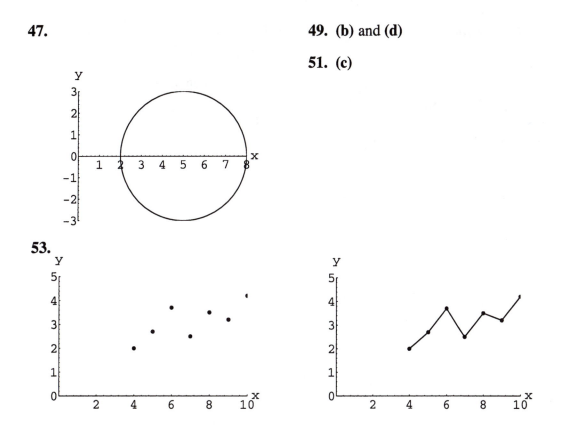

53.

55. Rewrite the equation in slope-intercept form.
$$y = x - \frac{x-2}{5} + \frac{3}{5}$$
$$y = x - \frac{1}{5}x + \frac{2}{5} + \frac{3}{5}$$
$$y = \frac{4}{5}x + 1$$
a. The y-intercept is 1. **b.** The slope is $\frac{4}{5}$.

57. Substitute 3 for m and $(2,-1)$ for (x_1,y_1) in the point-slope form.
$$y - (-1) = 3(x-2)$$
$$y + 1 = 3x - 6$$
$$y = 3x - 7$$

59. The line $2y - x = 5$, in slope-intercept form, becomes $y = \frac{1}{2}x + \frac{5}{2}$. Its slope is $\frac{1}{2}$. The slope of any perpendicular line is -2. Thus the required line has equation $y = -2x - b$. Since $(0,1)$ satisfies the equation, $1 = -2(0) + b$ and $b = 1$. Hence $y = -2x + 1$ is the required equation.

61. The line through $(1,3)$ and $(-4,2)$, has slope given by
$$\frac{y_2 - y_1}{x_2 - x_1} = \frac{2-3}{-4-1} = \frac{1}{5}$$
This is the slope of the required line. Substitute
$\frac{1}{5}$ for m and $(-4,5)$ for (x_1, y_1) in the point-slope form.
$$y - 5 = \frac{1}{5}\left(x - (-4)\right)$$
$$y - 5 = \frac{1}{5}(x + 4)$$
$$y - 5 = \frac{1}{5}x + \frac{4}{5}$$
$$y = \frac{1}{5}x + \frac{29}{5}$$

63. Since $\text{slope} = \dfrac{\text{change in } y}{\text{change in } x}$ we can

write $\dfrac{1}{5} = \dfrac{5000}{\text{change in } x}$

change in $x = 25{,}000$ feet

65. F   **67.** F   **69.** F   **71.** F

73. (d)   **75. (e)**

77. To find the y-intercept, set $x = 0$ and solve for y.
$$2(0) - 3y + 5 = 0$$
$$-3y = -5$$
$$y = \frac{5}{3}$$

79. Use the datapoints $(0, 70.8)$ and $(30, 76.8)$.

a. The slope is given by
$$\frac{y_2 - y_1}{x_2 - x_1} = \frac{76.8 - 70.8}{30 - 0} = 0.2$$
Substitute 0.2 for m and 70.8 for b in the slope-intercept form.
$$y = mx + b$$
$$y = 0.2x + 70.8$$

b. 1984 corresponds to $x = 14$.
$$y = 0.2(14) + 70.8$$
$$y = 73.6 \text{ years.}$$

c. Set $y = 80$ and solve for x.
$$80 = 0.2x + 70.8$$
$$9.2 = .02x$$
$$x = 46$$
This corresponds to the year 2016.

81. Graph **(c)** has slope $+75$.

83. Graph **(d)** has slope $+20$.

Chapter 2
Graphs and Technology

2.1 Graphs

1. Sketch:

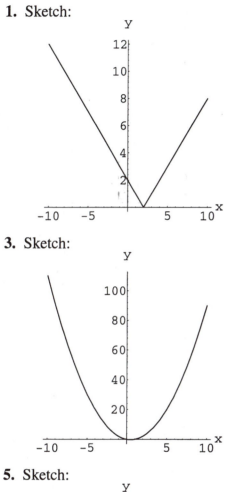

Calculator:

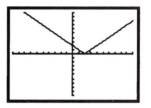

3. Sketch:

Calculator:

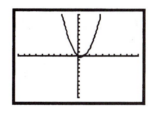

5. Sketch:

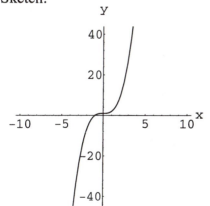

Calculator:

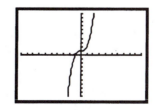

7. Graph $y = .5x - 3$

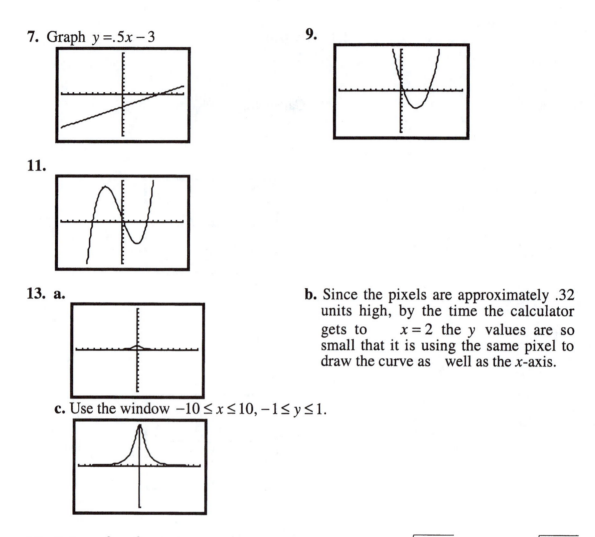

9.

11.

13. a.

b. Since the pixels are approximately .32 units high, by the time the calculator gets to $x = 2$ the y values are so small that it is using the same pixel to draw the curve as well as the x-axis.

c. Use the window $-10 \le x \le 10, -1 \le y \le 1$.

15. Solve $x^2 + y^2 = 9$ for y to obtain two solutions: $y = \sqrt{9 - x^2}$ and $y = -\sqrt{9 - x^2}$. Graph these in the standard window and choose ZSquare from the Zoom menu to get a square window.

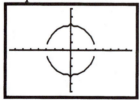

17. Solve to obtain two solutions: $y = \sqrt{\dfrac{48 - 3x^2}{2}}$ and $y = -\sqrt{\dfrac{48 - 3x^2}{2}}$.

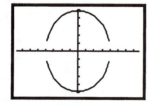

19. Solve to obtain two solutions: $y = -2 + \sqrt{25 - (x-4)^2}$ and $y = -2 - \sqrt{25 - (x-4)^2}$.

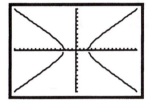

21. Solve to obtain two solutions: $y = \sqrt{\dfrac{4x^2 - 36}{9}}$ and $y = -\sqrt{\dfrac{4x^2 - 36}{9}}$.

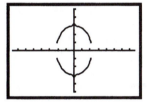

23. Solve to obtain two solutions: $y = \sqrt{\dfrac{45 - 9x^2}{5}}$ and $y = -\sqrt{\dfrac{45 - 9x^2}{5}}$.

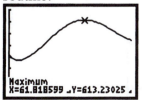

25. Graph the equation in the window specified and use the maximum and minimum routines.

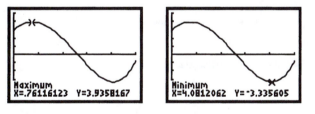

The highest point in the window is given by (.76116123, 3.9358167) and the lowest point is given by (4.0812062, –3.335605).

27. Graph the equation in the window $0 \le x \le 100, 0 \le y \le 700$ and use the maximum routine.

The highest point on the graph is given by $x = 61$. This corresponds to the year 1961.

29. Of all the choices given, **c** is the best, although a window which extends further in the positive x direction would be even better. **e**

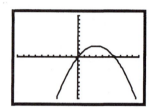

31. Of all the choices given, **d** is the best.

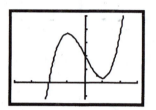

33. Of all the choices given, **d** is the best, although a window which extends further in the negative x direction would be even better. **e**

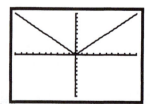

35. The two equations are not the same, as is confirmed by the following graphs:

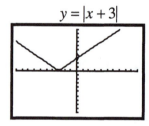

$$y = |x + 3|$$

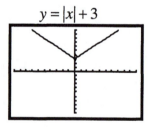

$$y = |x| + 3$$

37. The two equations are the same and have the same graph:

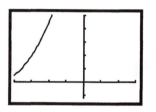

39. The two equations are not the same, as is confirmed by the following graphs:

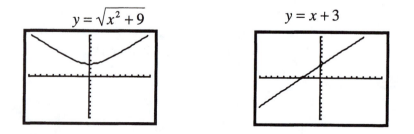

$$y = \sqrt{x^2 + 9}$$

$$y = x + 3$$

41. a. The two equations given have the same graph, hence the factorization can be considered confirmed. The window $-2 \le x \le 6, -1 \le y \le 10$ is shown.

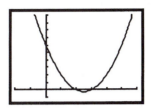

b. The two equations have different graphs, as is shown here in the window $-10 \le x \le 10, -10 \le y \le 100$. Thus the expressions are not equivalent.

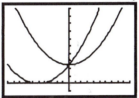

43. The two expressions give rise to the same graph, hence the factorization can be considered probably confirmed. The window $-2 \le x \le 4, 0 \le y \le 100$ is shown.

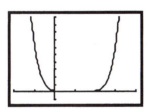

45. Of all the choices, **(c)** is the best, as it shows the rocket ascending, then descending to earth. The window $0 \le x \le 10, 0 \le y \le 250$ is shown.

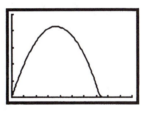

47. The window shown is $0 \le x \le 40$, $0 \le y \le 110,000$.

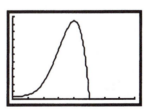

49. The window shown is $0 \le x \le 48$, $0 \le y \le 40$.

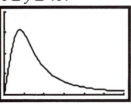

51. a. Values are approximately 12.3 and 24.2, corresponding to 1917 and 1929.
 b. Use the maximum routine to find a high point at $x = 20$, corresponding to 1925. There were approximately 104,000 deer at this time.

53. a. Use the maximum routine to find a high point at approximately $x = 6.3$ hours.
 b. The value is approximately $x = 18.7$ hours.

55. The window shown is $-3 \le x \le 3$, $-150 \le y \le 150$.

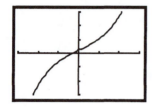

57. The standard window is shown. Note that $\sqrt{x^2} - x = 0$ if $x \ge 0$; $\sqrt{x^2} - x = -2x$ elsewhere (why?)

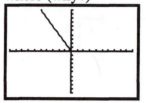

59. The window shown is $-6 \le x \le 12$, $-100 \le y \le 250$.

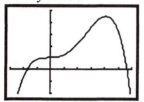

61. The window shown is $-4 \le x \le 4$, $-5 \le y \le 15$.

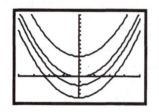

All four graphs have the same shape, but graph **(b)** is the result of shifting graph **(a)** vertically up 5 units, **(c)** of shifting vertically down 5 units, and **(d)** of shifting down 2 units.

63. The window shown is $0 \le x \le 10$, $0 \le y \le 10$.

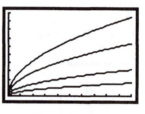

All four graphs have the same shape, but graph **(b)** is the result of stretching graph **(a)** away from the x-axis by a factor or 2, **(c)** by a factor of 3, and **(d)** by a factor of 1/2.

65. The window shown is $-5 \le x \le 5$, $-5 \le y \le 5$.

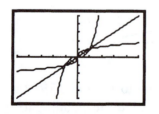

The graphs are reflections of each other in the line $y = x$.

67. The window shown is $-20 \le x \le 20$, $-20 \le y \le 20$.

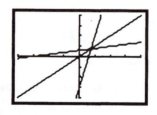

The graphs are reflections of each other in the line $y = x$.

2.2 Solving Equations Graphically and Numerically

1. Graph $y = x^5 + 5$ and $y = 3x^4 + x$ in the same viewing window to find the points of intersection; these will be the solutions of the equation. First use a window such as $-4 \leq x \leq 4, -5 \leq y \leq 15$. This shows two intersections. A second window such as $2 \leq x \leq 5, 10 \leq y \leq 500$ shows a third intersection. 3 solutions.

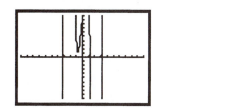

3. Graph $y = x^7 - 10x^5 + 15x + 10$ in the standard window. This suggests that there are three x-intercepts, that is, three solutions of the equation. A larger window such as $-20 \leq x \leq 20, -10 \leq y \leq 10$ does not indicate any more solutions.

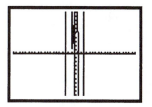

5. Graph $y = x^4 + 500x^2 - 8000x$ and $y = 16x^3 - 32000$ in the same viewing window to find the points of intersection. The window $-5 \leq x \leq 20, -50,000 \leq y \leq 50,000$ is shown. 2 solutions.

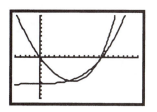

7. $x = -2.426453$ 9. $x = 1.1640351$ 11. $x = 1.2372475$ 13. $x = 1.1921235$

15. $x = -1.379414$ 17. $x = -1.600512$ 19. $x = 1.8608059$

21. Graph $y = 2x^3 - 4x^2 + x - 3$ and use the root (zero) routine to find the solution $x = 2.1016863$. The window $-2 \leq x \leq 3, -10 \leq y \leq 10$ is shown.

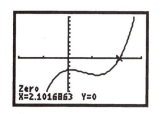

23. Graph $y = x^5 - 6x + 6$ and use the root (zero) routine to find the solution $x = -1.752119$. The window $-3 \leq x \leq 3, -5 \leq y \leq 15$ is shown.

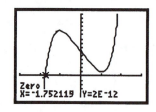

25. Graph $y = 10x^5 - 3x^2 + x - 6$ and use the root (zero) routine to find the solution $x = .95054589$. The window $-1 \leq x \leq 2, -20 \leq y \leq 10$ is shown.

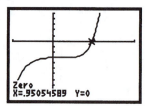

27. Rewrite the equation by multiplying both sides by 12 to obtain $24x - 6x^2 - x^4 = 0$. Graph $y = 24x - 6x^2 - x^4$ in the window $-1 \leq x \leq 3, -10 \leq y \leq 20$ to see two intercepts. One is obviously 0; the other is found by the root (zero) routine to be $x = 2.2073898$.

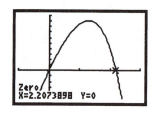

29. Graph $y = \dfrac{5x}{x^2 + 1} - 2x + 3$ and use the root (zero) routine to find the solution $x = 2.3901454$. The window $-3 \leq x \leq 3, -2 \leq y \leq 5$ is shown.

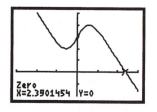

31. Graph $y = |x^2 - 4|$ and $y = 3x^2 - 2x + 1$ in the same window $-5 \leq x \leq 5, 0 \leq y \leq 10$ and use the intersection routine to locate the two solutions at $x = -.6513878$ and $x = 1.1513878$.

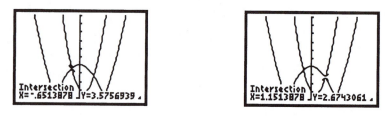

33. Graph $y = \sqrt{x^2 + 3}$ and $y = \sqrt{x - 2} + 5$ in the same window $0 \leq x \leq 10, 0 \leq y \leq 10$ and use the intersection routine to locate the only solution at $x = 7.033393$.

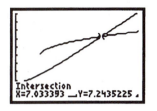

35. Graph $y = 3x^3 - 2x^2 + 3x - 2$ in the window $0 \leq x \leq 1, -2 \leq y \leq 2$ and use the root (zero) routine to find the solution approximated as $x = .66666667$. This suggests that the solution could be exactly $x = \frac{2}{3}$. Substituting into the original equation confirms this.

37. Graph $y = 12x^4 - x^3 - 12x^2 + 25x - 2$ in the window $0 \leq x \leq 1, -5 \leq y \leq 5$ and use the root (zero) routine to find the solution approximated as $x = .083333333$. This suggests that the solution could be exactly $x = \frac{1}{12}$. Substituting into the original equation confirms this.

39. Graph $y = 4x^4 - 13x^2 + 3$ in the window $-1 \leq x \leq 2, -10 \leq y \leq 10$ and use the root (zero) routine to find the solution approximated as $x = 1.7320508$. This suggests that the solution could possibly be exactly $x = \sqrt{3}$. Substituting into the original equation confirms this.

41. Graph $y = 10^x - \frac{1}{4}x$ and $y = 28$ in the same window $0 \leq x \leq 3, -10 \leq y \leq 40$ and use the intersection routine to locate the only solution at $x = 1.4527551$.

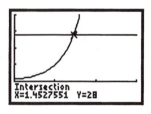

43. Graph $y = \ln x - x^2 + 3$ in the window $0 \leq x \leq 3, -3 \leq y \leq 3$ and use the root (zero) routine to locate the two solutions at $x = .04991125$ and $x = 1.9096976$.

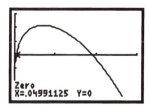
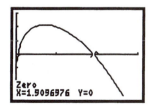

45. Graph $y = .024x^4 - .87x^3 + 9.6x^2 + 97.2x + 2196$ and $y = 4000$ in the same window $0 \le x \le 25, 0 \le y \le 5000$ and use the intersection routine to locate the only solution at $x = 14.2$. This corresponds to the year 2004 or early 2005.

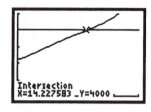

47. a. Graph the given function and the line $y = 750$ in the same window $0 \le x \le 25, 0 \le y \le 1500$ and use the intersection routine to locate the only solution at $x = 9.1$. This corresponds to the year 1999 or early 2000.

b. Graph the given function and the line $y = 1,000$ in the same window $0 \le x \le 25, 0 \le y \le 1500$ and use the intersection routine to locate the only solution at $x = 14.4$. This corresponds to the year 2004 or early 2005.

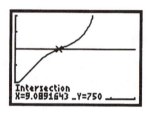

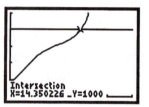

2.3 Applications of Equations

1.

English Language	Mathematical Language
The two numbers	x and y
Their sum is 15	$x + y = 15$
The difference of their squares is 5	$x^2 - y^2 = 5$

3.

English Language	Mathematical Language
The width and length of a rectangle	x and y
The perimeter is 45 centimeters	$2x + 2y = 45$
The area is 112.5 square centimeters	$xy = 112.5$

5.

English Language	Mathematical Language
Old salary	x
8% pay raise	$.08x$
New salary	$x + .08x$

$x + .08x = 1600$

7.

English Language	Mathematical Language
Area before decrease	$\frac{1}{4}\pi(16)^2$
Decrease radius	x
Diameter after decrease	$16 - 2x$
Area after decrease	$\frac{1}{4}\pi(16 - 2x)^2$

$$\frac{1}{4}\pi(16)^2 - \frac{1}{4}\pi(16 - 2x)^2 = 48\pi$$

9. Let x = amount invested at 12%, $1100 - x$ = amount invested at 6%

$\text{Return} = .12x + .06(1100 - x) + .11(550)$

$12x + .06(1100 - x) + .11(550) = .09(1650)$

$$.06x + 126.5 = 148.5$$

$$.06x = 22$$

$$x = \$366.67$$

11. Let x = amount drained

Then $.4(8 - x)$ = amount of antifreeze in drained radiator

x = amount of antifreeze added

$.6(8)$ = amount of antifreeze required

$$.4(8 - x) + x = .6(8)$$

$$3.2 + .6x = 4.8$$

$$x = 2\tfrac{2}{3} \text{ quarts}$$

13. Use $\text{distance} = \text{rate} \times \text{time}$

$\text{distance going} = 2.5(360 + r)$

$\text{distance returning} = 3.5(360 - r)$

Since the distances are equal,

$$2.5(360 + r) = 3.5(360 - r)$$

$$900 + 2.5r = 1260 - 3.5r$$

$$6r = 360$$

$$r = 60 \text{ mph}$$

15. Solve the equation in the example.

$$x + \frac{1683}{x} = 82.25$$

$$x^2 - 82.25x + 1683 = 0$$

$$x = \frac{82.25 \pm \sqrt{(82.25)^2 - 4(1683)}}{2(1)}$$

$$x = \frac{82.25 \pm 5.75}{2}$$

$x = 44$ and 38.25

17. Let x = rate of southbound car

50 = rate of northbound car

$3x$ = distance of southbound car

$3(50)$ = distance of northbound car

$$3x + 3(50) = 345$$

$$3x = 195$$

$$x = 65 \text{ miles per hour}$$

19. Before: After:

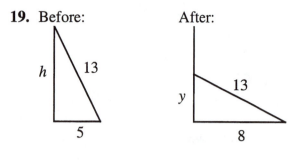

Before: $h^2 = 13^2 - 5^2$, hence $h = 12$

After: $y^2 = 13^2 - 8^2$, hence $y = \sqrt{105} \approx 10.2$

$h - y = 12 - 10.2 = 1.8$ feet.

21.   Let $x =$ width of walk

$12 =$ radius of pool

$x + 12 =$ outer radius of walk

Then, Area of walk $= \pi(x+12)^2 - \pi(12)^2$

$52\pi = \pi(x+12)^2 - \pi(12)^2$

$52 = x^2 + 24x + 144 - 144$

$0 = x^2 + 24x - 52$

$0 = (x-2)(x+26)$

$x = 2$ or $x = -26$

Only the positive answer makes sense: 2 meters.

23. Let $16t^2 =$ distance fallen

$1100(3-t) =$ distance covered by sound

Then $16t^2 = 1100(3-t)$

$16t^2 + 1100t - 3300 = 0$

$t = \dfrac{-1100 \pm \sqrt{(1100)^2 - 4(16)(-3300)}}{2(16)}$

$t = \dfrac{-1100 \pm 1192.14}{32}$

$t = -71.6$ or 2.88 sec

Only the positive answer makes sense. Then $d = 16t^2 = 16(2.88)^2 = 132.7$ feet.

25. Let $x =$ Red's speed

$x - 6 =$ Wolf's speed

$\dfrac{432}{x} =$ Red's time

$\dfrac{432}{x - 6} =$ Wolf's time

$\dfrac{432}{x - 6} - \dfrac{432}{x} = 1$

$432x - 432(x - 6) = x(x - 6)$

Then $0 = x^2 - 6x - 2592$

$0 = (x - 54)(x + 48)$

$x = 54$ or $x = -48$

Only the positive answer makes sense: Red drives 54 mph, Wolf drives 48 mph.

27. Let $x, x + 1, x + 2$ be the dimensions:

$x(x + 1)(x + 2) = 13800$

$x^3 + 3x^2 + 2x - 13800 = 0$

A calculator graph in the window $20 \le x \le 30, -1000 \le y \le 1000$:

shows the root at $x = 23$, dimensions $23\text{cm} \times 24\text{cm} \times 25\text{cm}$.

29. Use the given formula with $h = 5, S = 100$

$100 = \pi r \sqrt{r^2 + 25}$

A calculator graph in the window $0 \le x \le 10, 0 \le y \le 150$:

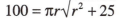

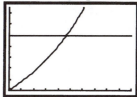

shows the root at 4.658 inches.

31. In Example 9, the dimensions of the box were found to be $x, 22 - 2x$, and $30 - 2x$
and the equation was found to be

$(30 - 2x)(22 - 2x)x = 1000$

$4x^3 - 104x^2 + 660x - 1000 = 0$

To have one dimension greater than 18, look for a solution with x less than 6, then $30 - 2x$ is greater than 18. The window $0 \le x \le 11, -10 \le y \le 10$ shown in the text yields the solution $x = 2.234$ inches.

33. The dimensions of the base of the box are given by x = width and $2x$ = length. Then the height is found from

$$\text{length} \times \text{width} \times \text{height} = \text{Volume}$$

$$2xxh = 38.72$$

$$h = \frac{19.36}{x^2}$$

The cost is given by

$$\text{Cost} = \text{area of base} \times 12 + \text{area of sides} \times 8$$

$$538.56 = 2x^2 \times 12 + 2\left(\frac{19.36}{x^2}\right)x \times 8 + 2\left(\frac{19.36}{x^2}\right)2x \times 8$$

which simplifies to

$$538.56 = \frac{929.28}{x} + 24x^2$$

A calculator graph of $y = \dfrac{929.28}{x} + 24x^2$ and $y = 538.56$ in the window $0 \leq x \leq 5$, $400 \leq y \leq 700$

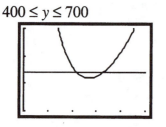

shows two roots, one at $x = 2.2$ and one at $x = 3.237$. Only the former leads to an acceptable height of 4 feet and the dimensions are 2.2 by 4.4 by 4 feet.

2.4 Optimization Applications

1.   Graph $y = 2x^3 - 3x^2 - 12x + 1$ and use the maximum routine to find the high point at $(-1, 8)$, The window $-3 \leq x \leq 3, -20 \leq y \leq 10$ is shown.

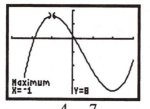

3.   Graph $y = \dfrac{4}{x^2} - \dfrac{7}{x} + 1$ and use the minimum routine to find the low point at $(1.1428553, -2.0625)$, The window $-10 \leq x \leq 10, -3 \leq y \leq 3$ is shown.

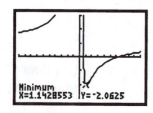

5. Graph $y = \dfrac{x^2(x+1)^3}{(x-2)^2(x-4)^2}$ and use the maximum routine to find the high point at $(-.3409026, .00032224)$, The window $-1 \le x \le 0, 0 \le y \le .0004$ is shown.

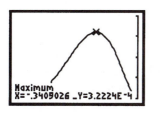

7. **a.** In the window $-2 \le x \le 0, -1 \le y \le 5$ the high point is at $(-1, 4)$.
    b. In the window $-2 \le x \le 2, -1 \le y \le 5$ there is a high point at $(-1, 4)$ and another at the endpoint $(2, 4)$ of the interval.
    c. In the window $-2 \le x \le 3, -1 \le y \le 25$ the high point occurs at the right endpoint of the interval $(3, 20)$.

9. The dimensions of the base of the box are given by x = width = length. Then the height is found from

$$\text{length} \times \text{width} \times \text{height} = \text{Volume}$$

$$xxh = 30,000$$

$$h = \frac{30,000}{x^2}$$

The total surface area is given by

$$\text{Area} = \text{Area of base} + 4 \times \text{Area of one side}$$

$$y = x^2 + 4 \cdot x \cdot \frac{30,000}{x^2}$$

$$y = x^2 + \frac{120,000}{x}$$

Graph this in the window $0 \le x \le 100, 0 \le y \le 20,000$ and use the minimum routine to find $x = 39.15 \, \text{cm}$ and $h = 30,000 \, / \, x^2 = 19.57 \, \text{cm}$

11. As in the example, the dimensions of the box are given by x = height, $20 - 2x$ = length = width. Then

$$\text{Volume} = \text{length} \times \text{width} \times \text{height}$$

$$y = (20 - 2x)(20 - 2x)x$$

a. Graph $y = (20 - 2x)(20 - 2x)x$ and $y = 550$ in the window $4 \le x \le 10$, $0 \le y \le 700$ and use the intersection routine to find $x = 4.43$ inches.
    b. Graph $y = (20 - 2x)(20 - 2x)x$ in the window $0 \le x \le 10, 0 \le y \le 700$ and use the maximum routine to find $x = 3.33$ inches.

13. The average cost per unit is given by

$$y = \frac{c(x)}{x} = \frac{.13x^3 - 70x^2 + 10,000x}{x}$$

$$y = .13x^2 - 70x + 10,000$$

a. Graph $y = .13x^2 - 70x + 10,000$ and $y = 1100$ in the window $0 \le x \le 400$, $0 \le y \le 5000$ and use the intersection routine to find $x = 206$. The other intersection shows x greater than 300 units and must be discarded.

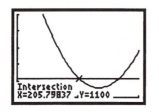

b. Graph $y = .13x^2 - 70x + 10,000$ in the window $0 \le x \le 400, 0 \le y \le 5000$ and use the minimum routine to find $x = 269$ units, which leads to an average cost of $577 per unit.

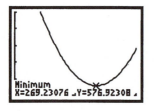

15. Use Revenue $= 142x$, Cost $= x^3 - 8x^2 + 20x + 40$. Then
Profit $=$ Revenue $-$ Cost

$$y = 142x - \left(x^3 - 8x^2 + 20x + 40\right)$$

$$y = -x^3 + 8x^2 + 122x - 40$$

a. Graph this equation in the window $0 \le x \le 6, -100 \le y \le 1000$. The high point occurs at the right hand endpoint $x = 6$, so 600 bookmarks should be made.
b. Graph this equation in the window $0 \le x \le 16, -100 \le y \le 1000$ and use the maximum routine to find $x = 9.5788127$, so 958 bookmarks should be made.

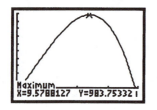

17. The figure of the garden consists of a rectangle of width 10, length $2x$ plus a triangle with base $2x$, height h. Since $h^2 + x^2 = 10^2$, we can write $h = \sqrt{100 - x^2}$. Then

Area = area of rectangle + area of triangle

$$y = 10(2x) + \tfrac{1}{2}(2x)\sqrt{100 - x^2}$$
$$y = 20x + x\sqrt{100 - x^2}$$

Graph this equation in the window $0 \le x \le 10, 0 \le y \le 300$ and use the maximum routine to find $x = 9.31$ feet, then area $= y = 220.18$ square feet.

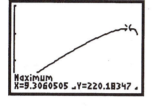

19. Any point on the graph of $y = 5 - x^2$ must have coordinates $(x, 5 - x^2)$. Then the distance from $(x, 5 - x^2)$ to $(0,1)$ is given by

$$y = \sqrt{(x - 0)^2 + ((5 - x^2) - 1)^2}$$
$$y = \sqrt{x^2 + (4 - x^2)^2}$$
$$y = \sqrt{x^4 - 7x^2 + 16}$$

Graph this equation in the window $0 \le x \le 5, 0 \le y \le 10$ and use the minimum routine to find $x = 1.871$.

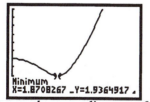

Hence the coordinates of the required point are given by $x = 1.871$, $y = 5 - x^2 = 5 - (1.871)^2 = 1.5$.

21. Let x = number of times ladders are ordered per year. Then

Cost = cost of ordering
+ cost of storage

$$y = 20x + 10\left(\frac{300}{x}\right)$$
$$y = 20x + \frac{3000}{x}$$

Graph this equation in the window $0 \le x \le 24, 0 \le y \le 3000$ and use the minimum routine to find $x = 12.2$. Since x must be an integer, the answer is 12 times per year.

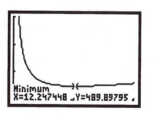

2.5 Linear Models

1. a. Here is a table of the residuals for each model.

$y = x$					$y = .5x + 1.5$			
Data Point (x,p)	Model Point (x,y)	Residual $p - y$	Squared Residual $(p-y)^2$		Data Point (x,p)	Model Point (x,y)	Residual $p - y$	Squared Residual $(p-y)^2$
(1,2)	(1,1)	1	1		(1,2)	(1,2)	0	0
(2,2)	(2,2)	0	0		(2,2)	(2,2.5)	.5	.25
(3,3)	(3,3)	0	0		(3,3)	(3,3)	0	0
(4,3)	(4,4)	−1	1		(4,3)	(4,3.5)	.5	.25
(5,5)	(5,5)	0	0		(5,5)	(5,4)	−1	1

b. The sum of the squares of the residuals in the first model is 2. The sum of the squares of the residuals in the second model is 1.5.

c. Clearly the second model is the better fit.

3. a. Here is a table of the residuals for each model.

$y = 11x + 121$					$y = 11.5x + 129$			
Data Point (x,p)	Model Point (x,y)	Residual $p - y$	Squared Residual $(p-y)^2$		Data Point (x,p)	Model Point (x,y)	Residual $p - y$	Squared Residual $(p-y)^2$
(0,129)	(0,121)	8	64		(0,129)	(0,129)	0	0
(5,161)	(5,176)	−15	225		(5,161)	(5,186.5)	−25.5	650.25
(9,215)	(9,220)	−5	25		(9,215)	(9,232.5)	−17.5	306.25
(10,244)	(10,231)	13	169		(10,244)	(10,244)	0	0

b. The sum of the squares of the residuals in the first model is 483. The sum of the squares of the residuals in the second model is 956.5.

c. Clearly the first model is the better fit.

5. There seems to be a positive correlation between the variables.

7. There seems to be little correlation between the variables. Certainly any relationship between them is not linear.

9. The scatter plot is shown below.

a. The data appears to be linear.

b. There appears to be a positive correlation.

11. The scatter plot is shown below.

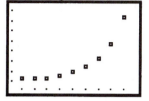

    a. The data does not appear to be linear.
    b. There appears to be a positive correlation.

13. The scatter plot is shown below.

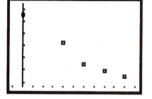

    a. The data appears to be linear.
    b. There appears to be a negative correlation.

15. a.

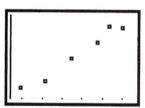

    b. The TI-83 yields the following model:

```
LinReg
 y=ax+b
 a=398.2421053
 b=-417.3157895
 r²=.9671168803
 r=.9834210087
```

    c. The Table feature of the TI-83 yields the following: when $x = 4$ (1994), $y = \$1175$ million. When $x = 8$ (1998), $y = \$2769$ million.
    d. The same Table predicts when $x = 14$ (2004), $y = \$5158$ million.

17. a. Entering the data into the TI-83 and using the linear regression routine yields:

```
LinReg
 y=ax+b
 a=67.68025415
 b=660.6403715
 r²=.9857658649
 r=.9928574242
```

    b. The slope of 67.68 represents the average rate of increase in billions of dollars per year. The y-intercept of 660.64 represents the model's estimate of expenditures in 1990.
    c. The Table feature of the TI-83 yields : when $x = 17$ (2007), $y = \$1811.2$ billion.

19. a. Entering the data into the TI-83 and using the linear regression routine twice yields:

Did not graduate: Graduated from high school:

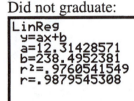

b. The result of using the Table feature of the TI-83 is shown below:

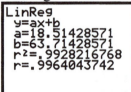

That is, the model predicts that when $x = 16$ (2006), those who did not graduate will have a median income of \$435.52, while those who did will have a median income of \$596.95.

21. a. Entering the data into the TI-83 and using the linear regression routine yields:

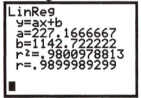

The Table feature of the TI-83 yields : when $x = 7$ (2004), $y = 193$ million users.

b. Since the model predicts that when $x = 12$ (2009), $y = 286$ million users, which is more than the present population and almost equal to the projected population, the model will have ceased to be valid before that.

23. a. Entering the data into the TI-83 and using the linear regression routine yields:

```
LinReg
 y=ax+b
 a=227.1666667
 b=1142.722222
 r²=.9800978813
 r=.9899989299
■
```

b. The model predicts that the maximum Pell Grant will reach \$5000 when $x = 17$, that is, in 2007. The fact that the maximum did not increase between 2002 and 2003, however, throws considerable doubt on the validity of the model.

Chapter 2 Review Exercises

1. a. Windows **a** and **d** give complete graphs.
  b. Windows **b** and **c** make the interesting part of the graph too close to the *x*-axis to see. Window **e** shows nothing sensible at all.
  c. The window $-4 \le x \le 6, -10 \le y \le 10$ (shown) would be a better choice.

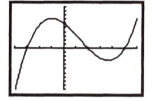

3. a. No window gives a complete graph.
  b. Windows **a** and **b** do not show the maxima and minima. Window **c** does not show the minima. Window **e** shows nothing at all, and window **d** is just a random stab.
  c. The window $-8 \le x \le 12, -1000 \le y \le 500$ (shown) would be a better choice.

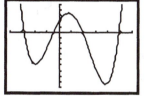

5. a. Windows **b** and **c** give a complete graph.
  b. Window **a** does not show the maxima and minima. Window **d** squeezes everything uselessly. Window **e** shows nothing at all.
  c. The window $-10 \le x \le 10, -150 \le y \le 150$ (shown) would be a better choice.

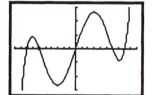

7.

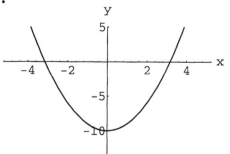

This shows all intercepts and the minimum.

9.

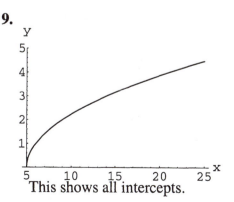

This shows all intercepts.

11. **13.**

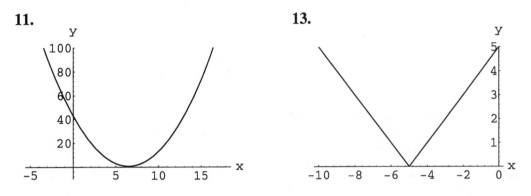

15. Graph $y = x^3 + 2x^2$ and $y = 11x + 6$ in the window $0 \le x \le 5, -10 \le y \le 60$ and use the intersection routine to obtain $x = 2.7644$.

17. Graph $y = x^4 + x^3 - 10x^2$ and $y = 8x + 16$ in the window $0 \le x \le 5, -20 \le y \le 60$ and use the intersection routine to obtain $x = 3.2678$.

19. The equation is equivalent to $x^3 + 2x^2 - 3x + 4 = 0$ as long as the denominator is defined, that is, $x \ne -5, 3$. Graph this equation in the window $-6 \le x \le 3$, $-10 \le y \le 11$ and use the root (zero) routine to obtain $x = -3.2843$.

21. The equation is equivalent to $x^3 + 2x^2 - 3x - 5 = 0$. Graph this equation in the window $0 \le x \le 3, -10 \le y \le 10$ and use the root (zero) routine to obtain $x = 1.6511$.

23. Let $x =$ amount of gold

Then $1 - x =$ amount of silver

$600x =$ value of gold

$50(1 - x) =$ value of silver

$600x + 50(1 - x) = 200$

$600x + 50 - 50x = 200$

$550x = 150$

$x = 3/11$ ounce of gold

$1 - x = 8/11$ ounce of silver

25. Let $x =$ time working

$\dfrac{x}{5} =$ Karen's rate

$\dfrac{x}{4} =$ Claire's rate

$\dfrac{x}{5} + \dfrac{x}{4} = 1$

$4x + 5x = 20$

$x = \dfrac{20}{9} = 2\dfrac{2}{9}$ hrs

27. Let x = length of smaller piece
  $4x$ = length of bigger piece
  $x + 4x = 12$
    $x = 2.4$
   $4x = 9.6$ feet

29.   Let x = original length
  $x + 2$ = new length
    $2x$ = new width
   x^2 = original area
  $2x(x + 2)$ = new area
  $2x(x + 2) = 3x^2$
  $2x^2 + 4x = 3x^2$
    $0 = x^2 - 4x$
  $x = 0$ (unreasonable) or $x = 4$ feet

31. Average cost $= \dfrac{\text{cost}}{x} = \dfrac{600x^2 + 600x}{x^2 + 1} \div x = \dfrac{600x + 600}{x^2 + 1}$

$\dfrac{600x + 600}{x^2 + 1} = 25$

$600x + 600 = 25x^2 + 25$

$0 = 25x^2 - 600x - 575$

$x = 12 \pm \sqrt{167}$

Only the positive answer makes sense, so $x = 12 + \sqrt{167} = 25$ caseloads

33. Let x = one of the two sides
  z = one of the three sides
  $2x + 3z = 120$

$z = 40 - \dfrac{2}{3}x$

Area $= xz = x\left(40 - \dfrac{2}{3}x\right)$

Graph $y = x\left(40 - \dfrac{2}{3}x\right)$ and use the maximum routine to find $x = 30$ yards, $z = 20$ yards.

35. Width of rectangle $= 2x$, length $= 9 - x^2$, hence area $= 2x(9 - x^2) = 18x - 2x^3$. Graph this equation and use the maximum routine to find $x = 1.7321$. This suggests that the exact value is $x = \sqrt{3}$, which can be proven.

37. a.

b. The data does not appear to be linear.

39. a. The TI-83 gives the following:

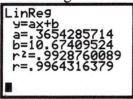

```
LinReg
 y=ax+b
 a=.3654285714
 b=10.67409524
 r²=.9928760089
 r=.9964316379
 ■
```

Thus $y = .365x + 10.674$ is the model.

b. For 1993, the model gives $11.77, slightly more than the actual. For 2000, the model gives $14.33, slightly less than the actual.

c. For 2004, the model gives $15.79.

41. a. The TI-83 gives the following:

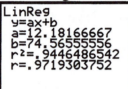

```
LinReg
 y=ax+b
 a=12.18166667
 b=74.56555556
 r²=.9446486542
 r=.9719303752
```

Thus $y = 12.18x + 74.57$ is the model.

b. For 2002, the model gives $220.7 billion, and for 2005, $257.3 billion.

c. The model gives $x = 11.9$, that is, 2002.

Chapter 3
Functions and Graphs

3.1 Functions

1. This could be a table of values of a function, because each input determines one and only one output.

3. This could not be a table of values of a function, because two output values are associated with the input –5.

5. 6

7. –2

9. –17

11. This defines y as a function of x.

13. Since $x = \dfrac{y^2 - 1}{4}$, this defines x as a function of y.

15. Since $y = \dfrac{12 - 3x}{2}$ and $x = \dfrac{12 - 2y}{3}$, this defines both y as a function of x and x as a function of y.

17. Neither

19.
x	–2	–1.5	–1	–.5	0	0.5	1
y	–2	–3.25	–4	–4.25	–4	–3.25	–2
x	1.5	2	2.5	3	3.5	4	
y	–.25	2	4.75	8	11.75	16	

21.
x	–2	–1.2	–.04	.04	1.2	2
y	0	1.6	1.99959996	1.99959996	1.6	0

23. $500 — $0
$1509 — $0
$3754 — $0
$6783 — $.02(6783 – 4500) = $45.66
$12,500 — $.04(12,500 – 10,000) + 110 = $210
$55,342 — $.06(55,342 – 50,000) + 1710 = $2030.52

25. A function may assign the same output to many different inputs.

27. Postage is a function of weight. From the word statement, it would be possible for the reverse to hold, but this is not true in reality.

29. **a.** Average Man

Drinks	2	3	4	5
Content	.03	.05	.07	.10

Average Woman

Drinks	2	3
Content	.05	.08

b. These tables define functions. For the average man, the domain is $[2,5]$ and the range is $[.03,.10]$; for the average woman, the domain is $[2,3]$ and the range is $[.05,.08]$.

31. a. $A = \pi r^2$ **b.** $A = \pi \left(\dfrac{d}{2}\right)^2$ or $A = \pi \dfrac{d^2}{4}$

33. Volume $=$ length $\times$ width $\times$ height

$$V = x \cdot x \cdot 4x$$
$$V = 4x^3$$

35. $D = 400 - 16t^2$. The range of this function is the possible values for D, that is, $0 \le D \le 400$.

37. a.

x	0	10	20	30	40
y	.1282	.13243	.17532	.20846	.23214

b. The values correspond to the accuracy of the data; the equation is a reasonable model.

c. 1995: $x = 35$ — 21.96%
2005: $x = 45$ — 25.38%

d. $x = 50$ corresponds to $y = 29.53\%$, thus year 2010 or 2011.

39. Domain: $[-3,3]$; Range: $[-4,3]$

41. Domain: $[-3,4]$; Range: $[-1.1,3]$

43. Domain: $[-2.5,2]$; Range: $[-3,3]$

45. In each case there is more than one y value associated with at least one domain value.

47. a. A calculator graph of $y = 10^{\tan x}$ is shown, $-10 \le x \le 10, 0 < y < 35$.

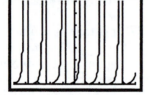

The largest possible domain is $\left(-\frac{\pi}{2}, \frac{\pi}{2}\right)$ together with displacements of this interval left and right by π units.

b. The range of the function is $(0,\infty)$.

3.2 Functional Notation

1. Using $f(x) = \dfrac{x-3}{x^2+4}$, obtain the following table:

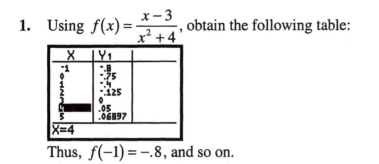

Thus, $f(-1) = -.8$, and so on.

3. $f(x) = \sqrt{x+3} - x + 1$
$f(0) = \sqrt{0+3} - 0 + 1 = \sqrt{3} + 1$

5. $f(x) = \sqrt{x+3} - x + 1$
$f(\sqrt{2}) = \sqrt{\sqrt{2}+3} - \sqrt{2} + 1$

7. $f(x) = \sqrt{x+3} - x + 1$
$f(-2) = \sqrt{-2+3} - (-2) + 1 = \sqrt{1} + 3 = 4$

9. $h(x) = x^2 + \dfrac{1}{x} + 2$
$h(3) = 3^2 + \dfrac{1}{3} + 2 = \dfrac{34}{3}$

11. $h(x) = x^2 + \dfrac{1}{x} + 2$
$h\left(\dfrac{3}{2}\right) = \left(\dfrac{3}{2}\right)^2 + \dfrac{1}{3/2} + 2 = \dfrac{9}{4} + \dfrac{2}{3} + 2 = \dfrac{59}{12}$

13. $h(x) = x^2 + \dfrac{1}{x} + 2$
$h(a+k) = (a+k)^2 + \dfrac{1}{a+k} + 2$

15. $h(x) = x^2 + \dfrac{1}{x} + 2$
$h(2-x) = (2-x)^2 + \dfrac{1}{2-x} + 2 = 4 - 4x + x^2 + \dfrac{1}{2-x} + 2 = 6 - 4x + x^2 + \dfrac{1}{2-x}$

17. $g(t) = t^2 - 1$
$g(3) = 3^2 - 1 = 8$

19. $g(t) = t^2 - 1$
$g(s+1) = (s+1)^2 - 1 = s^2 + 2s$

21. $g(t) = t^2 - 1$
$g(-t) = (-t)^2 - 1 = t^2 - 1$

23. $f(x) = 3x + 7$
a. $f(r) = 3r + 7$
b. $f(r) - f(x) = (3r+7) - (3x+7) = 3r - 3x$
c. $\dfrac{f(r) - f(x)}{r-x} = \dfrac{3r-3x}{r-x} = \dfrac{3(r-x)}{r-x} = 3$

25. $f(x) = x - x^2$
a. $f(r) = r - r^2$
b. $f(r) - f(x) = (r - r^2) - (x - x^2) = r - r^2 - x + x^2$
c. $\dfrac{f(r) - f(x)}{r-x} = \dfrac{r - r^2 - x + x^2}{r-x} = \dfrac{(r-x) - (r^2 - x^2)}{r-x} = \dfrac{(r-x)(1-r-x)}{r-x} = 1 - r - x$

27. $f(x) = x + 1$

$$\frac{f(x+h) - f(x)}{h} = \frac{(x+h+1) - (x+1)}{h} = \frac{h}{h} = 1$$

29. $f(x) = 3x + 7$

$$\frac{f(x+h) - f(x)}{h} = \frac{(3(x+h) + 7) - (3x+7)}{h} = \frac{3h}{h} = 3$$

31. $f(x) = x - x^2$

$$\frac{f(x+h) - f(x)}{h} = \frac{\left((x+h) - (x+h)^2\right) - (x - x^2)}{h} = \frac{x + h - x^2 - 2xh - h^2 - x + x^2}{h}$$

$$= \frac{h - 2xh - h^2}{h} = 1 - 2x - h$$

33. $f(x) = \sqrt{x}$

$$\frac{f(x+h) - f(x)}{h} = \frac{\sqrt{x+h} - \sqrt{x}}{h} = \frac{1}{\sqrt{x+h} + \sqrt{x}}$$

35. $f(x) = x^2 + 3$

$$\frac{f(x+h) - f(x)}{h} = \frac{\left((x+h)^2 + 3\right) - (x^2 + 3)}{h} = \frac{x^2 + 2xh + h^2 + 3 - x^2 - 3}{h}$$

$$= \frac{2xh + h^2}{h} = 2x + h$$

37. a. $f(x) = x^2$, $f(a) = a^2$, $f(b) = b^2$, $f(a+b) = (a+b)^2$
Since $(a+b)^2 \neq a^2 + b^2$, $f(a+b) \neq f(a) + f(b)$.
b. $f(x) = 3x$, $f(a) = 3a$, $f(b) = 3b$, $f(a+b) = 3(a+b)$
Since $3(a+b) = 3a + 3b$, $f(a+b) = f(a) + f(b)$.

39. Since $2 = f(1)$, we have

$$f(x) = x^3 + cx^2 + 4x - 1$$

$$2 = 1^3 + c \cdot 1^2 + 4 \cdot 1 - 1$$

$$2 = 4 + c$$

$$c = -2$$

41. a. $[-3, 4]$
b. $[-2, 3]$
c. -2
d. $.5$
e. 1
f. -1

43. a. $x \leq 20$
b. Since $-3 < 2$, $f(-3) = (-3)^2 + 2(-3) = 3$
c. Since $-1 < 2$, $f(-1) = (-1)^2 + 2(-1) = -1$
d. Since $2 \leq 2 \leq 20$, $f(2) = 3 \cdot 2 - 5 = 1$
e. Since $2 \leq 7/3 \leq 20$, $f(7/3) = 3(7/3) - 5 = 2$

45. All real numbers   **47.** All real numbers   **49.** $[0,\infty)$

51. All real numbers except 0   **53.** All real numbers

55. If $u^2 - u - 6 = 0$, then $(u-3)(u+2) = 0$, thus $u = 3$ or $u = -2$.
Domain: all real numbers exept 3 and -2.

57. We must have $9 - (x-9)^2 \geq 0$, thus $|x - 9| \leq 3$ and $6 \leq x \leq 12$. Domain: $[6,12]$

59. $f(x) = x^2$ and $g(x) = x^4$ are two of many examples.

61. $f(x) = g(x)$ **63.** $f(x) = g(x)$
$2x^2 + 4x - 3 = x^2 + 12x + 7$ $3x^2 - x + 5 = x^2 - 2x + 26$
$x^2 - 8x - 10 = 0$ $2x^2 + x - 21 = 0$
$x^2 - 8x + 16 = 26$ $(2x + 7)(x - 3) = 0$
$(x - 4)^2 = 26$ $x = -\dfrac{7}{2}$ or $x = 3$
$x = 4 \pm \sqrt{26}$

65. $f(x) = \sqrt{2x - 5}$ **67.** $f(x) = \dfrac{x^3 + 6}{5}$

69. a. $f(x) = 200 + 5x;\ g(x) = 10x$
    b. $f(20) = 200 + 5(20) = 300$ $g(20) = 10(20) = 200$
        $f(35) = 200 + 5(35) = 375$ $g(35) = 10(35) = 350$
        $f(50) = 200 + 5(50) = 450$ $g(50) = 10(50) = 500$
    c. Since $f(50) = 450,$ we must solve $450 = g(x)$, that is, $450 = 10x$ to obtain
       $x = 45$ suits.

71. a.
$$T(x) = \begin{cases} .1x \text{ if } 0 < x \leq 6000 \\ .15(x - 6000) + 600 \text{ if } 6000 < x \leq 27,950 \\ .27(x - 27,950) + 3892.5 \text{ if } 27,950 < x \leq 67,700 \\ .3(x - 67,700) + 14,625 \text{ if } 67,700 < x \leq 141,250 \\ .35(x - 141,250) + 36,690 \text{ if } 141,250 < x \leq 307,050 \\ .386(x - 307,050) + 94,720 \text{ if } x > 307,050 \end{cases}$$

    b. $T(24,000) = .15(24,000 - 6,000) + 600 = \$3300$
        $T(35,000) = .27(35,000 - 27,950) + 3892.5 = \$5796$
      $T(100,000) = .3(100,000 - 67,700) + 14,625 = \$24,315$

73.

$$d(t) = \begin{cases} 3t \text{ if } 0 \le t \le \dfrac{3}{4} \\[3mm] 3\left(\dfrac{3}{4}\right) + 5\left(t - \dfrac{3}{4}\right) \text{ if } \dfrac{3}{4} < t \le 2 \end{cases}$$

At $t = 2$ the distance is $d(2) = 3\left(\tfrac{3}{4}\right) + 5\left(2 - \tfrac{3}{4}\right) = 8.5$ miles. Continuing, therefore

$$d(t) = \begin{cases} 8.5 \text{ if } 2 < t \le \dfrac{5}{2} \\[3mm] 8.5 + 3\left(t - \dfrac{5}{2}\right) \text{ if } \dfrac{5}{2} < t \end{cases}$$

75. fourth side $= x$, $y =$ other length
$xy = 6000$
$$y = \frac{6000}{x}$$
$C = $ Cost of fencing $= 3.75y + 3.75y + 3.75x + 2.00x$
$C = 7.5y + 5.75x$
$$= 7.5\left(\frac{6000}{x}\right) + 5.75x$$
$$= \frac{45,000}{x} + 5.75x$$

77. a.

```
LinReg
 y=ax+b
 a=806.1714286
 b=8177.047619
 r²=.9887488406
 r=.9943585071

■
```

b. $f(6) = \$13,014;\ f(8) = \$14,626;\ f(10) = \$16,239$.
The first figure is lower than actual; others are higher.

c. $f(13) = \$18,657$

3.3 Graphs of Functions

1. This could be the graph of a function, since it passes the vertical line test. $f(3) = 0$.

3. This could not be the graph of a function, since it does not pass the vertical line test.

5.

7.

9.

11.

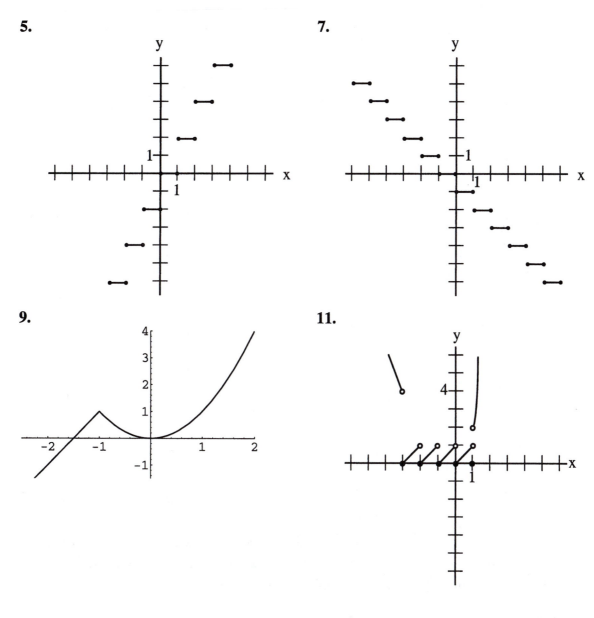

13. a. There are several possibilities, including $p(x) = -[-x], x > 0$.
   b.   **c.**

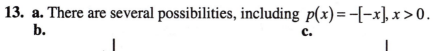

15. $f(x) = \begin{cases} x+2 \text{ if } x \geq 0 \\ -x+2 \text{ if } x < 0 \end{cases}$

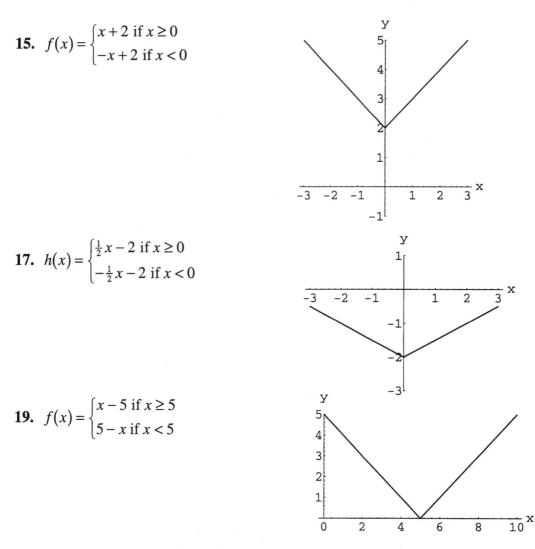

17. $h(x) = \begin{cases} \frac{1}{2}x-2 \text{ if } x \geq 0 \\ -\frac{1}{2}x-2 \text{ if } x < 0 \end{cases}$

19. $f(x) = \begin{cases} x-5 \text{ if } x \geq 5 \\ 5-x \text{ if } x < 5 \end{cases}$

21. Graph the function $f(x) = x^3 - x$ using the window $-2 \leq x \leq 2, -2 \leq y \leq 2$. Use the maximum routine to find the local maximum at the point $(-.57735,.3849)$ and the minimum routine to find the local minimum at the point $(.57735,-.3849)$.

23. Graph the function using the window $-5 \leq x \leq 5, -1 \leq y \leq 1$. Use the maximum routine to find the local maximum at the point $(1,.5)$ and the minimum routine to find the local minimum at the point $(-1,-.5)$.

25. Graph the function using the window $.3 \leq x \leq .9, 2.15 \leq y \leq 2.18$. Use the maximum routine to find the local maximum at the point $(.4367,2.1767)$ and the minimum routine to find the local minimum at the point $(.7633,2.1593)$.

27. The function is increasing on the intervals $(-2.5,0)$ and $(1.7,4)$. It is decreasing on the intervals $(-6,-2.5)$ and $(0,1.7)$.

29. Graph the function in the window $-5 \le x \le 5, -3 \le y \le 3$.

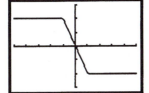

The function is constant on the intervals $(-\infty, -1)$ and $(1, \infty)$. The function is decreasing on the interval $(-1, 1)$.

31. Graph the function in the window $-10 \le x \le 5, -120 \le y \le 20$

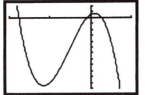

Using the minimum and maximum routines, we see that the function is decreasing on the intervals $(-\infty, -5.7936)$ and $(.4603, \infty)$. It is increasing on the interval $(-5.7936, .4603)$.

33. Graph the function in the window $-2 \le x \le 4, -4 \le y \le 1$.

Using the minimum and maximum routines, we see that the function is decreasing on the intervals $(-\infty, 0)$ and $(.867, 2.883)$. It is increasing on the intervals $(0, .867)$ and $(2.883, \infty)$.

35. a. $2x + 2z = 100$
    b. From the above equation, $z = 50 - x$, hence area $= f(x) = xz = x(50 - x)$.
    c. Graph the function in the window $0 \le x \le 50, 0 \le y \le 1000$ and use the maximum routine to find the local maximum at the point $(25, 625)$. Then the maximum area occurs when $x = 25$ inches, $z = 50 - x = 25$ inches.

37. a. The surface area $S = \text{top} + \text{bottom} + \text{four sides} = 2x^2 + 4xh$.
    b. Since Volume $=$ length $\times$ width $\times$ height, $867 = x^2 h$, hence $h = 867 / x^2$.
    c. Using parts **a** and **b**, we can write

$$S = 2x^2 + 4xh = 2x^2 + 4x\left(\frac{867}{x^2}\right) = 2x^2 + \frac{3468}{x}$$

d. Graph the function in the window $0 \le x \le 25, 0 \le y \le 1000$ and use the minimum routine to find the local minimum at the point $(9.5354, 545.545)$. Therefore the minimum volume occurs when $x = 9.5354$ inches. Then the height $h = 867 / (9.5354)^2 = 9.5354$ inches also.

39. a — iv
   b — i
   c — v
   d — iii
   e — ii

41. A possible graph is given below. The domain is actually $x \ge 0$, since one can buy any non-negative amount of gas, including none at all. The range then is $y \ge 0$.

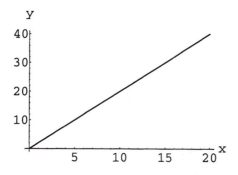

43. A possible graph is given below.

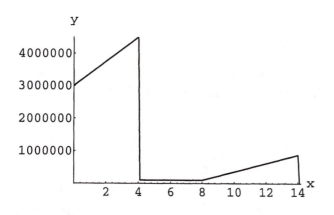

45. a. The point $(k, f(k))$ is on the function directly above k.

   b. The point $(-k, f(-k))$ is on the function directly above $-k$, which is on the x-axis opposite k..

   c. The point $(k, -f(k))$ is the mirror image of the point $(k, f(k))$ in the x-axis.

47. a. 9%   **b.** 16%   **c.** 8%   **d.** lowest: 1993, highest: 1990.
   e. between 1978 and 1981 the graph has the steepest change.

49. One of many possible graphs is shown.

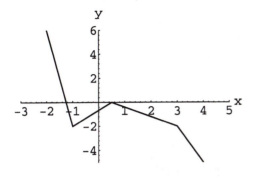

51. Graph $x = t^3 + 5t^2 - 4t - 5, y = t$ in the window
$-10 \le t \le 10, -10 \le x \le 30, -8 \le y \le 3$.

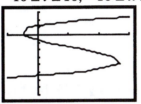

53. First, solve for x.

$$xy^2 + xy + x = y^3 - 2y^2 + 4$$

$$x(y^2 + y + 1) = y^3 - 2y^2 + 4$$

$$x = \frac{y^3 - 2y^2 + 4}{y^2 + y + 1}$$

Now graph $x = \frac{t^3 - 2t^2 + 4}{t^2 + t + 1}, y = t$
in the window $-10 \le t \le 10, -10 \le x \le 10, -10 \le y \le 10$.

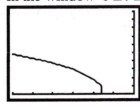

55. First, solve for x to obtain $x = \sqrt{y} - y^2 - 8$.

Now graph $x = \sqrt{t} - t^2 - 8, y = t$
in the window $0 \le t \le 10, -30 \le x \le 0, 0 \le y \le 10$.

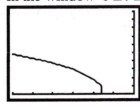

57. Use the window $-5 \le x \le 45, -65 \le y \le 65$.

59. The whole graph can be seen in the window $-2 \le x \le 32, -10 \le y \le 75$, but to see details near the origin use the window $-2 \le x \le 5, -10 \le y \le 10$.

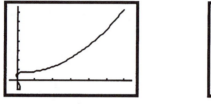 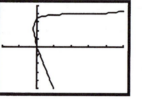

61. The whole graph can be seen in the window $-2 \le x \le 32, -10 \le y \le 75$, but to see details near the origin use the window $-2 \le x \le 5, -10 \le y \le 10$.

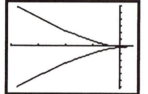 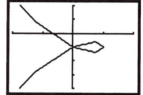

63. a.   **b.**

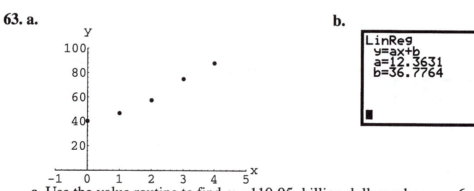

c. Use the value routine to find $y = 110.95$ billion dollars when $x = 6$

d. Use the intersection routine with the graph $y = 130$ to obtain $x = 7.54$, which corresponds to 2005 or 2006.

65. The graph can be seen in the window $-12 \le x \le 25, -20 \le y \le 20$. If a small value of T-step is used, the graphing is slow enough to count the 7 self-intersections.

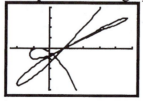

67. This equation does not define y as a function of x, however, solve for y to obtain

$$|y| = 1 - |x|$$

$y = 1 - |x|$ or $y = -(1 - |x|)$
Graph this in the window
$-1 \le x \le 1, -1 \le y \le 1$ to obtain:

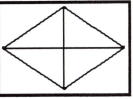

3.3A Graph Reading

1. Here is a possible calculator graph, using the standard window:

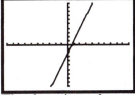

The domain and range appear to be all real numbers.

3. Here is a possible calculator graph, using the standard window:

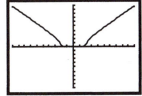

The domain appears to be $(-\infty, -2] \cup [2, \infty)$; the range appears to be the nonnegative numbers $[0, \infty)$.

5. $-3 \le x \le 5$ or $[-3, 5]$

7. $g(2t) = g(2 \cdot 1.5) = g(3) = 4$

9. $g(y + 1.5) = g(2 + 1.5) = g(3.5) = 3.5$

11. $g(y) + 1.5 = g(2) + 1.5 = 3 + 1.5 = 4.5$

13. $g(3v - 1.5) = g(3 \cdot 1.5 - 1.5) = g(3) = 4$

15. 1, 5

17. $-8 \le x \le 9$ or $[-8, 9]$

19. $-7, -3, 0, 3, 7$

21. 0, 1, 3 are some of the possibilities.

23. Since $f(2) = 0$, set $x + 1 = 2$. $x = 1$

25. The federal debt f is increasing during the period; however, the gross domestic product g is increasing faster, hence f/g is decreasing.

27. $T(4) = 39°$; $T(7 + 11) = T(18) = 75°$

29. $12 \le x \le 20$ is a possibility.

31. domain of f: $-6 \le x \le 7$;

domain of g: $-8 \le x \le 9$

33. -1.6 and 0

35. 3

37. $-2 \le x \le -1, 3 \le x \le 7$

39. If 5000 hammers are manufactured, expenses are \$18,000 and income is \$5,000.
Profit $= 5000 - 18000 = -\$13,000$.

41. 12,000 hammers (or slightly more).

43. $h(x) = f(x) - 1 = \sqrt{4 - x^2} + 1 - 1 = \sqrt{4 - x^2}$

45. $h(x) = g(x) - f(x) = x - x^3$

47. During the next 5 minutes she rested then continued her run at her original pace for 10 minutes. She then ran home at this pace, a 25 minute run.

3.4 Graphs and Transformations

1. H   **3.** F   **5.** K   **7.** C

9.   **11.**

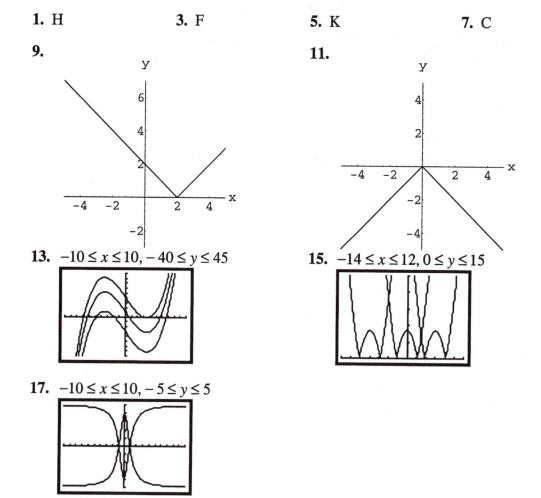

13. $-10 \le x \le 10, -40 \le y \le 45$ **15.** $-14 \le x \le 12, 0 \le y \le 15$

17. $-10 \le x \le 10, -5 \le y \le 5$

19. Shift the graph of f 3 units horizontally to the right, then 2 units vertically up.

21. Reflect the graph of f with respect to the x-axis, stretch vertically by a factor of $\frac{1}{2}$, then shift vertically 6 units down.

23. First note that g can be written as

$$g(x) = \frac{-2 \cdot 3(x-2)}{(x-2)^2 + 10} = -2f(x-2)$$

Therefore to obtain the graph of g, shift the graph of f horizontally 2 units to the right, then reflect with respect to the x-axis, then stretch vertically by a factor of 2.

25. $g(x) = (x+5)^2 + 6$ **27.** $g(x) = 2\sqrt{x-6} + 3$

29. a. $g(x) = x^2 + 3x + 2$
   b. $f(x) = x^2 + 3x$

$$\frac{f(x+h) - f(x)}{h} = \frac{\left((x+h)^2 + 3(x+h)\right) - \left(x^2 + 3x\right)}{h}$$

$$= \frac{x^2 + 2xh + h^2 + 3x + 3h - x^2 - 3x}{h}$$

$$= \frac{2xh + h^2 + 3h}{h}$$

$$= 2x + h + 3$$

$$\frac{g(x+h) - g(x)}{h} = \frac{\left((x+h)^2 + 3(x+h) + 2\right) - \left(x^2 + 3x + 2\right)}{h}$$

$$= \frac{x^2 + 2xh + h^2 + 3x + 3h + 2 - x^2 - 3x - 2}{h}$$

$$= \frac{2xh + h^2 + 3h}{h}$$

$$= 2x + h + 3$$

Thus, the difference quotients are identical.

31.   **33.**

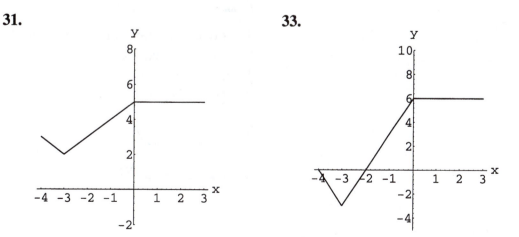

35.

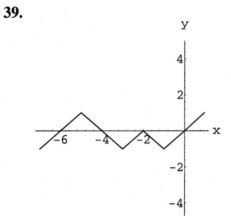

37.

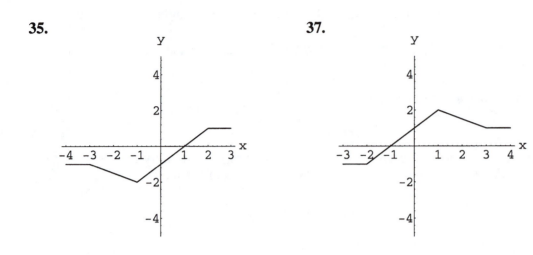

39.

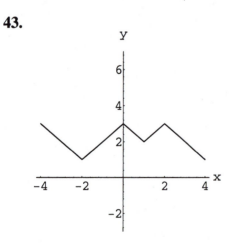

41.

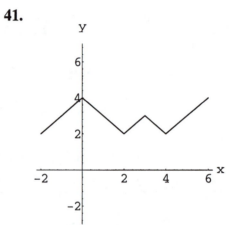

43.

45.

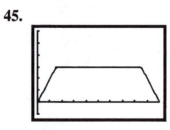

To reflect the graph in the horizontal line $y = 6$, set $g(x) = 12 - f(x)$. Then $g(x) = |x - 3| + |x - 17| - 8$ has the required appearance.

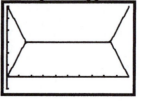

47.

49.

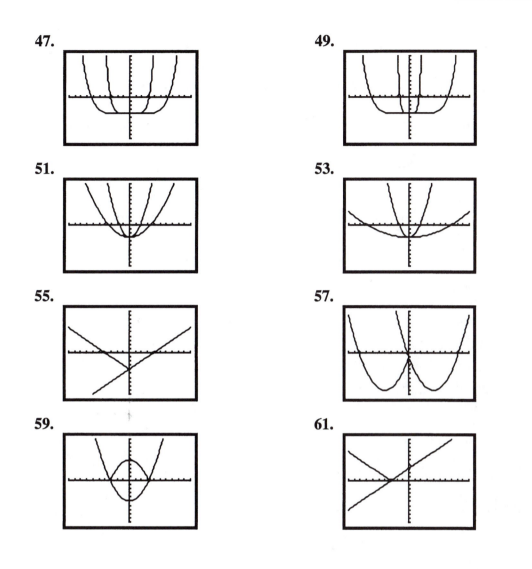

51.

53.

55.

57.

59.

61.

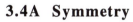
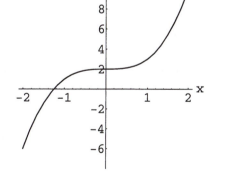

3.4A Symmetry

1.

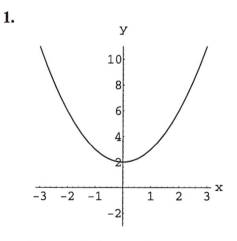

The graph has *y*-axis symmetry.

3.

The graph has none of the three symmetries.

5. $f(-x) = 4(-x) = -4x = -f(x).$ Therefore, the function is odd.

7. $f(-x) = (-x)^2 - |-x| = x^2 - |x| = f(x).$ Therefore, the function is even.

9. $k(-t) = (-t)^4 - 6(-t)^2 + 5 = t^4 - 6t^2 + 5 = k(t).$ Therefore, the function is even.

11. $f(-t) = \sqrt{(-t)^2 - 5} = \sqrt{t^2 - 5} = f(t).$ Therefore, the function is even.

13. $f(-x) = \dfrac{(-x)^2 + 2}{(-x) - 7} = -\dfrac{x^2 + 2}{x + 7} \neq f(x).$ Therefore, the function is not even.

  $-f(x) = -\dfrac{x^2 + 2}{x - 7} \neq f(-x).$ Therefore, the function is not odd.
  Neither even nor odd.

15. Replace y with $-y$:

  $x^2 - 6x + (-y)^2 + 8 = 0$

  $x^2 - 6x + y^2 + 8 = 0$

  Equation is unchanged, graph has x-axis symmetry.

17. Replace y with $-y$:

  $x^2 - 2x + (-y)^2 + 2(-y) = 2$

  $x^2 - 2x + y^2 - 2y = 2$

  Equation changed, graph has no y-axis symmetry.

19. The graph has origin symmetry.

21. The graph has origin symmetry.

23. The graph has y-axis symmetry.

25.

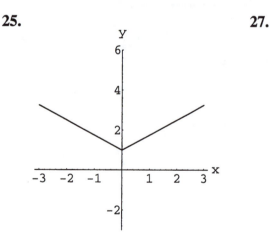

27.

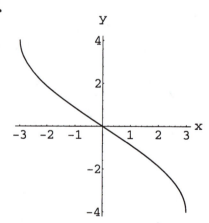

29. a.   **b.**

31. a.   **b.**

c. One possibility is shown below:

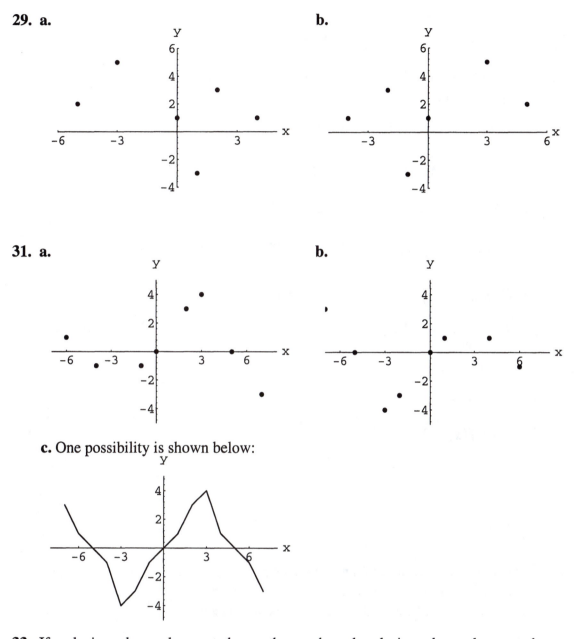

33. If replacing x by $-x$ does not change the graph, and replacing y by $-y$ does not change the graph, then doing both does not change the graph. Thus x-axis plus y-axis symmetry guarantees origin symmetry.
If replacing x by $-x$ does not change the graph, and replacing y by $-y$ DOES change the graph, then doing both must change the graph.
Thus y-axis symmetry and origin symmetry without x-axis symmetry is impossible.
The argument is analogous for x-axis and origin symmetry.

3.5 Operations on Functions

1. $(f+g)(x) = f(x) + g(x) = -3x + 2 + x^3 = x^3 - 3x + 2$
    $(f-g)(x) = f(x) - g(x) = -3x + 2 - x^3 = -x^3 - 3x + 2$
    $(g-f)(x) = g(x) - f(x) = x^3 - (-3x + 2) = x^3 + 3x - 2$

3. $(f+g)(x) = f(x) + g(x) = \dfrac{1}{x} + x^2 + 2x - 5$
    $(f-g)(x) = f(x) - g(x) = \dfrac{1}{x} - (x^2 + 2x - 5) = \dfrac{1}{x} - x^2 - 2x + 5$
    $(g-f)(x) = g(x) - f(x) = (x^2 + 2x - 5) - \dfrac{1}{x} = x^2 + 2x - 5 - \dfrac{1}{x}$

5. $(fg)(x) = f(x)g(x) = (-3x + 2)x^3 = -3x^4 + 2x^3$
    $\left(\dfrac{f}{g}\right)(x) = \dfrac{f(x)}{g(x)} = \dfrac{-3x + 2}{x^3}$
    $\left(\dfrac{g}{f}\right)(x) = \dfrac{g(x)}{f(x)} = \dfrac{x^3}{-3x + 2}$

7. $(fg)(x) = f(x)g(x) = (x^2 - 3)\sqrt{x - 3}$
    $\left(\dfrac{f}{g}\right)(x) = \dfrac{f(x)}{g(x)} = \dfrac{x^2 - 3}{\sqrt{x - 3}} = \dfrac{(x^2 - 3)\sqrt{x - 3}}{x - 3}$
    $\left(\dfrac{g}{f}\right)(x) = \dfrac{g(x)}{f(x)} = \dfrac{\sqrt{x - 3}}{x^2 - 3}$

9. The domain of f consists of all real numbers.
    The domain of g consists of all real $x \neq 0$. $g(x)$ is never 0.
    Therefore the domain of both fg and f/g is all real $x \neq 0$.

11. The domain of f is $-2 \leq x \leq 2$.
    The domain of g is $x \geq -4/3$.
    Therefore the domain of fg is $-4/3 \leq x \leq 2$.
    $g\left(-\frac{4}{3}\right) = 0$, therefore the domain of f/g is $-4/3 < x \leq 2$.

13.
$$f(x) = 1 + x$$
$$f(0) = 1 + 0$$
$$= 1$$
$$g(t) = t^2 - t$$
$$g(f(0)) = g(1)$$
$$= 1^2 - 1$$
$$= 0$$

15.
$$f(x) = 1 + x$$
$$f(2) = 1 + 2$$
$$= 3$$
$$g(t) = t^2 - t$$
$$g(f(2) + 3) = g(3 + 3)$$
$$= g(6)$$
$$= 6^2 - 6$$
$$= 30$$

17.
$$f(x) = 3x - 2$$
$$f(3) = 3 \cdot 3 - 2$$
$$= 7$$
$$g(x) = x^2$$
$$(g \circ f)(3) = g(f(3))$$
$$= g(7)$$
$$= 7^2$$
$$= 49$$

$$g(1) = 1^2$$
$$= 1$$
$$(f \circ g)(1) = f(g(1))$$
$$= f(1)$$
$$= 3 \cdot 1 - 2$$
$$= 1$$

$$f(0) = 3 \cdot 0 - 2$$
$$= -2$$
$$(f \circ f)(0) = f(f(0))$$
$$= f(-2)$$
$$= 3(-2) - 2$$
$$= -8$$

19.
$$f(x) = x$$
$$f(3) = 3$$
$$g(x) = -3$$
$$(g \circ f)(3) = g(f(3))$$
$$= g(3)$$
$$= -3$$

$$g(1) = -3$$
$$(f \circ g)(1) = f(g(1))$$
$$= f(-3)$$
$$= -3$$

$$f(0) = 0$$
$$(f \circ f)(0) = f(f(0))$$
$$= f(0)$$
$$= 0$$

21.
$$f(x) = x^2$$
$$g(x) = x + 3$$
$$(f \circ g)(x) = f(g(x))$$
$$= (g(x))^2$$
$$= (x + 3)^2$$
Domain: all real numbers

$$(g \circ f)(x) = g(f(x))$$
$$= f(x) + 3$$
$$= x^2 + 3$$
Domain: all real numbers

23.

$$f(x) = \frac{1}{x}$$

$$g(x) = \sqrt{x}$$

$$(f \circ g)(x) = f(g(x))$$

$$= \frac{1}{g(x)}$$

$$= \frac{1}{\sqrt{x}} \text{ or } \frac{\sqrt{x}}{x}$$

Domain: all $x > 0$

$$(g \circ f)(x) = g(f(x))$$

$$= \sqrt{f(x)}$$

$$= \sqrt{\frac{1}{x}} \text{ or } \frac{\sqrt{x}}{x}$$

Domain: all $x > 0$

25. $f(x) = x^3$

$$(ff)(x) = f(x)f(x) = x^3 \cdot x^3 = x^6$$

$$(f \circ f)(x) = f(f(x)) = (f(x))^3 = (x^3)^3 = x^9$$

27. $f(x) = \frac{1}{x}$

$$(ff)(x) = f(x)f(x) = \frac{1}{x} \cdot \frac{1}{x} = \frac{1}{x^2}$$

$$(f \circ f)(x) = f(f(x)) = \frac{1}{f(x)} = \frac{1}{1/x} = x$$

29. $(f \circ g)(x) = f(g(x))$

$$= 9g(x) + 2$$

$$= 9\left(\frac{x-2}{9}\right) + 2$$

$$= x - 2 + 2$$

$$= x$$

$$(g \circ f)(x) = g(f(x))$$

$$= \frac{f(x) - 2}{9}$$

$$= \frac{9x + 2 - 2}{9}$$

$$= \frac{9x}{9}$$

$$= x$$

31. $(f \circ g)(x) = f(g(x))$

$$= \sqrt[3]{g(x)} + 2$$

$$= \sqrt[3]{(x-2)^3} + 2$$

$$= x - 2 + 2$$

$$= x$$

$$(g \circ f)(x) = g(f(x))$$

$$= (f(x) - 2)^3$$

$$= \left(\sqrt[3]{x} + 2 - 2\right)^3$$

$$= \left(\sqrt[3]{x}\right)^3$$

$$= x$$

33.

x	$f(x)$	$g(x)=f(f(x))$
-4	-3	-1
-3	-1	1/2
-2	0	1
-1	1/2	5/4
0	1	3/2
1	3/2	2
2	1	3/2
3	-2	0
4	-2	0

35.

x	$(g \circ f)(x)$
1	4
2	2
3	5
4	4
5	4

37.

x	$(f \circ f)(x)$
1	1
2	3
3	3
4	5
5	1

39. Let $g(x) = x^2 + 2$, $h(t) = \sqrt[3]{t}$, then $(h \circ g)(x) = h(g(x)) = \sqrt[3]{g(x)} = \sqrt[3]{x^2 + 2}$.

41. Let $g(x) = 7x^3 - 10 + 17$, $f(t) = t^7$,
then $(f \circ g)(x) = f(g(x)) = (g(x))^7 = (7x^3 - 10x + 17)^7$.

43. Let $g(x) = 3x^2 + 5x - 7$, $h(t) = \dfrac{1}{t}$, then $(h \circ g)(x) = h(g(x)) = \dfrac{1}{g(x)} = \dfrac{1}{3x^2 + 5x - 7}$.

45. Let $h(x) = x + 2$, $k(t) = (t-1)^2$,
then $(k \circ h)(x) = k(h(x)) = (h(x) - 1)^2 = (x + 2 - 1)^2 = x^2 + 2x + 1$.

47. $(f \circ g)(x) = (g(x))^3 = (\sqrt{x})^3$ Defined, domain $x \geq 0$

$(g \circ f)(x) = \sqrt{f(x)} = \sqrt{x^3}$ Defined, domain $x \geq 0$

49. $(f \circ g)(x) = \sqrt{g(x) + 10} = \sqrt{5x + 10}$ Defined, domain $x \geq -2$

$(g \circ f)(x) = 5f(x) = 5\sqrt{x + 10}$ Defined, domain $x \geq -10$

51. $f(x) = 2x^3 + 5x - 1$
a. $f(x^2) = 2(x^2)^3 + 5x^2 - 1 = 2x^6 + 5x^2 - 1$
b. $(f(x))^2 = (2x^3 + 5x - 1)^2 = 4x^6 + 20x^4 - 4x^3 + 25x^2 - 10x + 1$
c. The answers are not the same. In general, $f(x^2) \neq (f(x))^2$.

53. $(f \circ g)(x) = f(g(x)) = (x-2)^5 - (x-2)^3 - (x-2);$
$(g \circ f)(x) = g(f(x)) = x^5 - x^3 - x - 2$

Graphing in the standard window shows that the two functions are not the same.

55. $f(x) = x + 3 \quad g(x) = x^2 + 1$

$(g \circ f)(x) = g(f(x)) = (f(x))^2 + 1 = (x+3)^2 + 1 = x^2 + 6x + 10$

$\dfrac{(g \circ f)(x+h) - (g \circ f)(x)}{h} = \dfrac{\left((x+h)^2 + 6(x+h) + 10\right) - \left(x^2 + 6x + 10\right)}{h}$

$= \dfrac{x^2 + 2xh + h^2 + 6x + 6h + 10 - x^2 - 6x - 10}{h}$

$= \dfrac{2xh + h^2 + 6h}{h} = 2x + h + 6$

57. $f(x) = x + 1, \quad g(x) = \dfrac{1}{x}, \quad (g \circ f)(x) = \dfrac{1}{f(x)} = \dfrac{1}{x+1}$

$\dfrac{(g \circ f)(x+h) - (g \circ f)(x)}{h} = \dfrac{\dfrac{1}{x+h+1} - \dfrac{1}{x+1}}{h}$

$= \dfrac{(x+1) - (x+h+1)}{h(x+1)(x+h+1)}$

$= \dfrac{-h}{h(x+1)(x+h+1)}$

$= \dfrac{-1}{(x+1)(x+h+1)}$

59. a. $A = \pi\left(\dfrac{18}{2t+3}\right)^2$

After 1 day, $t = 1440$ min, $A = \pi\left(\dfrac{18}{2 \cdot 1440 + 3}\right)^2 = 1.2 \times 10^{-4}$ sq. in.

After 1 week, $t = 10{,}080$ min, $A = \pi\left(\dfrac{18}{2 \cdot 10{,}080 + 3}\right)^2 = 2.5 \times 10^{-6}$ sq. in.

After 1 month $= 30$ days, $t = 43{,}200$ min,

$$A = \pi\left(\dfrac{18}{2 \cdot 43{,}200 + 3}\right)^2 = 1.4 \times 10^{-7} \text{ sq. in.}$$

b. In this model, the puddle never totally evaporates. This is not realistic, however, the model may be valid over a certain time period.

61. a. $A = \dfrac{\pi}{4} d^2 = \dfrac{\pi}{4}\left(6 - \dfrac{50}{t^2 + 10}\right)^2$

b. At $t = 0$, $A = \dfrac{\pi}{4}\left(6 - \dfrac{50}{0^2 + 10}\right)^2 = .7854$ sq. in.

At $t = 8$, $A = \dfrac{\pi}{4}\left(6 - \dfrac{50}{8^2 + 10}\right)^2 = 22.265$ sq. in.

When $A = 25$, we must solve

$$25 = \dfrac{\pi}{4}\left(6 - \dfrac{50}{t^2 + 10}\right)^2$$

Using the calculator intersection routine, we obtain $t = 11.4$ weeks.

63. Since $r = 4t$ and $V = 4\pi r^3 / 3$, we have

$$V(t) = \dfrac{4\pi(4t)^3}{3} = \dfrac{256\pi t^3}{3}$$

At $t = 4$,

$$V(4) = \dfrac{256\pi 4^3}{3} = 17{,}157 \text{ cm}^3$$

65. From similar triangles, we have

$$\dfrac{s}{6} = \dfrac{s + d}{15}$$

$$15s = 6s + 6d$$

$$9s = 6d$$

$$s = 2d/3$$

Using distance $= $ rate $\times$ time, $d = 5t$, hence

$$s(t) = 2(5t)/3$$

$$s(t) = 10t/3$$

67. The function $f(x) = \dfrac{x - 3}{x - 2}$, $x \neq 1$ is an example. Note that the domain of this function excludes 1 and 2. In fact, all functions of form $f(x) = \dfrac{x + a}{bx - 2}$, where $ab = -3$, satisfy the condition, if the necessary values are excluded from the domain.

3.6 Rates of Change

1. average speed $= \dfrac{\text{distance traveled}}{\text{time interval}}$

    a. $\dfrac{d(10)-d(0)}{10-0} = \dfrac{140-0}{10-0} = 14$ ft per sec

    b. $\dfrac{d(20)-d(10)}{20-10} = \dfrac{680-140}{20-10} = 54$ ft per sec

    c. $\dfrac{d(30)-d(20)}{30-20} = \dfrac{1800-680}{30-20} = 112$ ft per sec

    d. $\dfrac{d(30)-d(15)}{30-15} = \dfrac{1800-400}{30-15} = 93.3$ ft per sec

3. **a.** average rate of change $= \dfrac{f(60)-f(20)}{60-20} = \dfrac{845-761.58}{40} \approx 2.086$
    The population was increasing at the rate of 2086 people per year.

    b. average rate of change $= \dfrac{f(100)-f(90)}{100-90} = \dfrac{476-503.56}{10} \approx -2.756$
    The population was decreasing at the rate of 2756 people per year.

5. **a.** average rate of change $= \dfrac{f(1985)-f(1980)}{1985-1980} = \dfrac{44,979-46,208}{5} = -245.8$
    Decreasing at the rate of 245,800 per year.

    b. average rate of change $= \dfrac{f(1995)-f(1985)}{1995-1985} = \dfrac{50,503-44,979}{10} = 552.4$
    Increasing at the rate of 552,400 per year.

    c. average rate of change $= \dfrac{f(2005)-f(1995)}{2005-1995} = \dfrac{53,465-50,503}{10} = 296.2$
    Increasing at the rate of 296,200 per year.

    d. average rate of change $= \dfrac{f(2010)-f(2005)}{2010-2005} = \dfrac{53,017-53,465}{5} = -89.6$
    Decreasing at the rate of 89,600 per year.
    e. Fastest: 1985 - 1995. Slowest: 1995 - 2005.
    f. Fastest: 1980 - 1985. Slowest: 2005 - 2010.

7. **a.** average rate of change $= \dfrac{f(1995)-f(1991)}{1995-1991} = \dfrac{87.2-45.3}{4} = 10.475$
    Increasing at the rate of 10.475 billion shares per year.

    b. average rate of change $= \dfrac{f(1999)-f(1995)}{1999-1995} = \dfrac{203.9-87.2}{4} = 29.175$
    Increasing at the rate of 29.175 billion shares per year.

    c. average rate of change $= \dfrac{f(2001)-f(1999)}{2001-1999} = \dfrac{307.5-203.9}{2} = 51.8$
    Increasing at the rate of 51.8 billion shares per year.

d. average rate of change $= \dfrac{f(2001) - f(1991)}{2001 - 1991} = \dfrac{307.5 - 45.3}{10} = 26.22$

Increasing at the rate of 26.22 billion shares per year.

e. 1999 - 2001

9. a. average rate of change $= \dfrac{f(20) - f(10)}{20 - 10} = \dfrac{100 - 50}{10} = 5$: $5000 per page

b. average rate of change $= \dfrac{f(60) - f(20)}{60 - 20} = \dfrac{175 - 100}{40} = 1.875$: $1875 per page

c. average rate of change $= \dfrac{f(100) - f(60)}{100 - 60} = \dfrac{200 - 175}{40} = .625$: $625 per page

d. average rate of change $= \dfrac{f(100) - f(0)}{100 - 0} = \dfrac{200 - 30}{100} = 1.7$: $1700 per page

e. The average rate of change between 70 and 80 pages is:

$\dfrac{f(80) - f(70)}{80 - 70} = \dfrac{185 - 175}{10} = 1$: $1000 per page.

Therefore, it is only worthwhile to buy more than 70 pages of ads if the cost of a page is less than $1000.

11. $\dfrac{f(2) - f(0)}{2 - 0} = \dfrac{(-2) - 2}{2} = -2$

13. $\dfrac{f(3) - f(-1)}{3 - (-1)} = \dfrac{0 - 4}{4} = -1$

15. $\dfrac{f(2) - f(1)}{2 - 1} = \dfrac{3 - 1.4142}{1} = 1.5858$

17. $\dfrac{f(x + h) - f(x)}{h} = \dfrac{(x + h + 5) - (x + 5)}{h}$

$\qquad\qquad\qquad\quad = \dfrac{x + h + 5 - x - 5}{h}$

$\qquad\qquad\qquad\quad = \dfrac{h}{h} = 1$

19. $\dfrac{f(x + h) - f(x)}{h} = \dfrac{\left((x + h)^2 + 3\right) - \left(x^2 + 3\right)}{h}$

$\qquad\qquad\qquad = \dfrac{x^2 + 2xh + h^2 + 3 - x^2 - 3}{h}$

$\qquad\qquad\qquad = \dfrac{2xh + h^2}{h}$

$\qquad\qquad\qquad = 2x + h$

21. $\dfrac{f(t + h) - f(t)}{h} = \dfrac{\left(160,000 - 8000(t + h) + (t + h)^2\right) - \left(160,000 - 8000t + t^2\right)}{h}$

$\qquad\qquad\qquad = \dfrac{160,000 - 8000t - 8000h + t^2 + 2th + h^2 - 160,000 + 8000t - t^2}{h}$

$\qquad\qquad\qquad = \dfrac{-8000h + 2th + h^2}{h}$

$\qquad\qquad\qquad = -8000 + 2t + h$

23. $\dfrac{A(r+h)-A(r)}{h} = \dfrac{\pi(r+h)^2 - \pi r^2}{h}$

$\qquad\qquad\qquad = \dfrac{\pi r^2 + 2\pi rh + \pi h^2 - \pi r^2}{h}$

$\qquad\qquad\qquad = \dfrac{2\pi rh + \pi h^2}{h}$

$\qquad\qquad\qquad = 2\pi r + \pi h$

25. From Exercise **21**, average rate of change between t and $t+h$ is $-8000 + 2t + h$
   a. Here $t=10$, $h=.1$, thus, average rate $= -8000 + 2(10)+.1 = -7979.9$
       average rate $= 7979.9$ gallons per minute draining
   b. Here $t=10$, $h=.01$, thus, average rate $= -8000 + 2(10)+.01 = -7979.99$
       average rate $= 7979.99$ gallons per minute draining
   c. 7980 gallons per minute draining

27. From Exercise **23**, average rate of change between r and $r+h$ is $2\pi r + \pi h$
   a. Here $r=3$, $h=.5$, thus, average rate $= 2\pi(3) + \pi(.5) = 6.5\pi$
   b. Here $r=3$, $h=.2$, thus, average rate $= 2\pi(3) + \pi(.2) = 6.2\pi$
   c. Here $r=3$, $h=.1$, thus, average rate $= 2\pi(3) + \pi(.1) = 6.1\pi$
   d. $2\pi r = 2\pi(3) = 6\pi$
   e. The two are identical.

29. average speed $= \dfrac{\text{distance traveled}}{\text{time interval}}$

   a. Car C: average speed $= \dfrac{1000}{16} = 62.5$ ft per sec

       Car D: average speed $= \dfrac{1200}{16} = 75$ ft per sec

   b. Car C: average speed $\approx \dfrac{800-200}{10-2} \approx 75$ ft per sec

       Car D: average speed $= \dfrac{d-80}{t-4} = 75$. if $t=9$, $d=480$; the average speed is then
       80 ft per sec on this interval, approximately that of car C.
   c. In this interval, the slopes of the secant lines (average rate of change) are
       approximately:

       Car C: $\dfrac{800-500}{10-4} = 50$ ft per sec

       Car D: $\dfrac{600-90}{10-4} = 85$ ft per sec

   Thus, the statement is justified.

31. a. day 0 to day 50.
   b. day 0 to day 100.

c. average growth rate $= \dfrac{900 - 2200}{100 - 50} = -26$

The number of chipmunks is decreasing at the rate of 26 per day.

d. $\dfrac{2200 - 2100}{50 - 45} = 20$ chipmunks per day.

$\dfrac{2100 - 2200}{55 - 50} = -20$ chipmunks per day.

Since there is no change in population, the average growth rate from day 49 to day 51 is zero.

33. a. Entering this data into the calculator gives:

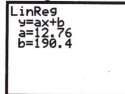

b. The average rate of change is the slope of the line: 12.76.

c. 1980-1990: $\dfrac{f(10) - f(0)}{10 - 0} = \dfrac{319 - 191}{10} = 12.8;$

1990-2000: $\dfrac{f(20) - f(10)}{20 - 10} = \dfrac{454 - 319}{10} = 13.5$

These rates are somewhat larger than the overall rate found in **b.**

d. Solve $12.76x + 190.4 = 550$ to obtain $x = 28.2$, which would correspond to 2008 or early 2009.

3.7 Inverse Functions

1.

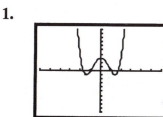

This fails the horizontal line test.
The function is not one-to-one.

3.

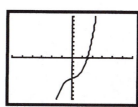

This passes the horizontal line test.
The function is one-to-one.

5.

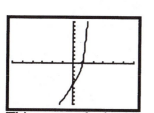

This passes the horizontal line test.
The function is one-to-one.

7.

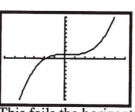

This fails the horizontal line test.
The function is not one-to-one.

9. Set $y = -x$

$$x = -y$$

$$g(y) = -y$$

Thus $g(x) = -x$

11. Set $y = 5x - 4$

$$y + 4 = 5x$$

$$x = \frac{y+4}{5}$$

$$g(y) = \frac{y+4}{5}$$

$$g(x) = \frac{x+4}{5}$$

13. Set $y = 5 - 2x^3$

$$y - 5 = -2x^3$$

$$\frac{5-y}{2} = x^3$$

$$x = \sqrt[3]{\frac{5-y}{2}}$$

$$g(y) = \sqrt[3]{\frac{5-y}{2}}$$

$$g(x) = \sqrt[3]{\frac{5-x}{2}}$$

15. Set $y = \sqrt{4x - 7} \quad (y \geq 0)$

$$y^2 = 4x - 7$$

$$y^2 + 7 = 4x$$

$$x = \frac{y^2 + 7}{4}$$

$$g(y) = \frac{y^2 + 7}{4} \quad (y \geq 0)$$

$$g(x) = \frac{x^2 + 7}{4} \quad (x \geq 0)$$

17. Set $y = \frac{1}{x}$

$$x = \frac{1}{y}$$

$$g(y) = \frac{1}{y}$$

$$g(x) = \frac{1}{x}$$

19. Set $y = \frac{1}{2x + 1}$

$$2xy + y = 1$$

$$2xy = 1 - y$$

$$x = \frac{1-y}{2y}$$

$$g(y) = \frac{1-y}{2y}$$

$$g(x) = \frac{1-x}{2x}$$

21. Set $y = \frac{x^3 - 1}{x^3 + 5}$

$$x^3 y + 5y = x^3 - 1$$

$$5y + 1 = x^3 - x^3 y$$

$$5y + 1 = x^3(1 - y)$$

$$x^3 = \frac{5y + 1}{1 - y}$$

$$x = \sqrt[3]{\frac{5y + 1}{1 - y}}$$

$$g(y) = \sqrt[3]{\frac{5y + 1}{1 - y}}$$

$$g(x) = \sqrt[3]{\frac{5x + 1}{1 - x}}$$

23. $g(f(x)) = g(x+1) = x+1-1 = x$
$f(g(x)) = f(x-1) = x-1+1 = x$
Thus g is the inverse of f.

25. $g(f(x)) = g\left(\dfrac{1}{x+1}\right) = \dfrac{1-\dfrac{1}{x+1}}{\dfrac{1}{x+1}} = \dfrac{x+1-1}{1} = x$

$f(g(x)) = f\left(\dfrac{1-x}{x}\right) = \dfrac{1}{\dfrac{1-x}{x}+1} = \dfrac{x}{1-x+x} = x$

Thus g is the inverse of f.

27. $g(f(x)) = g(x^5) = \sqrt[5]{x^5} = x$
$f(g(x)) = f(\sqrt[5]{x}) = (\sqrt[5]{x})^5 = x$
Thus g is the inverse of f.

29. Calculate $f(f(x))$:

$$f(f(x)) = \frac{2f(x)+1}{3f(x)-2} = \frac{2\cdot\dfrac{2x+1}{3x-2}+1}{3\cdot\dfrac{2x+1}{3x-2}-2} = \frac{4x+2+3x-2}{6x+3-6x+4} = \frac{7x}{7} = x$$

Since $f(f(x)) = x$, f is the inverse of itself.

31.

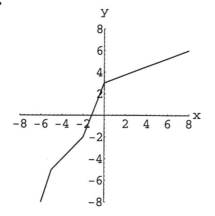

33. Sketch $x = \sqrt{t+3}$, $y = t$, $-3 \le t \le 13$
in parametric mode:

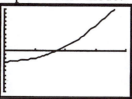

35. Sketch $x = .3t^5 + 2$, $y = t$, $-3 \le t \le 3$
in parametric mode:

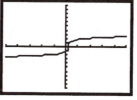

37. Sketch $x = \sqrt[5]{x^3 + x - 2}$, $y = t$, $-10 \le t \le 10$ in parametric mode:

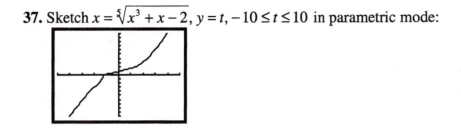

39. $h(x) = |x|$, $x \ge 0$ is equivalent to $h(x) = x$, $x \ge 0$. This function is its own inverse.

41. Let $h(x) = -x^2$, $x \ge 0$. Then set
$$y = -x^2,\ x \ge 0,\ y \le 0$$
$$-y = x^2$$
$$x = \sqrt{-y}$$
$$g(y) = \sqrt{-y},\ y \le 0$$
$$g(x) = \sqrt{-x},\ x \le 0$$
   is the inverse function.

43. Let $h(x) = \dfrac{x^2 + 6}{2}$, $x \ge 0$. Then set
$$y = \frac{x^2 + 6}{2},\ x \ge 0,\ y \ge 3$$
$$2y = x^2 + 6$$
$$x = \sqrt{2y - 6}$$
$$g(y) = \sqrt{2y - 6},\ y \ge 3$$
$$g(x) = \sqrt{2x - 6},\ x \ge 3$$
   is the inverse function.

45. Let $h(x) = \dfrac{1}{x^2 + 1}$, $x \ge 0$. Then set
$$y = \frac{1}{x^2 + 1},\ x \ge 0,\ 0 < y \le 1$$
$$x^2 + 1 = \frac{1}{y}$$
$$x^2 = \frac{1}{y} - 1$$
$$x = \sqrt{\frac{1}{y} - 1}$$
$$g(y) = \sqrt{\frac{1}{y} - 1},\ 0 < y \le 1$$
$$g(x) = \sqrt{\frac{1}{x} - 1},\ 0 < x \le 1 \text{ is the inverse function.}$$

47. **a.**   Set $y = 3x + 2$.

$$3x = y - 2$$

$$x = \frac{y - 2}{3}$$

$$f^{-1}(y) = \frac{y - 2}{3}$$

$$f^{-1}(x) = \frac{x - 2}{3}$$

b. $f^{-1}(1) = \dfrac{1-2}{3} = -\dfrac{1}{3}$

$$\frac{1}{f(1)} = \frac{1}{3(1) + 2} = \frac{1}{5}$$

Therefore f^{-1} is not the same as $\dfrac{1}{f}$.

49. First note that f is one-to-one, since if $a \neq b$, $ma \neq mb$ and $ma + b \neq mb + b$. To find the inverse function, set

$$y = mx + b$$

$$y - b = mx$$

$$x = \frac{y - b}{m}$$

$$g(y) = \frac{y - b}{m}$$

$$g(x) = \frac{x - b}{m} \text{ is the inverse function.}$$

51. **a.** slope $= \dfrac{a - b}{b - a} = -1$

b. Since the slope of $y = x$ is 1, and $(-1)1 = -1$, the two lines are perpendicular.

c. Length of $PR = \sqrt{(c - a)^2 + (c - b)^2}$

Length of $RQ = \sqrt{(b - c)^2 + (a - c)^2}$

Clearly, these are equal, since $(b - c)^2 = (c - b)^2$ and $(c - a)^2 = (a - c)^2$.
Thus, using the result from part **b**, $y = x$ is perpendicular to PQ and bisects PQ.
$y = x$ is the perpendicular bisector of PQ and thus P and Q are symmetric with respect to $y = x$.

53. **a.** If $f(a) = f(b)$, then $g(f(a)) = g(f(b))$, hence $a = b$. Thus if $a \neq b$, $f(a) \neq f(b)$.

b. If $g(y) = x$, then $y = f(g(y)) = f(x)$.

c. If $f(x) = y$, then $x = g(f(x)) = g(y)$.

Chapter 3 Review Exercises

1. **a.** $[-5/2] = [-2.5] = -3$

c. $[18.7] + [-15.7] = 18 + (-16) = 2$

b. $[1755] = 1755$

d. $[-7] - [7] = -7 - 7 = -14$

3.

x	0	1	2	-4	t	k	$b-1$	$1-b$	$6-2u$
$f(x)$	7	5	3	15	$7-2t$	$7-2k$	$9-2b$	$5+2b$	$4u-5$

5. Many examples are possible:

    a. Let $f(x) = x+1$, $a=1$, $b=2$ **b.** Let $f(x) = x+1$, $a=1$, $b=2$

$$f(a+b) = f(3) = 4 \qquad\qquad f(ab) = f(2) = 3$$

$$f(a)+f(b) = 2+3 = 5 \qquad\qquad f(a)f(b) = 2\cdot 3 = 6$$

7. $r \geq 4$.

9. $h(x) = x^2 - 3x$

$$h(t+2) = (t+2)^2 - 3(t+2) = t^2 + 4t + 4 - 3t - 6 = t^2 + t - 2$$

11. $f(x) = 2x^3 + x + 1$

$$f\left(\frac{x}{2}\right) = 2\left(\frac{x}{2}\right)^3 + \frac{x}{2} + 1 = \frac{x^3}{4} + \frac{x}{2} + 1 = \frac{x^3 + 2x + 4}{4}$$

13. a. $f(t) = 50\sqrt{t}$

    b. $g(t) = \pi r^2 = \pi\left(f(t)\right)^2 = \pi\left(50\sqrt{t}\right)^2 = 2500\pi t$

    c. radius: $f(9) = 50\sqrt{9} = 150$ meters

        area: $g(9) = 2500\pi(9) = 22{,}500\pi$ square meters

    d. Solve $100{,}000 = 2500\pi t$

$$t = \frac{100{,}000}{2500\pi}$$

$$t \approx 12.7$$

    After approximately 12.7 hours.

15.

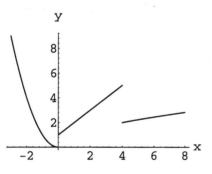

17. **a** fails the vertical line test. **b** is the graph of a function of x.

19. Graph the function using the window $-10 \leq x \leq 10, -1 \leq y \leq 10$ and use the minimum routine to find the local minimum at $x = -.5$. Then the function is decreasing on the interval $(-\infty, -.5)$ and increasing on the interval $(-.5, \infty)$.

21. Graph the function using the window $-8 \le x \le 3, -10 \le y \le 60$ and use the minimum routine to find the local minimum at $x = -.263$ and the maximum routine to find the local maximum at $x = -5.07$. Then the function is increasing on the intervals $(-\infty, -5.07)$ and $(-.26, \infty)$ and decreasing on the interval $(-5.07, -.26)$.

23.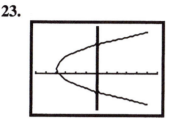

25. Here is one possibility:

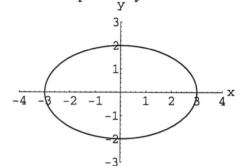

27. Replace x with $-x$:
$$(-x)^2 = y^2 + 2$$
$$x^2 = y^2 + 2$$
Equation unchanged; graph has y-axis symmetry.

Replace y with $-y$:
$$x^2 = (-y)^2 + 2$$
$$x^2 = y^2 + 2$$
Equation unchanged; graph has x-axis symmetry.

Since the graph has both x- and y-axis symmetry, it has origin symmetry.

29. $g(-x) = 9 - (-x)^2 = 9 - x^2 = g(x)$
Therefore, the function is even.

31. $h(-x) = 3(-x)^5 - (-x)((-x)^4 - (-x)^2)$
$h(-x) = -3x^5 + x(x^4 - x^2) \ne h(x)$
Therefore, the function is not even.
$-h(x) = -(3x^5 - x(x^4 - x^2)) = -3x^5 + x(x^4 - x^2) = h(-x)$
Therefore, the function is odd.

33. Replace x with $-x$:
$$(-x)^2 + y^2 + 6y = -5$$
$$x^2 + y^2 + 6y = -5$$
Equation is unchanged; graph has y-axis symmetry.

Replace y with $-y$:
$$x^2 + (-y)^2 + 6(-y) = -5$$
$$x^2 + y^2 - 6y = -5$$
Equation is changed; graph has no x-axis symmetry.

Since the graph has y-axis symmetry but not x-axis symmetry, it cannot have origin symmetry.

35.

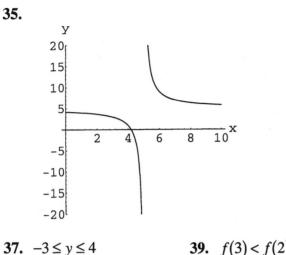

37. $-3 \le y \le 4$

39. $f(3) < f(2)$, thus 2 is a possible answer.

41. 1

43. $f(-1) = 0$, $f(1) = -3$
$f(-1) + f(1) = -3$

45. True. $3f(2) = -3$
$= -f(4)$

47. 4

49. $x \le 3$

51. $x < 3$

53. a. King Richard made 2 pitstops while Fireball Bob made 1.
b. King Richard started out faster (greater slope).
c. Fireball Bob won (finished at an earlier time).

55. Shrink vertically toward the x-axis by a factor of .25, then shift vertically 2 units up.

57. Shift horizontally 7 units to the right, then stretch vertically away from the x-axis by a factor of 3, then reflect in the x-axis, then shift vertically 2 units up.

59. e (only)

61. a. $f(2) = \dfrac{1}{2-1} = 1$ $g(2) = \sqrt{2^2 + 5} = \sqrt{9} = 3$. Hence $(f/g)(2) = \dfrac{f(2)}{g(2)} = \dfrac{1}{3}$

b. $(g/f)(x) = \dfrac{g(x)}{f(x)} = \dfrac{\sqrt{x^2 + 5}}{1/(x-1)} = (x-1)\sqrt{x^2 + 5}$ $(x \ne 1)$

c. $(fg)(c+1) = f(c+1)g(c+1) = \dfrac{1}{c+1-1}\sqrt{(c+1)^2 + 5} = \dfrac{\sqrt{(c+1)^2 + 5}}{c}$

63.

x	−4	−3	−2	−1	0	1	2	3	4
$g(x)$	1	4	3	1	−1	−3	−2	−4	−3
$h(x) = g(g(x))$	−3	−3	−4	−3	1	4	3	1	4

65. $f(2) = \dfrac{1}{2+1} = \dfrac{1}{3}$. $(g \circ f)(2) = g(f(2)) = g\left(\dfrac{1}{3}\right) = \left(\dfrac{1}{3}\right)^3 + 3 = \dfrac{82}{27}$

67. $f(x-1) = \dfrac{1}{x-1+1} = \dfrac{1}{x}.$ $(g \circ f)(x-1) = g(f(x-1)) = (f(x-1))^3 + 3 = \left(\dfrac{1}{x}\right)^3 + 3$

69. $g(1) = 1^3 + 3 = 4.$ $f(g(1)-1) = f(4-1) = f(3) = \dfrac{1}{3+1} = \dfrac{1}{4}$

71. $(f \circ g)(x) = f(g(x)) = \dfrac{1}{g(x)} = \dfrac{1}{x^2 - 1}$

$(g \circ f)(x) = g(f(x)) = (f(x))^2 - 1 = \left(\dfrac{1}{x}\right)^2 - 1$

73. The domain of $f \circ g$ is those numbers in the domain of g for which $f(g(x))$ is defined. The domain of g is $x \geq 0.$ $f(g(x))$ defined for all of these except 1. The domain of $f \circ g$ is $x \geq 0, x \neq 1.$

75. a. $g(-1) = \dfrac{(-1)^3 - (-1) + 1}{(-1) + 2} = \dfrac{1}{1} = 1$ $g(1) = \dfrac{(1)^3 - 1 + 1}{1 + 2} = \dfrac{1}{3}$

average rate of change $= \dfrac{g(1) - g(-1)}{1 - (-1)} = \dfrac{\frac{1}{3} - 1}{2} = -\dfrac{1}{3}$

b. $g(2) = \dfrac{2^3 - 2 + 1}{2 + 2} = \dfrac{7}{4}$ $g(0) = \dfrac{0^3 - 0 + 1}{0 + 2} = \dfrac{1}{2}$

average rate of change $= \dfrac{g(2) - g(0)}{2 - 0} = \dfrac{\frac{7}{4} - \frac{1}{2}}{2} = \dfrac{5}{8}$

77. $(f \circ g)(x) = 2g(x) + 1 = 2(3x - 2) + 1 = 6x - 3$
$(f \circ g)(5) = 6 \cdot 5 - 3 = 27$ $(f \circ g)(3) = 6 \cdot 3 - 3 = 15$

average rate of change $= \dfrac{(f \circ g)(5) - (f \circ g)(3)}{5 - 3} = \dfrac{27 - 15}{2} = 6$

79. $\dfrac{f(x+h) - f(x)}{h} = \dfrac{(3(x+h) + 4) - (3x + 4)}{h} = \dfrac{3x + 3h + 4 - 3x - 4}{h} = \dfrac{3h}{h} = 3$

81. $\dfrac{g(x+h) - g(x)}{h} = \dfrac{((x+h)^2 - 1) - (x^2 - 1)}{h}$

$= \dfrac{x^2 + 2xh + h^2 - 1 - x^2 + 1}{h} = \dfrac{2xh + h^2}{h} = 2x + h$

83. $P(4) = .2(4)^2 + .5(4) - 1 = 4.2$

   a. $P(8) = .2(8)^2 + .5(8) - 1 = 15.8$

     average rate of change $= \dfrac{P(8) - P(4)}{8 - 4} = \dfrac{15.8 - 4.2}{4} = 2.9$

     $290 per ton.

   b. $P(5) = .2(5)^2 + .5(5) - 1 = 6.5$

     average rate of change $= \dfrac{P(5) - P(4)}{5 - 4} = \dfrac{6.5 - 4.2}{1} = 2.3$

     $230 per ton.

   c. $P(4.1) = .2(4.1)^2 + .5(4.1) - 1 = 4.412$

     average rate of change $= \dfrac{P(4.1) - P(4)}{4.1 - 4} = \dfrac{4.412 - 4.2}{.1} = 2.12$

     $212 per ton.

85. **a.** 40 to 45   **b.** 25 to 35   **c.** 30 to 44 (approximately)

87. Let t = time, P = price

   a. $\dfrac{P(1998) - P(1985)}{1998 - 1985} = \dfrac{164,800 - 95,400}{13} \approx \$5338$ per year

     $\dfrac{P(2001) - P(1998)}{2001 - 1998} = \dfrac{194,500 - 164,800}{3} = \$9900$ per year

     $\dfrac{P(2001) - P(1985)}{2001 - 1985} = \dfrac{194,500 - 95,400}{16} = \$6194$ per year

   b. Four years of change at $9900 per year = $39,600. Adding this to the price in
     2001 yields $194,500 + \$39,600 = \$234,100$ as the (projected) median price.

89. Set $y = \sqrt{5 - x} + 7 \quad (y \ge 7)$

     $\sqrt{5 - x} = y - 7$

     $5 - x = (y - 7)^2$

     $x = 5 - (y - 7)^2$

     $f^{-1}(y) = 5 - (y - 7)^2 \quad (y \ge 7)$

     $f^{-1}(x) = 5 - (x - 7)^2 \quad (x \ge 7)$

91.

93. Graph *f* in the window
$-5 \le x \le 5, -5 \le y \le 5$:

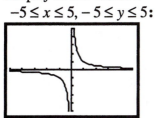

f appears to be one-to-one, therefore to have an inverse. In fact it is its own inverse, hence the graph of the inverse function is identical.

95. Graph *f* in the window
$-5 \le x \le 25, -500 \le y \le 500$:

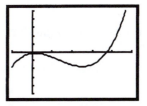

f is not a one-to-one function. Therefore it does not have an inverse function.

Chapter 4
Polynomial and Rational Functions

4.1 Quadratic Functions

1. The parabola opens upward and has vertex $(0,2)$. Graph **I.**

3. The parabola opens upward and has vertex $(2,0)$. Graph **K.**

5. The parabola opens upward and has vertex $(2,2)$. Graph **J.**

7. The parabola opens downward and has vertex $(-2,2)$. Graph **F.**

9. Vertex: $(5,2)$. Since $a = 3$, opens upward.

11. Vertex: $(1,2)$. Since $a = -1$, opens downward.

13. Complete the square:
$$f(x) = x^2 - 6x + 3 = x^2 - 6x + 9 - 6 = (x-3)^2 - 6.$$
Vertex: $(3,-6)$. Since $a = 1$, opens upward.

15. Complete the square:
$$h(x) = x^2 + 3x + 6 = x^2 + 3x + \frac{9}{4} + \frac{15}{4} = \left(x + \frac{3}{2}\right)^2 + \frac{15}{4}$$
Vertex: $\left(-\frac{3}{2}, \frac{15}{4}\right)$. Since $a = 1$, opens upward.

17. Complete the square:
$$y = 2x^2 + 12x - 3 = 2\left(x^2 + 6x + 9\right) - 21 = 2(x+3)^2 - 21.$$
Vertex: $(-3,-21)$. Since $a = 2$, opens upward.

19. Complete the square:
$$f(x) = -x^2 + 8x - 2 = -\left(x^2 - 8x + 16\right) + 14 = -(x-4)^2 + 14.$$
Vertex: $(4,14)$. Since $a = -1$, opens downward.

21. a.
$$\frac{f(x+h) - f(x)}{h} = \frac{\left(-(x+h)^2 + (x+h)\right) - \left(-x^2 + x\right)}{h}$$
$$= \frac{-x^2 - 2xh - h^2 + x + h + x^2 - x}{h}$$
$$= \frac{-2xh - h^2 + h}{h}$$
$$= -2x - h + 1$$

b. Complete the square:

$$f(x) = -x^2 + x = -\left(x^2 - x + \frac{1}{4}\right) + \frac{1}{4} = -\left(x - \frac{1}{2}\right)^2 + \frac{1}{4}$$

Vertex: $\left(\frac{1}{2}, \frac{1}{4}\right)$

c. Substitute $\frac{1}{2}$ for x in $-2x - h + 1$ to obtain $-2\left(\frac{1}{2}\right) - h + 1 = -h.$

23. a. $\dfrac{f(x+h) - f(x)}{h} = \dfrac{\left(-2(x+h)^2 + 2(x+h) - 1\right) - \left(-2x^2 + 2x - 1\right)}{h}$

$$= \frac{-2x^2 - 4xh - 2h^2 + 2x + 2h - 1 + 2x^2 - 2x + 1}{h}$$

$$= \frac{-4xh - 2h^2 + 2h}{h}$$

$$= -4x - 2h + 2$$

b. Complete the square:

$$f(x) = -2x^2 + 2x - 1 = -2\left(x^2 - x + \frac{1}{4}\right) - \frac{1}{2} = -2\left(x - \frac{1}{2}\right)^2 - \frac{1}{2}$$

Vertex: $\left(\frac{1}{2}, -\frac{1}{2}\right)$

c. Substitute $\frac{1}{2}$ for x in $-4x - 2h + 2$ to obtain $-4\left(\frac{1}{2}\right) - 2h + 2 = -2h.$

25. Rule: $g(x) = 2x^2 - 5$ Vertex: $(0, -5)$

27. Since $\frac{1}{2}\left[h(x+3) - 4\right] = f(x) = x^2$, we can write

$$h(x+3) - 4 = 2x^2$$
$$h(x+3) = 2x^2 + 4$$
$$h(x) = 2(x-3)^2 + 4$$

Vertex: $(3, 4)$

29. Since the vertex is at $(0, 0)$, the rule must be of form $f(x) = ax^2$.
Since $(2, 12)$ is on the graph, $f(2) = 12$, thus

$$12 = a(2)^2$$
$$a = 3$$

The rule is $f(x) = 3x^2.$

31. Since the vertex is at $(2, 5)$, the rule must be of form $f(x) = a(x-2)^2 + 5$.
Since $(-3, 80)$ is on the graph, $f(-3) = 80$, thus

$$80 = a(-3-2)^2 + 5$$
$$75 = 25a$$
$$a = 3$$

The rule is $f(x) = 3(x-2)^2 + 5$ or $f(x) = 3x^2 - 12x + 17.$

33. Since the x-coordinate of the vertex satisfies $-b/2 = 0$, $b = 0$.

35. Since the x-coordinate of the vertex is $x = -b/2a$, we have

$$2 = -\frac{b}{2(1)}$$
$$b = -4$$

Since $f(2) = 4$, we have

$$f(x) = x^2 - 4x + c$$
$$4 = 2^2 - 4(2) + c$$
$$c = 8$$

37. a. $p(50) = 1600(50) - 4(50)^2 - 50,000 = \$20,000$

$p(250) = 1600(250) - 4(250)^2 - 50,000 = \$100,000$

b. The maximum profit occurs at the vertex of the parabola that is the graph of $p(x)$. Since $p(x) = -4x^2 + 1600x - 50,000$ has vertex at $x = -b/2a = -\frac{1600}{2(-4)} = 200$, 200 units will maximize the profit. Then

$$p(200) = 1600(200) - 4(200)^2 - 50,000 = \$110,000$$

39. The maximum height occurs at the vertex of the parabola that is the graph of $p(x)$. Since the vertex is at $x = -b/2a = -\frac{.23}{2(-.0000167)} = 6886$ feet, the maximum height is given by $p(6886) = -.0000167(6886)^2 + .23(6886) + 50 = 842$ feet.

The shell hits the water when $p(x) = 0$. Solve

$$-.0000167x^2 + .23x + 50 = 0$$

$$x = \frac{-.23 \pm \sqrt{(.23)^2 - 4(-.0000167)(50)}}{2(-.0000167)}$$

$$x = \frac{-.23 \pm .2371497}{-.0000334}$$

$$x = -214 \quad \text{or} \quad 13,987$$

The shell hits the water 13,987 feet, or 2.65 miles, away.

41. a. $B(30) = .01(30)^2 + .7(30) = 30$ meters; $B(100) = .01(100)^2 + .7(100) = 170$ meters

b. Substitute 60 for $B(s)$ and solve for s.

$$60 = .01s^2 + .7s$$

$$0 = .01s^2 + .7s - 60$$

$$s = \frac{-.7 \pm \sqrt{(.7)^2 - 4(.01)(-60)}}{2(.01)}$$

$$s = -120 \quad \text{or} \quad 50$$

50 kilometers per hour

43. Substitute 80 for v_0 and 96 for h_0.

$$h = -16t^2 + 80t + 96$$

The maximum height occurs when $t = -b/2a = -\frac{80}{2(-16)} = 2.5$ sec.

At that time

$$h = -16(2.5)^2 + 80(2.5) + 96 = 196 \text{ feet.}$$

45. Substitute 32 for v_0 and 6 for h_0.

$$h = -16t^2 + 32t + 6$$

The maximum height occurs when $t = -b/2a = -\frac{32}{2(-16)} = 1$ sec.

At that time

$$h = -16(1)^2 + 32(1) + 6 = 22 \text{ feet.}$$

47. Set $h = 30 - b$. The area is given by

$$A = \frac{1}{2}bh = \frac{1}{2}b(30 - b)$$

$$A = -\frac{1}{2}b^2 + 15b$$

This will be maximum when $b = -\dfrac{15}{2\left(-\frac{1}{2}\right)} = 15.$

Thus $b = 15, 30 - b = h = 15,$ for maximum area.

49. Set x = length perpendicular to river and y = length parallel to river. Then

$$2x + y = 200$$

$$y = 200 - 2x$$

$$A = xy$$

$$A = x(200 - 2x)$$

$$A = -2x^2 + 200x$$

This will be maximum when $x = -\frac{200}{2(-2)} = 50$ feet. Then $y = 200 - 2(50) = 100$ feet.

51. Set x = length and y = width. Then

$$2x + 3y = 130$$

$$y = \frac{130 - 2x}{3}$$

$$A = xy$$

$$A = x\left(\frac{130 - 2x}{3}\right)$$

$$A = -\frac{2}{3}x^2 + \frac{130}{3}x$$

This will be maximum when $x = -\frac{130}{3} \div 2\left(-\frac{2}{3}\right) = 32.5$ feet.

Then $A = -\frac{2}{3}(32.5)^2 + \frac{130}{3}(32.5) = 704.2$ square feet.

53. a. Since the vertex is at the origin, the equation has form $y = ax^2$. Since the end of the board is half of 8 feet, or 48 inches horizontally from the vertex, and y is 2 inches there, we have

$$2 = a(48)^2$$

$$a = \frac{1}{1152}$$

$$y = \frac{1}{1152}x^2$$

b. Substitute 1 for y and solve.

$$1 = \frac{1}{1152}x^2$$

$$x = \sqrt{1152} \approx 34 \text{ inches.}$$

55. Let x = the number of 50 cent decreases. Then

price per bowl $= 4 - .5x$

number of bowls $= 120 + 20x$

Income $= (120 + 20x)(4 - .5x)$

$\qquad = -10x^2 + 20x + 480$

This will be maximum when $x = -\frac{20}{2(-10)} = 1$.

The price should be $4 - .5(1) = \$3.50$.

57. Let x = the number of 20 cent decreases. Then

price per ticket $= 4 - .2x$

attendance $= 500 + 30x$

Income $= (500 + 30x)(4 - .2x)$

$\qquad = -6x^2 + 20x + 2000$

This will be maximum when $x = -\frac{20}{2(-6)} = \frac{5}{3}$.

The price will be $4 - .2\left(\frac{2}{3}\right) = \$3.67$.

4.2 Polynomial Functions

1. Write the expression as $x^3 + 1$.
This is a polynomial with leading coefficient 1, constant term 1, and degree 3.

3. Write the expression as $x^3 - x^2 + x - 1$.
This is a polynomial with leading coefficient 1, constant term -1, and degree 3.

5. Write the expression as $x^2 - 3$.
This is a polynomial with leading coefficient 1, constant term -3, and degree 2.

7. This is not a polynomial.

9.

$$x+1 \overline{)\,3x^4 \qquad\;\; +2x^2-6x+1\,}$$

Quotient above: $3x^3 - 3x^2 + 5x - 11$

$$\underline{3x^4 + 3x^3}$$
$$-3x^3 + 2x^2$$
$$\underline{-3x^3 - 3x^2}$$
$$5x^2 - 6x$$
$$\underline{5x^2 + 5x}$$
$$-11x + 1$$
$$\underline{-11x - 11}$$
$$12$$

Quotient: $3x^3 - 3x^2 + 5x - 11$

Remainder: 12

Check: $\left(3x^3 - 3x^2 + 5x - 11\right)(x+1) + 12 =$
$$3x^4 + 2x^2 - 6x + 1$$

11.

$$x^3+1 \overline{)\,x^5 + 2x^4 - 6x^3 + x^2 - 5x + 1\,}$$

Quotient above: $x^2 + 2x - 6$

$$\underline{x^5 \qquad\qquad\; + x^2}$$
$$2x^4 - 6x^3 \qquad -5x$$
$$\underline{2x^4 \qquad\qquad +2x}$$
$$-6x^3 \qquad -7x + 1$$
$$\underline{-6x^3 \qquad\qquad -6}$$
$$-7x + 7$$

Quotient: $x^2 + 2x - 6$

Remainder: $-7x + 7$

Check: $\left(x^2 + 2x - 6\right)\left(x^3 + 1\right) + \left(-7x + 7\right) =$
$$x^5 + 2x^4 - 6x^3 + x^2 - 5x + 1$$

13.

$$
\begin{array}{r}
5x^2 + 5x + 5 \\
x^2 - x + 1\overline{)5x^4 \qquad + 5x^2 \qquad + 5} \\
\underline{5x^4 - 5x^3 + 5x^2} \\
5x^3 \\
\underline{5x^3 - 5x^2 + 5x} \\
5x^2 - 5x + 5 \\
\underline{5x^2 - 5x + 5} \\
0
\end{array}
$$

Quotient: $5x^2 + 5x + 5$ Remainder: 0

Check: $\left(5x^2 + 5x + 5\right)\left(x^2 - x + 1\right) + 0 =$

$$5x^4 + 5x^2 + 5$$

15. Divide:

$$
\begin{array}{r}
x - 1 \\
x^2 + 3x - 1\overline{)x^3 + 2x^2 - 5x - 6} \\
\underline{x^3 + 3x^2 - \ x} \\
-x^2 - 4x - 6 \\
\underline{-x^2 - 3x + 1} \\
-x - 7
\end{array}
$$

Since the remainder is non-zero, the first polynomial is not a factor of the second.

17. Divide:

$$
\begin{array}{r}
x^2 - 1 \\
x^2 + 3x - 1\overline{)x^4 + 3x^3 - 2x^2 - 3x + 1} \\
\underline{x^4 + 3x^3 - \ x^2} \\
-x^2 - 3x + 1 \\
\underline{-x^2 - 3x + 1} \\
0
\end{array}
$$

Since the remainder is zero, the first polynomial is a factor of the second.

19. Substitute:

$g(2) = 2^4 + 6 \cdot 2^3 - 2^2 - 30 \cdot 2 = 0.$ 2 is a root.

$g(3) = 3^4 + 6 \cdot 3^3 - 3^2 - 30 \cdot 3 = 144.$ 3 is not a root.

$g(0) = 0^4 + 6 \cdot 0^3 - 0^2 - 30 \cdot 0 = 0.$ 0 is a root.

$g(-1) = (-1)^4 + 6 \cdot (-1)^3 - (-1)^2 - 30 \cdot (-1) = 24.$ -1 is not a root.

21. Substitute:

$h\left(2\sqrt{2}\right) = \left(2\sqrt{2}\right)^3 + \left(2\sqrt{2}\right)^2 - 8\left(2\sqrt{2}\right) - 8 = 0.$ $2\sqrt{2}$ is a root.

$h\left(\sqrt{2}\right) = \left(\sqrt{2}\right)^3 + \left(\sqrt{2}\right)^2 - 8\left(\sqrt{2}\right) - 8 = -6\sqrt{2} - 6.$ $\sqrt{2}$ is not a root.

$h\left(-\sqrt{2}\right) = \left(-\sqrt{2}\right)^3 + \left(-\sqrt{2}\right)^2 - 8\left(-\sqrt{2}\right) - 8 = 6\sqrt{2} - 6.$ $-\sqrt{2}$ is not a root.

$h(1) = (1)^3 + (1)^2 - 8(1) - 8 = -14.$ 1 is not a root.

$h(-1) = (-1)^3 + (-1)^2 - 8(-1) - 8 = 0.$ -1 is a root.

23. Apply the Remainder Theorem with $c = 1$. The remainder is
$$f(1) = 1^{10} + 1^8 = 2$$

25. Apply the Remainder Theorem with $c = -1$. The remainder is
$$f(-1) = 3(-1)^4 - 6(-1)^3 + 2(-1) - 1 = 6$$

27. Apply the Remainder Theorem with $c = -2$. The remainder is
$$f(-2) = (-2)^3 - 2(-2)^2 + 5(-2) - 4 = -30$$

29. Apply the Remainder Theorem with $c = 10$. The remainder is
$$f(10) = 2(10)^5 - 3(10)^4 + (10)^3 - 2(10)^2 + 10 - 8 = 170,802$$

31. Apply the Remainder Theorem with $c = 20$. The remainder is
$$f(20) = 2(20)^5 - 3(20)^4 + 2(20)^3 - 8(20) - 8 = 5,935,832$$

33. $h(x)$ is a factor of $f(x)$ if 1 is a root of $f(x)$. Evaluating $f(x)$ at 1 shows that
$$f(1) = 1^5 + 1 = 2$$
Therefore, 1 is not a root and $h(x)$ is not a factor.

35. $h(x)$ is a factor of $f(x)$ if -2 is a root of $f(x)$. Evaluating $f(x)$ at -2 shows that
$$f(-2) = (-2)^3 - 3(-2)^2 - 4(-2) - 12 = -24$$
Therefore, -2 is not a root and $h(x)$ is not a factor.

37. $h(x)$ is a factor of $f(x)$ if 1 is a root of $f(x)$. Evaluating $f(x)$ at 1 shows that
$$f(1) = 14 \cdot 1^{99} - 65 \cdot 1^{56} + 51 = 0$$
Therefore, 1 is a root and $h(x)$ is a factor.

39. After some trial and error, or a calculator plot, we find that -4 is a root, since
$f(-4) = 6(-4)^3 - 7(-4)^2 - 89(-4) + 140 = 0$. By the Factor Theorem, $x + 4$ is a
factor. Division by $x + 4$ yields
$$f(x) = (x + 4)(6x^2 - 31x + 35)$$
$$f(x) = (x + 4)(2x - 7)(3x - 5)$$

41. After some trial and error, or a calculator plot, we find that 3 is a root, since
$f(3) = 0$. By the Factor Theorem, $x - 3$ is a factor. Division by $x - 3$ yields
$$f(x) = (x - 3)(4x^3 + 16x^2 + 13x + 3) = (x - 3)g(x)$$
Similarly, -3 is a root, since $g(-3) = 4(-3)^3 + 16(-3)^2 + 13(-3) + 3 = 0$
By the Factor Theorem, $x + 3$ is a factor. Division by $x + 3$ yields
$$f(x) = (x - 3)(x + 3)(4x^2 + 4x + 1)$$
$$f(x) = (x - 3)(x + 3)(2x + 1)^2$$

43. The graph indicates that $-2, -1, 1, 2$, and 3 are roots. By the Factor Theorem,
$$f(x) = (x+2)(x+1)(x-1)(x-2)(x-3)$$
$$f(x) = x^5 - 3x^4 - 5x^3 + 15x^2 + 4x - 12$$

45. The graph indicates that $-1, 0, 1, 2$, and 3 are roots. By the Factor Theorem,
$$f(x) = (x+1)x(x-1)(x-2)(x-3)$$
$$f(x) = x^5 - 5x^4 + 5x^3 + 5x^2 - 6x$$

47. $f(x) = (x-1)(x-7)(x+4) = x^3 - 4x^2 - 25x + 28$

49. There are many possible polynomials, such as $f(x) = (x-1)^4(x-2)(x-\pi)$ or
$f(x) = (x-1)^2(x-2)^2(x-\pi)^2$, and so on.

51. Let $f(x) = a(x-8)(x-5)x$. Then
$$17 = f(10) = a(10-8)(10-5)10$$
$$17 = 100a$$
$$a = .17$$
$$f(x) = .17(x-8)(x-5)x$$

53. Since $x+2$ is a factor, -2 is a root. Hence
$$(-2)^3 + 3(-2)^2 + k(-2) - 2 = 0$$
$$-2k + 2 = 0$$
$$k = 1$$

55. Since $x-1$ is a factor, 1 is a root. Hence
$$k^2 \cdot 1^4 - 2k \cdot 1^2 + 1 = 0$$
$$k^2 - 2k + 1 = 0$$
$$(k-1)^2 = 0$$
$$k = 1$$

57. If $x-c$ were a factor, by the Factor Theorem, c would be a root. Then $c^4 + c^2 + 1 = 0$. This equation has discriminant given by $1^2 - 4(1)(1) = -3$, hence it has no real roots. Thus $x-c$ cannot be a factor for any real c.

59. a. An example would be $n = 1$. $x + c$ is not a factor of $x - c$.
b. $x + c$ is a factor of $x^n + c^n$ if n is odd, because $-c$ is a root, which follows because $(-c)^n + c^n = -c^n + c^n = 0$.

61. The difference quotient for $f(x)$ is given by

$$\frac{f(x+h)-f(x)}{h} = \frac{\left((x+h)^2 + k(x+h)\right) - \left(x^2 + kx\right)}{h}$$

$$= \frac{x^2 + 2xh + h^2 + kx + kh - x^2 - kx}{h}$$

$$= \frac{2xh + h^2 + kh}{h}$$

$$= 2x + h + k$$

Thus we must have $k = 5$.

4.2A   Synthetic Division

1.
$$\begin{array}{r|rrrrr} 2 & 3 & -8 & 0 & 9 & 5 \\ & & 6 & -4 & -8 & 2 \\ \hline & 3 & -2 & -4 & 1 & 7 \end{array}$$

Quotient: $3x^3 - 2x^2 - 4x + 1$

Remainder: 7

3.
$$\begin{array}{r|rrrrr} -3 & 2 & 5 & 0 & -2 & -8 \\ & & -6 & 3 & -9 & 33 \\ \hline & 2 & -1 & 3 & -11 & 25 \end{array}$$

Quotient: $2x^3 - x^2 + 3x - 11$

Remainder: 25

5.
$$\begin{array}{r|rrrrr} 7 & 5 & 0 & -3 & -4 & 6 \\ & & 35 & 245 & 1694 & 11830 \\ \hline & 5 & 35 & 242 & 1690 & 11836 \end{array}$$

Quotient: $5x^3 + 35x^2 + 242x + 1690$

Remainder: 11836

7.
$$\begin{array}{r|rrrrr} 2 & 1 & -6 & 4 & 2 & -7 \\ & & 2 & -8 & -8 & -12 \\ \hline & 1 & -4 & -4 & -6 & -19 \end{array}$$

Quotient: $x^3 - 4x^2 - 4x - 6$

Remainder: -19

Details of the synthetic division are omitted in the following.

9. Quotient: $3x^3 + \dfrac{3}{4}x^2 - \dfrac{29}{16}x - \dfrac{29}{64}$ Remainder: $\dfrac{483}{256}$

11. Quotient: $2x^3 - 6x^2 + 2x + 2$ Remainder: 1

13. $3x^3 + 9x^2 - 11x + 4 = (x+4)(3x^2 - 3x + 1)$

15. $2x^5 - 7x^4 + 15x^3 - 16x^2 - 10x + 5 = \left(x - \dfrac{1}{2}\right)(2x^4 - 6x^3 + 12x - 10)$

17. Quotient: $x^2 - 2.15x + 4$ Remainder: 2.25

19.

$$\underline{-2}|\;1 \quad 0 \qquad c \qquad\quad 4$$
$$\underline{\qquad -2 \qquad 4 \quad -2c-8\qquad}$$
$$1 \quad -2 \quad c+4 \quad -2c-4$$

Since the remainder is given as 4, we must have

$$-2c-4=4$$
$$-2c=8$$
$$c=-4$$

4.3 Real Roots of Polynomials

1. Graph the function in the window $-4 \le x \le 2, -10 \le y \le 10$. The graph has intercepts -3, -1, 1. These are the rational roots.

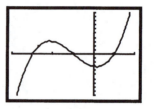

3. Graph the function in the window $-6 \le x \le 3, -10 \le y \le 20$. The graph has intercepts -5, -1, 1. These are the rational roots.

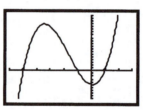

5. $f(x) = 2x^5 + 5x^4 - 11x^3 + 4x^2$

$$= x^2\left(2x^3 + 5x^2 - 11x + 4\right)$$

$$= x^2 g(x)$$

0 is a rational root of f. Graph g in the window $-5 \le x \le 3, -5 \le y \le 35$. The graph has intercepts -4, 1, and $\frac{1}{2}$. These are the remaining rational roots.

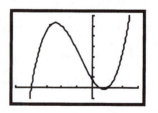

7. $12f(x) = g(x) = x^3 - x^2 - 8x + 12$. The roots of f are the same as the roots of g. Graph g in the window $-4 \le x \le 4$, $-5 \le y \le 20$. The graph has intercepts -3 and 2. These are the rational roots.

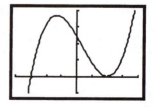

Details of the synthetic division are omitted in the following.

9. $3f(x) = g(x) = x^4 - 3x^3 - 3x^2 + 13x - 6$.
The roots of f are the same as the roots of
g. Graph g in the window $-6 \le x \le 6$,
$-5 \le y \le 5$. The graph has an intercept at
2; since this is the only integer root, it is
the only rational root, since all the
rational roots must be integers in this
case.

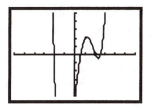

11. $10f(x) = g(x) = x^3 - 19x + 30$. The roots
of f are the same as the roots of g. Graph
g in the window $-6 \le x \le 6$,
$-10 \le y \le 30$. The graph has intercepts at
–5, 2, and 3; these are the rational roots.

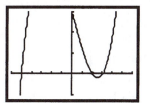

13. Graph the function in the window
$-3 \le x \le 3, -20 \le y \le 20$ and observe
that 2 is a root, hence $x - 2$ is a factor.
Synthetically divide by $x - 2$ to obtain
$2x^2 + 1$, which has no further rational
roots. $(x - 2)(2x^2 + 1)$

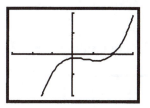

15. First note that the function can be fac-
tored into $x^3(x^3 + 2x^2 + 3x + 6)$. Graph
$x^3 + 2x^2 + 3x + 6$ in the window
$-6 \le x \le 6, -20 \le y \le 20$ and observe
that –2 is a root, hence $x + 2$ is a factor.
Synthetically divide by $x + 2$ to obtain
$x^2 + 3$ which has no further rational
roots. $x^3(x + 2)(x^2 + 3)$.

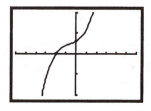

17. Graph the function in the window
$0 \le x \le 3, -3 \le y \le 3$ and observe: the
graph is tangent to the x-axis at 1, hence
1 is a root and $x - 1$ is a factor twice;
also, 2 is a root, hence $x - 2$ is a factor.
Synthetically divide by $(x - 1)^2(x - 2)$ to
obtain $x^2 + 3$ which has no further
rational roots. $(x - 1)^2(x - 2)(x^2 + 3)$.

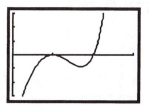

19. Synthetic division by $x - 2$ yields a bottom row which is all positive, hence 2 is an
upper bound. Synthetic division by $x + 5$ yields a bottom row which alternates in
sign, hence –5 is a lower bound.

21. Following the hint, analyze $-f(x) = x^5 + 5x^4 - 9x^3 - 18x^2 + 68x - 176,$ which has the same roots as f. Synthetic division by $x - 3$ yields a bottom row which is all positive, hence 3 is an upper bound for the roots of $-f$ and f. Synthetic division by $x + 7$ yields a bottom row which alternates in sign, hence -7 is a lower bound for the roots of $-f$ and f.

23. Graph the function in the window $-1 \le x \le 3, -5 \le y \le 5$ and observe that the roots are $-\frac{1}{2}, 1,$ and 2.

25. Graph the function in the window $0 \le x \le 1.5, -.2 \le y \le .1$ and observe that the roots are $\frac{1}{3}, \frac{1}{2},$ and 1.

27. Graph the function in the window $-6 \le x \le 6, -10 \le y \le 10$ and observe: the graph is tangent to the x-axis at 2, hence 2 is a root and $x - 2$ is a factor twice. Synthetic division twice by $x - 2$ yields $x^2 + 5x - 3$. Applying the quadratic formula yields $\dfrac{-5 \pm \sqrt{37}}{2}$ as well as the root 2.

29. Graph the function in the window $-3 \le x \le 3, -3 \le y \le 3$ (shown).

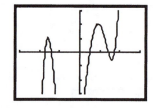

Five roots are seen, of which the only rational root is $\frac{1}{2}$. Use the root (zero) routine to locate the other four roots at ± 1.414 and ± 1.732. This suggests that these roots are exactly $\pm\sqrt{2}$ and $\pm\sqrt{3}$, which can be confirmed by substitution.

31. Graph the function in the window $-10 \le x \le 10, -200 \le y \le 200$ (shown).

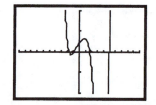

Four roots are seen, including the two integer roots -1 and 5. Use the root (zero) routine to locate the other two roots at ± 1.732. This suggests that these roots are exactly $\pm\sqrt{3}$, which can be confirmed by substitution.

33. Graph the function in the window $-3 \le x \le 3, -10 \le y \le 10$ (shown).

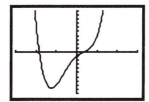

Use the root (zero) routine to locate the roots at -1.8393 and $.3333$ ($\frac{1}{3}$).

35. First graph the function in the window $-60 \le x \le 60, -10{,}000 \le y \le 10{,}000$ (left) and observe that 50 is a root, but -50 is not. Synthetically divide by $x - 50$ to obtain $x^3 + 2x^2 - x - 1$. Then graph this in the window $-3 \le x \le 3, -3 \le y \le 3$ (right).

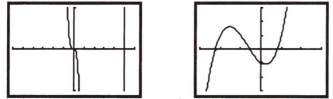

Use the root (zero) routine to locate the roots at -2.2470, $-.5550$, and $.8019$, as well as 50.

37. a. By the Rational Roots Theorem, the only possible rational roots of $x^2 - 2$ are ± 1 and ± 2. Since $\sqrt{2}$ is a root of the polynomial, and it is not one of those four numbers, it is not rational.
   b. By the Rational Roots Theorem, the only possible rational roots of $x^2 - 3$ are ± 1 and ± 3. Since $\sqrt{3}$ is a root of the polynomial, and it is not one of those four numbers, it is not rational.

39. a. In 1995, $x = 5$, hence the murder rate was $f(5) = 8.2$. In 2000, $x = 10$, hence the murder rate was $f(10) = 5.5$.
   b. Graph the function and the horizontal line $y = 7$ in the window $0 \le x \le 11$, $0 \le y \le 15$ and use the intersection routine to find $x = 6.75$, that is, 1996.
   c. Graph the function in the same window and use the maximum routine to find $x = 1$, that is, 1991.

41. Since
   height $= x$
   length $= 36 - 4x$
   width $= 12 - 2x$
   we can write: volume $= V(x) = x(36 - 4x)(12 - 2x) = 8x^3 - 120x^2 + 432x$. Graph this function and the horizontal line $y = 448$ in the window $0 \le x \le 3, 0 \le y \le 600$ and use the intersection routine to find the only root less than 2.5 exactly at $x = 2$.

43. a. At $t = 0$, $F(0) = 6$ degrees per day. At $t = 11$, $F(11) = 6.6435$ degrees per day.

    b. Graph the function and the horizontal line $y = 4$ in the window $0 \le x \le 12$, $-10 \le y \le 30$ and use the intersection routine to find $t = 2.0330$ and $t = 10.7069$.

    c. Graph the function and the horizontal line $y = -3$ (the temperature is decreasing, hence the rate is negative) in the same window and use the intersection routine to find $t = 5.0768$ and $t = 9.6126$.

    d. Graph the function and use the minimum routine to find $t = 7.6813$.

4.4 Graphs of Polynomial Functions

1. This could be a graph of a polynomial function (of a very large degree).

3. This could be a graph of a polynomial function (of first degree).

5. This is discontinuous and could not be the graph of a polynomial function.

7. This could be the graph of a polynomial function. Since the ends go in opposite directions, the polynomial has odd degree. Since there are 3 intercepts and 2 local extrema, it must have degree at least 3 and possibly 5.

9. This could not be the graph of a polynomial function, since the domain is not all real numbers.

11. This could be the graph of a polynomial function. Since the ends go in opposite directions, the polynomial has odd degree. Since there are 5 intercepts and 4 local extrema, it must have degree at least 5.

13. In a very large window like $-50 \le x \le 50$, $-100,000 \le y \le 1,000,000$ the graph looks like the graph of $y = x^4$.

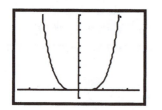

15. Roots –2, 1, 3; each has multiplicity 1.

17. Root –2 has multiplicity 1, root –1 has multiplicity 1, and root 2 has multiplicity 2 (or possibly higher).

19. Graph **(e)**   **21.** Graph **(f)**   **23.** Graph **(c)**

25.

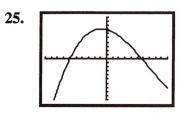

27.

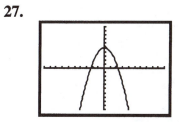

Since this is a polynomial function of degree 3, the complete graph must have another *x*-intercept.

Since this is a polynomial function of degree 4 with positive leading coefficient, both of the ends should point up; here they are pointing down.

29. $-9 \leq x \leq 3, -20 \leq y \leq 40$

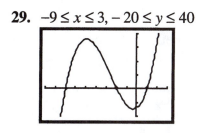

31. $-6 \leq x \leq 6, -60 \leq y \leq 320$

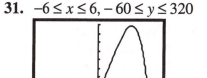

33. $-3 \leq x \leq 4, -35 \leq y \leq 20$

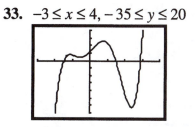

35. $-33 \leq x \leq -2, -50,000 \leq y \leq 260,000$ then $-2 \leq x \leq 3, -20 \leq y \leq 30$

37. $-90 \leq x \leq 120, -15,000 \leq y \leq 5000$

39. $-3 \le x \le 3, -20 \le y \le 20$ then $-.1 \le x \le .2, 4.997 \le y \le 5.001$

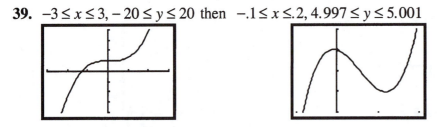

41. a. The graph of a cubic polynomial can have no more than two local extrema. If only one, the ends would go in the same direction. Hence it can have two or none.
   b. When the end behavior and the number of extrema are both accounted for, these four shapes are the only possible ones.

43. a. The ends go in opposite directions, so the degree is odd.
   b. Since if x is large and positive, so is $f(x)$, the leading coefficient is positive.
   c. Root –2 has multiplicity 1, root 0 has multiplicity at least 2, root 4 has multiplicity 1, root 6 has multiplicity 1.
   d. Adding the multiplicities from **c**, the polynomial must have degree at least 5.

45. Pattern (**d**).

47. Use the window $-2 \le x \le 4, -9 \le y \le 9$.
   x-intercepts: $-1, 2$
   local maximum: $(0,4)$
   local minimum: $(2,0)$

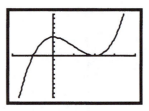

49. Use the window $-2 \le x \le 6, -1 \le y \le 8$.
   x-intercepts: $0, 4$
   local maximum: $(2,4)$
   local minima: $(0,0), (4,0)$

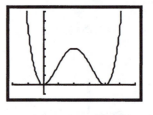

51. Use the window $-4 \le x \le 8, -200 \le y \le 50$.
   x-intercept: 6.72
   local maximum: $(-.12, -44.73)$
   local minimum: $(4.23, -167.99)$

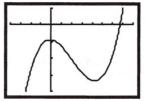

53. Use the window $-2 \le x \le 2, -1 \le y \le 3$.
   x-intercepts: $-1.69, 1, 1.51$
   local maxima: $(-1.30, 2.58), (.34, 1.23)$
   local minima: $(-.34, .77), (1.30, -.58)$

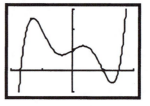

55. a.

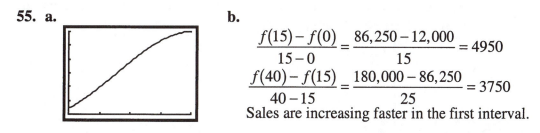

b.
$$\frac{f(15)-f(0)}{15-0} = \frac{86,250-12,000}{15} = 4950$$
$$\frac{f(40)-f(15)}{40-15} = \frac{180,000-86,250}{25} = 3750$$
Sales are increasing faster in the first interval.

c. The inflection point separates intervals on which the rate of increase is increasing from those on which it is decreasing. When the rate of increase is decreasing, the returns are diminishing.

57. a. Revenue: $R(x) = 347x$

Profit: $P(x) = R(x) - C(x) = .0001x^5 + .29x^3 - 8x^2 + 347x - 1600$

Graph this function using the window $0 \le x \le 60, -1000 \le y \le 9000$ to answer the questions below.

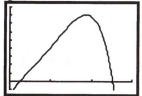

b. Use the zero (root) routine to locate the first positive root at $x = 5.101$, that is, 5101 radiators.

c. Use the zero (root) routine to locate the next positive root at $x = 50.517$, that is, 50,517 radiators.

d. Use the maximum routine to locate the maximum profit at $x = 37.167$, that is, 37,167 radiators.

59. Use revenue $R(x) = 29x$,

profit $P(x) = R(x) - C(x) = -.001x^3 - .06x^2 + 30.5x$.

Graph this function using the window $0 \le x \le 160, -100 \le y \le 2000$

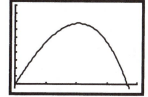

a. Use the maximum routine to locate the maximum at $x = 82.794$, $y = 1546.39$. Thus the maximum profit is $1546.39, if 82,794 name tags are sold.

b. Use the zero (root) routine to locate the right-hand x-intercept at $x = 147.2$. Thus the largest number of tags that can be produced without losing money is 147,200.

61. Use the function derived in the text: $V(x) = x(12 - 2x)(36 - 4x)$. Graph the function in the window $0 \le x \le 6, 0 \le y \le 700$ and use the maximum routine.

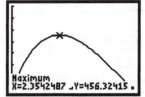

The maximum volume is achieved when squares of side 2.3542 inches are removed.

63. a.

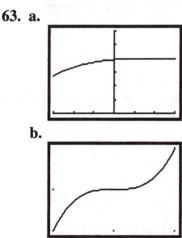

If the graph had the horizontal portion appearing in the window, then the equation $g(x) = 4$ would have infinite solutions, which is impossible since a polynomial equation of degree 3 can have at most 3 solutions.

b.

When graphed in the window shown, which is $1 \le x \le 3, 3.99 \le y \le 4.01$, the graph has a much smaller incorrect horizontal portion.

c. If the graph of a polynomial of degree n had a horizontal portion appearing in the window, it would be a portion also of the line $y = k$. Then the equation $f(x) = k$ would have infinite solutions, which is impossible since a polynomial equation of degree n can have at most n solutions.

65. a. In the standard window the graphs appear similar only in the region between 0 and 2.

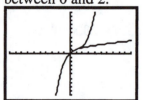

b. In this window the graphs appear similar.

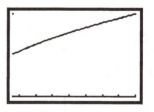

c. The maximum error is .00073994.

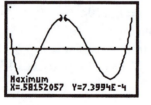

67. a. The general shape of the graph would be something like this, with an intercept   and a minimum at $x = 2$.

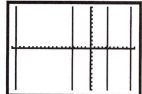

b. The graph is not correct; only 4 intercepts are shown, not including 2.

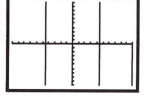

c. The graph is still not correct and 2 is still missed.

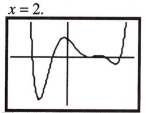

d. Different windows would be needed for different portions of the graph.

$-20 \le x \le -3, -5 \times 10^6 \le y \le 10^6$

$-3 \le x \le 2, -5000 \le y \le 60,000$

$1 \le x \le 5, -5000 \le y \le 5000$

$5 \le x \le 11, -100,000 \le y \le 100,000$

The Table feature of the TI-83 can be very useful in estimating the size of the y-values.

4.4A Polynomial Models

1. A curve that would fit this data well would have one local maximum and one local minimum and therefore be a polynomial of degree 3. A cubic model seems best.

3. A curve that would fit this data well would have one local maximum and therefore be a polynomial of degree 2. A quadratic model seems best.

5. a. Entering the data into the TI-83 and applying the cubic regression routine yields the following:

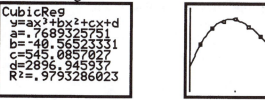

b. Evaluating the function in **(a)** gives:

1987 $(x = 7)$—4988.6, 1993 $(x = 13)$—4816.9.

c. Evaluating the function in **(a)** gives 4541.2 for 1995 $(x = 15)$.

d. Noting that the data shows an uptick for 2001 while the graph continues to decrease suggests that the model will immediately cease to be valid.

7. a. Entering the data into the TI-83 yields the following:

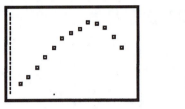 

   b. Evaluating the function in **(a)** gives:
   noon $(x = 12)$ 80.4°, 9 A.M. $(x = 9)$ 69°, 2 P.M. $(x = 14)$ 82.8°

9. a. Entering the data into the TI-83 yields the following:

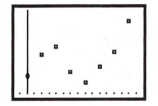

   b. Of the two choices offered, a quartic model seems better, although an optimist might prefer a cubic model. The TI-83 gives the following quartic model:

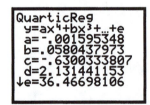

   c. Evaluating the function in **(b)** gives $42,546.
   d. Since the quartic model has a negative leading coefficient, at a certain point the graph will achieve a maximum and then head down to zero and below (2009 yields negative income). We can hope that this is inaccurate.

11. a. The scatter plot is shown:

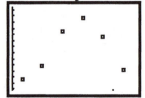

b. Quadratic, cubic, and quartic models are shown and graphed:

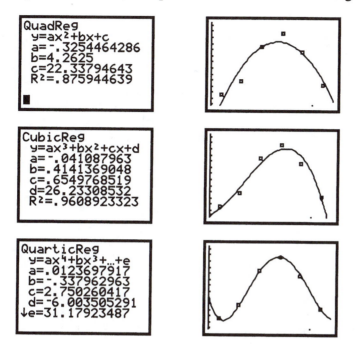

```
QuadReg
  y=ax²+bx+c
  a=⁻.3254464286
  b=4.2625
  c=22.33794643
  R²=.875944639
■
```

```
CubicReg
  y=ax³+bx²+cx+d
  a=⁻.041087963
  b=.4141369048
  c=.6549768519
  d=26.23308532
  R²=.9608923323
```

```
QuarticReg
  y=ax⁴+bx³+...+e
  a=.0123697917
  b=⁻.337962963
  c=2.750260417
  d=⁻6.003505291
↓e=31.17923487
```

c. The quartic model both fits the data best, and, with its indication that the decreasing trend is levelling off, probably seems most reasonable for the immediate future.

13. a. Cubic and quartic models are shown and graphed:

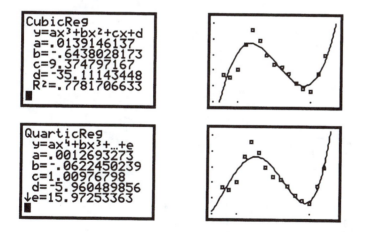

```
CubicReg
  y=ax³+bx²+cx+d
  a=.0139146137
  b=⁻.6438028173
  c=9.374797167
  d=⁻35.11143448
  R²=.7781706633
■
```

```
QuarticReg
  y=ax⁴+bx³+...+e
  a=.0012693273
  b=⁻.0622450239
  c=1.00976798
  d=⁻5.960489856
↓e=15.97253363
■
```

b. Both models seem to rise disturbingly steeply and only a pessimist would expect them to be accurate very far into the future.

4.5 Rational Functions

1. $f(x)$ is defined except where $2x + 5 = 0$, or $x = -\frac{5}{2}$.

3. $h(x)$ is defined except where $x^2 - 6x + 4 = 0$, or $x = 3 \pm \sqrt{5}$.
   Domain: $\left(-\infty, 3 - \sqrt{5}\right) \cup \left(3 - \sqrt{5}, 3 + \sqrt{5}\right) \cup \left(3 + \sqrt{5}, \infty\right)$

5. $f(x)$ is defined except where $x^3 - x^2 - 2x + 2 = 0$.
$$x^3 - x^2 - 2x + 2 = 0$$
$$(x - 1)(x^2 - 2) = 0$$
$$(x - 1)(x - \sqrt{2})(x + \sqrt{2}) = 0$$
$$x = 1, \sqrt{2}, -\sqrt{2}$$
   Domain: $\left(-\infty, -\sqrt{2}\right) \cup \left(-\sqrt{2}, 1\right) \cup \left(1, \sqrt{2}\right) \cup \left(\sqrt{2}, \infty\right)$

7. The graph of f has vertical asymptote $x = 2$. **F**.

9. The graph of h has vertical asymptote $x = -1$. **A**.

11. $f(x) = \dfrac{x^2 + 4}{x^2 - 5x - 6} = \dfrac{x^2 + 4}{(x - 6)(x + 1)}$
   vertical asymptotes: $x = 6, x = -1$
   holes: none

13. $f(x) = \dfrac{x}{x^3 + 2x^2 + x} = \dfrac{x}{x(x + 1)^2} = \dfrac{1}{(x + 1)^2}$ if $x \neq 0$.
   vertical asymptote: $x = -1$
   holes: $x = 0$

15. $f(x) = \dfrac{x^2 - 4x + 4}{(x + 2)(x - 2)^3} = \dfrac{(x - 2)^2}{(x + 2)(x - 2)^3} = \dfrac{1}{(x + 2)(x - 2)}$ if $x \neq 2$.
   vertical asymptotes: $x = -2, x = 2$
   holes: none

17. $f(x) = \dfrac{3x - 2}{x + 3} = \dfrac{3 - \frac{2}{x}}{1 + \frac{3}{x}} \to 3$ as $x \to \infty$.
   horizontal asymptote: $y = 3$
   window: $-100 \leq x \leq 100, -8 \leq y \leq 8$

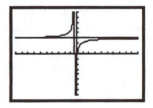

19. $h(x) = \dfrac{5 - x}{x - 2} = \dfrac{-1 + \frac{5}{x}}{1 - \frac{2}{x}} \to -1$ as $x \to \infty$.
   horizontal asymptote: $y = -1$
   window: $-40 \leq x \leq 40, -8 \leq y \leq 8$

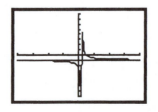

21. $g(x) = \dfrac{5x^3 - 8x^2 + 4}{2x^3 + 2x} = \dfrac{5 - \frac{8}{x} + \frac{4}{x^3}}{2 + \frac{2}{x^2}} \to \dfrac{5}{2}$ as $x \to \infty$.

horizontal asymptote: $y = \dfrac{5}{2}$

window: $-40 \le x \le 40, -8 \le y \le 8$

23. $f(x) = \dfrac{1}{x+5}$

vertical asymptote: $x = -5$
horizontal asymptote: $y = 0$
holes: none

25. $f(x) = \dfrac{3x}{x-1} = \dfrac{3}{1 - \frac{1}{x}}, \; x \neq 0$

vertical asymptote: $x = 1$
horizontal asymptote: $y = 3$
holes: none

27. $f(x) = \dfrac{2-x}{x-3} = \dfrac{-1 + \frac{2}{x}}{1 - \frac{3}{x}}, \; x \neq 0$

vertical asymptote: $x = 3$
horizontal asymptote: $y = -1$
holes: none

29. $f(x) = \dfrac{1}{x(x+1)^2}$

vertical asymptotes: $x = 0, \; x = -1$
horizontal asymptote: $y = 0$
holes: none

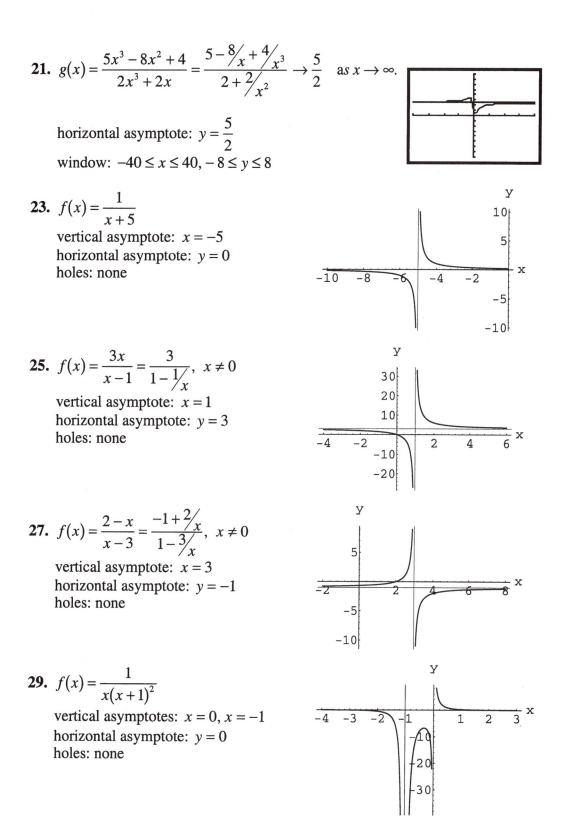

31. $f(x) = \dfrac{x-3}{x^2+x-2} = \dfrac{\frac{1}{x} - \frac{3}{x^2}}{1 + \frac{1}{x} - \frac{2}{x^2}}, \quad x \neq 0$

vertical asymptotes: $x = -2, x = 1$
horizontal asymptote: $y = 0$
holes: none

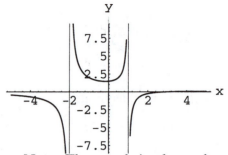

Note: The graph is above the x-axis for $x > 2$ (not seen in this small picture).

33. $h(x) = \dfrac{(x^2+6x+5)(x+5)}{(x+5)^3(x-1)} = \dfrac{(x+1)(x+5)^2}{(x+5)^3(x-1)}$

$= \dfrac{x+1}{(x+5)(x-1)}, \quad x \neq -5$

vertical asymptotes: $x = -5, x = 1$
horizontal asymptote: $y = 0$
holes: none

35. $h(x) = \dfrac{-4x^2+1}{x^2} = -4 + \dfrac{1}{x^2}$

vertical asymptote: $x = 0$
horizontal asymptote: $y = -4$
holes: none

37. $q(x) = \dfrac{x^2+2x}{x^2-4x-5} = \dfrac{1 + \frac{2}{x}}{1 - \frac{4}{x} - \frac{5}{x^2}}, \quad x \neq 0$

$q(x) = \dfrac{x(x+2)}{(x-5)(x+1)}$

vertical asymptotes: $x = 5, x = -1$
horizontal asymptote: $y = 1$
holes: none

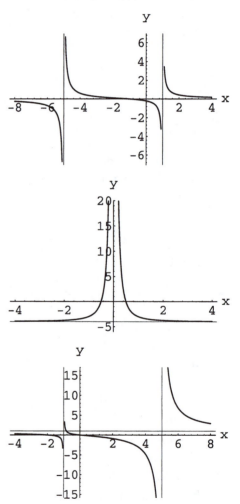

39. $p(x) = \dfrac{(x+3)(x-3)}{(x-5)(x+4)(x+3)}$

$= \dfrac{x-3}{(x-5)(x+4)}, \quad x \neq -3$

vertical asymptotes: $x = 5, x = -4$

horizontal asymptote: $y = 0$

hole: $x = -3, y = .75$ (not shown here)

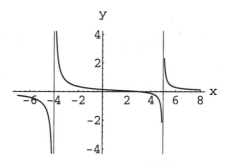

41. vertical asymptotes: $x = \pm 2, x = \pm 3$

horizontal asymptote: $y = 0$

holes: none

Three windows are needed.

$-5 \leq x \leq 4.4, -8 \leq y \leq 4$

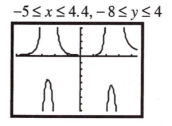

$-2 \leq x \leq 2, -.5 \leq y \leq .5$

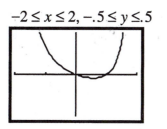

$-15 \leq x \leq -3, -.07 \leq y \leq .02$

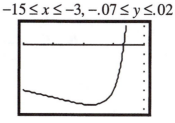

43. $h(x) = \dfrac{(2x+3)(x-2)}{x(x-2)(x+3)}$

$= \dfrac{2x+3}{x(x+3)}, \quad x \neq 2$

vertical asymptotes: $x = 0, x = -3$

horizontal asymptote: $y = 0$

holes: $x = 2, y = .7$

Window: $-9.4 \leq x \leq 9.4, -4 \leq y \leq 4$

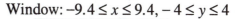
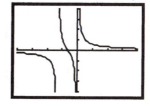

45. $h(x) = \dfrac{(3x+4)(x-1)}{x(2x-5)}$

vertical asymptotes: $x = 0, x = \frac{5}{2}$

horizontal asymptote: $y = \frac{3}{2}$

holes: none

Two windows are needed.

$-4.7 \leq x \leq 4.7, -8 \leq y \leq 8$

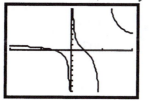

$-40 \leq x \leq 35, -2 \leq y \leq 3$

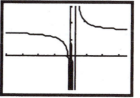

47. $h(x) = \dfrac{(x-4)}{(x-2)^2(2x+3)}$

vertical asymptotes: $x = 2$, $x = -\frac{3}{2}$
horizontal asymptote: $y = 0$
holes: none
Two windows are needed.

$-4.7 \le x \le 4.7, -2 \le y \le 2$

$3 \le x \le 14, -.02 \le y \le .01$

49. $f(x) = \dfrac{1}{x}$

$$\dfrac{f(x+h)-f(x)}{h} = \dfrac{\dfrac{1}{x+h}-\dfrac{1}{x}}{h} = \dfrac{x-(x+h)}{hx(x+h)} = \dfrac{-h}{hx(x+h)} = \dfrac{-1}{x(x+h)}$$

51. $g(x) = \dfrac{1}{x^2}$

$$\dfrac{g(x+h)-g(x)}{h} = \dfrac{\dfrac{1}{(x+h)^2}-\dfrac{1}{x^2}}{h} = \dfrac{x^2-(x+h)^2}{hx^2(x+h)^2} = \dfrac{-2xh-h^2}{hx^2(x+h)^2} = \dfrac{-2x-h}{x^2(x+h)^2}$$

53. a. From 2 to 2.1: $x = 2$, $h = .1$ $\dfrac{-1}{x(x+h)} = \dfrac{-1}{2(2+.1)} = -\dfrac{1}{4.2} = -.2381$

From 2 to 2.01: $x = 2$, $h = .01$ $\dfrac{-1}{x(x+h)} = \dfrac{-1}{2(2+.01)} = -\dfrac{1}{4.02} = -.2488$

From 2 to 2.001: $x = 2$, $h = .001$ $\dfrac{-1}{x(x+h)} = \dfrac{-1}{2(2+.001)} = -\dfrac{1}{4.002} = -.2499$

Instantaneous rate of change: $-\dfrac{1}{4}$

b. From 3 to 3.1: $x = 3$, $h = .1$ $\dfrac{-1}{x(x+h)} = \dfrac{-1}{3(3+.1)} = -\dfrac{1}{9.3} = -.1075$

From 3 to 3.01: $x = 3$, $h = .01$ $\dfrac{-1}{x(x+h)} = \dfrac{-1}{3(3+.01)} = -\dfrac{1}{9.03} = -.1107$

From 3 to 3.001: $x = 3$, $h = .001$ $\dfrac{-1}{x(x+h)} = \dfrac{-1}{3(3+.001)} = -\dfrac{1}{9.003} = -.1111$

Instantaneous rate of change: $-\dfrac{1}{9}$

c. They are identical.

55. a. $g(x) = \dfrac{x-1}{x-2}$ **b.** $g(x) = \dfrac{x-1}{-x-2}$

   c. Clearly, $x = 2$ and $x = -2$ are vertical asymptotes for the two "halves "of the graph. Graph using the window $-4.7 \le x \le 4.7, -6 \le y \le 6$.

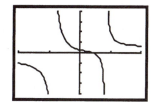

57. For the top, area $= x^2$, cost $= 2.5x^2$

   For the bottom, area $= x^2$, cost $= 2.5x^2$

   For the sides, area $= 4 \cdot \dfrac{1000}{x}$, cost $= 1.5 \cdot 4 \cdot \dfrac{1000}{x} = \dfrac{6000}{x}$

   Cost $= c(x) = 2.5x^2 + 2.5x^2 + \dfrac{6000}{x}$

   $c(x) = 5x^2 + \dfrac{6000}{x}$

   Graph in the window $0 \le x \le \sqrt{500}, \ 0 \le y \le 5000$ to obtain minimum $x = 8.4343$ inches. Then $h = \dfrac{1000}{8.4343^2} = 14.057$ inches.

59. a. Cost $= c(x) = 1.40(500)\dfrac{800 + x^2}{200x} = \dfrac{2800 + 3.5x^2}{x}$

   b. Solve $250 = \dfrac{2800 + 3.5x^2}{x}$

   $250x = 2800 + 3.5x^2$

   $0 = 3.5x^2 - 250x + 2800$

   From the quadratic formula, $x = 13.9$ mph and $x = 57.5$ mph are both solutions.

   c. Graph $c(x)$ in the window $0 \le x \le 100, 0 \le y \le 1000$ to find the minimum at 28.3 mph.

61. a. Let $y =$ the length of the non-parallel sides. Then $xy = 250, \ y = \dfrac{250}{x}$.

   Total length $= P(x) = x + 2y = x + 2\left(\dfrac{250}{x}\right) = x + \dfrac{500}{x}$

   b. Graph $P(x)$ and $y = 60$ in the window $0 \le x \le 60, 0 \le y \le 100$. Then use the intersection routine to find: if $10 < x < 50$, then $P(x) < 60$.

   c. Graph $P(x)$ in the same window, then use the minimum routine to find $x \approx 22.4$. The dimensions are then 22.4 meters by 11.2 meters.

63. a. $h_1 = h - 2$

b. Since volume $= 150 = \pi r^2 h$, $h = \dfrac{150}{\pi r^2}$. Hence $h_1 = \dfrac{150}{\pi r^2} - 2$.

c. Set $r_1 = r - 1$. Then $V = \pi r_1^2 h_1$, hence

$$V = \pi(r-1)^2 \left(\frac{150}{\pi r^2} - 2 \right)$$

d. Since the walls are 1 foot thick, r must be greater than 1.

e. Graph V in the window $0 \le r \le 5, 0 \le V \le 150$, then use the maximum routine to find $r = 2.88$ feet. Hence

$$h = \frac{150}{\pi r^2} = 5.76 \text{ feet.}$$

65. a. $g(0) = \dfrac{3.987 \times 10^{14}}{\left(6.378 \times 10^6\right)^2} = 9.801$ meters per second^2.

b. Use the window $0 \le r \le 9{,}000{,}000, \ 0 \le g(r) \le 10$.

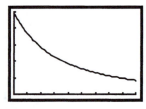

c. Since the r-axis is a horizontal asymptote, and the graph never crosses the axis, as $r \to \infty$, g approaches 0 but never reaches 0. You can never escape the pull of gravity.

4.5A Other Rational Functions

1. Divide to obtain $f(x) = x + \dfrac{4x-1}{x^2-4}$. Then the nonvertical asymptote is $y = x$.

Use viewing window $-14 \le x \le 14, -14 \le y \le 14$.

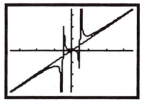

(vertical segments are erroneous; the calculator incorrectly connects points on opposite sides of the vertical asymptotes.)

3. Divide to obtain $h(x) = x^2 - x + \dfrac{1}{x+4}$. Then the nonvertical asymptote is $y = x^2 - x$.

Use viewing window $-6 \le x \le 6, -20 \le y \le 40$.

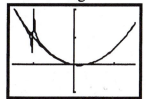

(vertical segments are erroneous; the calculator incorrectly connects points on opposite sides of the vertical asymptotes.)

5. $f(x) = \dfrac{x^2 - x - 6}{x - 2} = \dfrac{(x-3)(x+2)}{x-2}$

$= x + 1 + \dfrac{-4}{x - 2}$

vertical asymptote: $x = 2$
nonvertical asymptote: $y = x + 1$
holes: none

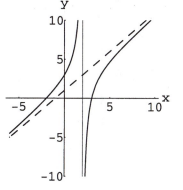

7. $Q(x) = \dfrac{4x^2 + 4x - 3}{2x - 5} = \dfrac{(2x-1)(2x+3)}{2x-5}$

$= 2x + 7 + \dfrac{32}{2x - 5}$

vertical asymptote: $x = \frac{5}{2}$
nonvertical asymptote: $y = 2x + 7$
holes: none

9. $f(x) = \dfrac{x^3 - 2}{x - 1} = x^2 + x + 1 + \dfrac{-1}{x - 1}$.

vertical asymptote: $x = 1$
nonvertical asymptote: $y = x^2 + x + 1$
holes: none

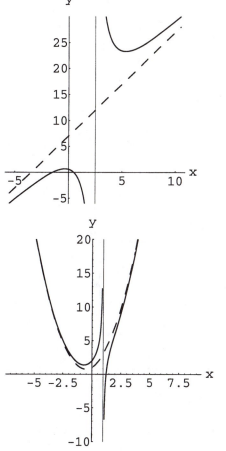

11. $q(x) = \dfrac{x^3 - 1}{x - 2} = x^2 + 2x + 4 + \dfrac{7}{x - 2}$.

vertical asymptote: $x = 2$
nonvertical asymptote:
$\quad y = x^2 + 2x + 4$
holes: none

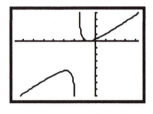

13. $f(x) = \dfrac{2x^2 + 5x + 2}{2x + 7} = x - 1 + \dfrac{9}{2x + 7}$

vertical asymptote: $x = -\frac{7}{2}$
nonvertical asymptote: $y = x - 1$
$-15.5 \le x \le 8.5, -16 \le y \le 8$

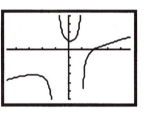

15. $h(x) = \dfrac{x^3 - 2x^2 + x - 2}{x^2 - 1} = x - 2 + \dfrac{2x - 4}{x^2 - 1}$

vertical asymptotes: $x = \pm 1$
nonvertical asymptote: $y = x - 2$
$-4.7 \le x \le 4.7, -12 \le y \le 8$

17. $g(x) = \dfrac{2x^4 + 7x^3 + 7x^2 + 2x}{x^3 - x + 50} = 2x + 7 + \dfrac{9x^2 - 91x - 350}{x^3 - x + 50}$

To find the vertical asymptote, solve $x^3 - x + 50 = 0$ to obtain $x = -3.774494$.
nonvertical asymptote: $y = 2x + 7$
Two windows are needed:
$-13 \le x \le 7, -20 \le y \le 20$ $\qquad\qquad -2.5 \le x \le 1, -.02 \le y \le .02$

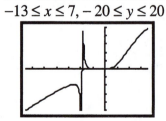   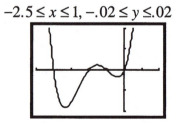

19. a.

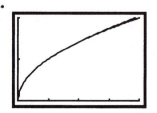

b. Shown is the graph of $r(x) - s(x)$ in the window $0 \le x \le 4, -.01 \le y \le .02$ together with the horizontal line $y = .01$. The graph is below the line, and thus $r(x)$ differs from $s(x)$ by less than .01, in the interval $[.06, 2.78]$.

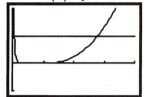

4.6 Polynomial and Rational Inequalities

1. $2x + 4 \le 7$

$\quad 2x \le 3$

$\quad x \le \dfrac{3}{2}$

$\quad \left(-\infty, \dfrac{3}{2}\right]$

3. $3 - 5x < 13$

$\quad -5x < 10$

$\quad x > -2$

$\quad (-2, \infty)$

5. $6x + 3 \le x - 5$

$\quad 5x + 3 \le -5$

$\quad 5x \le -8$

$\quad x \le -\dfrac{8}{5}$

$\quad \left(-\infty, -\dfrac{8}{5}\right]$

7. $5 - 7x < 2x - 4$

$\quad 5 - 9x < -4$

$\quad -9x < -9$

$\quad x > 1$

$\quad (1, \infty)$

9. $2 < 3x - 4 < 8$

$\quad 6 < 3x < 12$

$\quad 2 < x < 4$

$\quad (2, 4)$

11. $0 < 5 - 2x \le 11$

$\quad -5 < -2x \le 6$

$\quad \dfrac{5}{2} > x \ge -3$

$\quad \left[-3, \dfrac{5}{2}\right)$

13. $2x + 7(3x - 2) < 2(x - 1)$

$\quad 2x + 21x - 14 < 2x - 2$

$\quad 21x - 14 < -2$

$\quad 21x < 12$

$\quad x < \dfrac{4}{7}$

$\quad \left(-\infty, \dfrac{4}{7}\right)$

15. $\dfrac{x+1}{2} - 3x \le \dfrac{x+5}{3}$

$\quad 3(x+1) - 18x \le 2(x+5)$

$\quad 3x + 3 - 18x \le 2x + 10$

$\quad -15x + 3 \le 2x + 10$

$\quad -17x \le 7$

$\quad x \ge -\dfrac{7}{17}$

$\quad \left[-\dfrac{7}{17}, \infty\right)$

17. Solve $2x + 3 \leq 5x + 6$ and $5x + 6 < -3x + 7$

$\qquad\qquad -3x \leq 3 \qquad\qquad\qquad\quad 8x < 1$

$\qquad\qquad\quad x \geq -1 \qquad\qquad\qquad\quad x < \dfrac{1}{8}$

$[-1, 1/8)$

19. Solve $3 - x < 2x + 1$ and $2x + 1 \leq 3x - 4$

$\qquad\qquad -3x \leq -2 \qquad\qquad\qquad -x \leq -5$

$\qquad\qquad\quad x > \dfrac{2}{3} \qquad\qquad\qquad\quad x \geq 5$

Combining these yields $x \geq 5 \colon [5, \infty)$

21. $ax - b < c$

$\quad ax < b + c$

$\quad x < \dfrac{b + c}{a}$

23. $0 < x - c < a$

$\quad c < x < a + c$

25. The solutions of $x^2 - 4x + 3 \leq 0$ are the numbers x at which the graph of $f(x) = x^2 - 4x + 3 = (x - 1)(x - 3)$ lies on or below the x-axis. This condition holds when x is between the roots 1 and 3. Solution: $[1, 3]$

27. $x^2 + 9x + 15$ has roots $x = \dfrac{-9 \pm \sqrt{21}}{2}$. The solutions of $x^2 + 9x + 15 > 0$ are the values that are not between thsese roots: $\left[\dfrac{-9 - \sqrt{21}}{2}, \dfrac{-9 + \sqrt{21}}{2} \right]$.

29. $8 + x - x^2 \leq 0$

$x^2 - x - 8 \geq 0$

$x^2 - x - 8$ has roots $\dfrac{1 \pm \sqrt{33}}{2}$. The solutions of $x^2 - x - 8 \geq 0$ are the values of x that are not between these roots: $\left[\dfrac{1 - \sqrt{33}}{2}, \dfrac{1 + \sqrt{33}}{2} \right]$.

31. $x^3 - x \geq 0$

$x(x - 1)(x + 1) \geq 0$

Interval	$x < -1$	$-1 < x < 0$	$0 < x < 1$	$1 < x$
Test number	-2	$-\frac{1}{2}$	$\frac{1}{2}$	2
Sign	$-$	$+$	$-$	$+$
Graph	Below x-axis	Above x-axis	Below x-axis	Above x-axis

$[-1, 0] \cup [1, \infty]$

33. $x^3 - 2x^2 - 3x < 0$

$x(x-3)(x+1) < 0$

Interval	$x < -1$	$-1 < x < 0$	$0 < x < 3$	$3 < x$
Test number	-2	$-\frac{1}{2}$	1	4
Sign	$-$	$+$	$-$	$+$
Graph	Below x-axis	Above x-axis	Below x-axis	Above x-axis

$(-\infty, -1) \cup (0, 3)$

35. $x^4 - 5x^2 + 4 < 0$ $(x-1)(x-2)(x+1)(x+2) < 0$

Interval	$x < -2$	$-2 < x < -1$	$-1 < x < 1$	$1 < x < 2$	$2 < x$
Test number	-3	-1.5	0	1.5	3
Sign	$+$	$-$	$+$	$-$	$+$
Graph	Above x-axis	Below x-axis	Above x-axis	Below x-axis	Above x-axis

$(-2, -1) \cup (1, 2)$

37. $x^3 - 2x^2 - 5x + 7 \geq 2x + 1$

$x^3 - 2x^2 - 7x + 6 \geq 0$

Graph this equation in the standard window and apply the zero (root) routine to find that the graph is above the x-axis in the intervals $[-2.264, .756] \cup [3.508, \infty)$.

39. $2x^4 + 3x^3 < 2x^2 + 4x - 2$

$2x^4 + 3x^3 - 2x^2 - 4x + 2 < 0$

Graph this equation in the window $-5 \leq x \leq 5, -1 \leq y \leq 10$ and apply the zero (root) routine to find that the graph is below the x-axis in the interval $(.5, .839)$.

41. $\dfrac{3x+1}{2x-4} > 0$

Interval	$x < -\frac{1}{3}$	$-\frac{1}{3} < x < 2$	$2 < x$
Test number	-1	0	3
Sign	$+$	$-$	$+$
Graph	Above x-axis	Below x-axis	Above x-axis

$\left(-\infty, -\frac{1}{3}\right) \cup (2, \infty)$

43. $\dfrac{x^2 + x - 2}{x^2 - 2x - 3} < 0$ $\dfrac{(x+2)(x-1)}{(x-3)(x+1)} < 0$

Interval	$x < -2$	$-2 < x < -1$	$-1 < x < 1$	$1 < x < 3$	$3 < x$
Test number	-3	-1.5	0	2	4
Sign	$+$	$-$	$+$	$-$	$+$
Graph	Above x-axis	Below x-axis	Above x-axis	Below x-axis	Above x-axis

$(-2, -1) \cup (1, 3)$

45. $\dfrac{x-2}{x-1} < 1$

$\dfrac{x-2}{x-1} - 1 < 0$

$\dfrac{-1}{x-1} < 0$

Interval	$x < 1$	$1 < x$
Test number	0	2
Sign	+	−
Graph	Above x-axis	Below x-axis

$(1, \infty)$

47. $\dfrac{x-3}{x+3} \le 5$

$\dfrac{x-3}{x+3} - 5 \le 0$

$\dfrac{x-3-5(x+3)}{x+3} \le 0$

$\dfrac{-4x-18}{x+3} \le 0$

Interval	$x < -\frac{18}{4} = -\frac{9}{2}$	$-\frac{18}{4} < x < -3$	$-3 < x$
Test number	−5	−4	0
Sign	−	+	−
Graph	Below x-axis	Above x-axis	Below x-axis

$\left(-\infty, -\frac{9}{2}\right] \cup (-3, \infty)$

49. $\dfrac{2}{x+3} \ge \dfrac{1}{x-1}$

$\dfrac{2(x-1)-(x+3)}{(x+3)(x-1)} \ge 0$

$\dfrac{x-5}{(x+3)(x-1)} \ge 0$

Interval	$x < -3$	$-3 < x < 1$	$1 < x < 5$	$5 < x$
Test number	−4	0	2	6
Sign	−	+	−	+
Graph	Below x-axis	Above x-axis	Below x-axis	Above x-axis

$(-3, 1) \cup [5, \infty)$

51. Here a complete solution is necessarily approximate. Graph the function

$$f(x) = \frac{x^3 - 3x^2 + 5x - 29}{x^2 - 7}$$

and the horizontal line $y = 3$ in the window $-10 \le x \le 10, -20 \le y \le 20$.

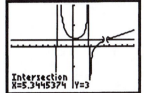

Observe that the graph of the function is above the horizontal line between the vertical asymptotes $x = \pm\sqrt{7}$ as well as for $x > 5.3445$.

$$\left(-\sqrt{7}, \sqrt{7}\right) \cup (5.3445, \infty)$$

53. $\dfrac{2x^2 + 6x - 8}{2x^2 + 5x - 3} < 1$

$$\frac{2x^2 + 5x - 3 + x - 5}{2x^2 + 5x - 3} < 1$$

$$1 + \frac{x - 5}{2x^2 + 5x - 3} < 1$$

$$\frac{x - 5}{(2x - 1)(x + 3)} < 0$$

Interval	$x < -3$	$-3 < x < \frac{1}{2}$	$\frac{1}{2} < x < 5$	$5 < x$
Test number	-4	0	1	6
Sign	$-$	$+$	$-$	$+$
Graph	Below x-axis	Above x-axis	Below x-axis	Above x-axis

$$(-\infty, -3) \cup \left(\tfrac{1}{2}, 5\right)$$

55. $x > -1.43$ **57.** $x < -3.79$ or $x > .79$

59. Let x = rate per kwh. Then

Cost of first freezer = $62,395 + 144(90x)$ over 12 years (144 months)

Cost of second freezer = $50,000 + 144(100x)$

Solve $62,395 + 144(90x) < 50,000 + 144(100x)$

$$-1440x < -12,395$$

$$x > 8.6 \text{ cents per kwh.}$$

61. Let x = amount of sales. Then

Earnings of first = $1000 + .02x$

Earnings of second = $.1x$

$.1x > 1000 + .02x$

$.08x > 1000$

$x > 12,500$

At least $12,500.

63. Let x = investment at 10%. Then

$.1x$ = income at 10%

$.08(5000)$ = income at 8%

$$800 < .1x + .08(5000) < 940$$

$$800 < .1x + 400 < 940$$

$$400 < .1x < 540$$

$4000 < x < 5400$
Between $4000 and $5400.

65. Let x = one number, $20 - x$ = other number. Then

$$x^2 + (20 - x)^2 < 362$$

$$2x^2 - 40x + 400 < 362$$

$$2x^2 - 40x + 38 < 0$$

$$x^2 - 20x + 19 < 0$$

The solutions are the numbers x for which the graph of $f(x) = x^2 - 20x + 19 = (x-1)(x-19)$ lies below the x-axis. This condition holds when x is between the roots 1 and 19. Thus any two numbers between 1 and 19 whose sum is 20, e.g. 2 and 18, 3 and 17, π and $20 - \pi$, etc., satisfy the conditions.

67. Cost $= 350 + 5x$ Revenue $= x(50 - x)$

Profit = Revenue − Cost
For the profit to be positive, then:

$$x(50 - x) - (350 + 5x) > 0$$

$$50x - x^2 - 350 - 5x > 0$$

$$x^2 - 45x + 350 < 0$$

The solutions are the numbers x for which the graph of $f(x) = x^2 - 45x + 350 = (x-10)(x-35)$ lies below the x-axis. This condition holds when x is between the roots 10 and 35. More than 10 but fewer than 35 medallions.

69. $h = -16t^2 + 80t$. Solve

$$-16t^2 + 80t \geq 64$$

$$-16t^2 + 80t - 64 \geq 0$$

$$t^2 - 5t + 4 \leq 0$$

The solutions are the numbers x for which the graph of $f(t) = t^2 - 5t + 4 = (t-1)(t-4)$ lies below the t-axis. This condition holds when t is between the roots 1 and 4.

$1 \leq t \leq 4$

71. $h = -16t^2 + 120t$. Solve $39 < -16t^2 + 120 < 56$

Graph the equation $h = -16t^2 + 120t$ in the window $0 \leq x \leq 5, 0 \leq y \leq 120$ together with the horizontal lines $y = 39$ and $y = 56$ and apply the intersection routine twice to obtain $2 < t < 2.25$.

73. a.

$$x^2 < x \qquad\qquad x^2 > x$$

$$x^2 - x < 0 \qquad\qquad x^2 - x > 0$$

$$x(x-1) < 0 \qquad\qquad x(x-1) > 0$$

Interval	$x < 0$	$0 < x < 1$	$1 < x$
Test number	-1	$\frac{1}{2}$	2
Sign	$+$	$-$	$+$
Graph	Above x-axis	Below x-axis	Above x-axis

Thus $x^2 < x$: if $0 < x < 1$

$\qquad x^2 > x$: if $x < 0$ or $x > 1$

b. If $0 < c < 1$, then $c^2 < c$. Since $|c| = c$, $c^2 < |c|$.

If $-1 < c < 0$, then $-c > c^2$. Since $|c| = -c$, $c^2 < |c|$.

Thus if $-1 < c < 1$, $c \neq 0$, $c^2 < |c|$.

c. If $c > 1$, then $c^2 > c$.

If $c < -1$, then $c^2 > -c$. Since $c < 0$, $-c > c$, hence $c^2 > c$.

Thus if $|c| > 1$, $c^2 > c$.

4.6A Absolute Value Inequalities

1. $|3x + 2| \leq 2$

$\qquad -2 \leq 3x + 2 \leq 2$

$\qquad -4 \leq 3x \leq 0$

$\qquad -4/3 \leq x \leq 0$

3. $|3 - 2x| < 2/3$

$\qquad -2/3 < 3 - 2x < 2/3$

$\qquad -2 < 9 - 6x < -2$

$\qquad -11 < -6x < -7$

$\qquad 11/6 > x > 7/6$

$\qquad 7/6 < x < 11/6$

5. $|2x + 3| > 1$

$\qquad 2x + 3 > 1 \quad$ or $\quad 2x + 3 < -1$

$\qquad\quad 2x > -2 \qquad\qquad 2x < -4$

$\qquad\quad\ \ x > -1 \qquad\qquad\ \ x < -2$

7. $|5x + 2| \geq 3/4$

$\qquad 5x + 2 \geq 3/4 \quad$ or $\quad 5x + 2 \leq -3/4$

$\qquad\quad 5x \geq -5/4 \qquad\qquad 5x \leq -11/4$

$\qquad\quad\ \ x \geq -1/4 \qquad\qquad\ \ x \leq -11/20$

9. $|12/5 + 2x| > 1/4$

$\qquad 12/5 + 2x > 1/4 \quad$ or $\quad 12/5 + 2x < -1/4$

$\qquad 48 + 40x > 5 \qquad\qquad 48 + 40x < -5$

$\qquad\quad 40x > -43 \qquad\qquad\quad 40x < -53$

$\qquad\quad\ \ x > -43/40 \qquad\qquad\ \ x < -53/40$

11. $\left|\dfrac{x-1}{x+2}\right| \le 3.$ Graph $\left|\dfrac{x-1}{x+2}\right| - 3$ using the window $-10 \le x \le 5, -5 \le y \le 10.$
The graph is at or below the x-axis on the intervals $(-\infty, -7/2]$ and $(-5/4, \infty]$.

13. $\left|\dfrac{2x-1}{x+5}\right| > 1.$ Graph $\left|\dfrac{2x-1}{x+5}\right| - 1$ using the window $-10 \le x \le 10, -2 \le y \le 6.$
The graph is above the x-axis and the function is defined on the intervals $(-\infty, -5) \cup (-5, -4/3) \cup (6, \infty).$

15. $\left|\dfrac{1-4x}{2+3x}\right| < 1.$ Graph $\left|\dfrac{1-4x}{2+3x}\right| - 1$ using the window $-3 \le x \le 8, -2 \le y \le 2.$
The graph is below the x-axis on the interval $(-1/7, 3).$

17. $\left|x^2 - 2\right| < 1.$ Graph $\left|x^2 - 2\right| - 1$ using the window $-5 \le x \le 5, -2 \le y \le 5.$ Since the roots of the function are the solutions of $x^2 - 2 = 1$ and $x^2 - 2 = -1$, that is, $\pm\sqrt{3}$ and $\pm\sqrt{1}$, the graph is below the x-axis on the interval $\left(-\sqrt{3}, -1\right) \cup \left(1, \sqrt{3}\right).$

19. $\left|x^2 - 2\right| > 4.$ Graph $\left|x^2 - 2\right| - 4$ using the window $-5 \le x \le 5, -2 \le y \le 10.$ Since the roots of the function are the solutions of $\left|x^2 - 2\right| = 4$, that is, $\pm\sqrt{6}$, the graph is above the x-axis on the interval $\left(-\infty, -\sqrt{6}\right) \cup \left(\sqrt{6}, \infty\right).$

21. $\left|x^2 + x - 1\right| \ge 1.$ Graph $\left|x^2 + x - 1\right| - 1.$ using the window $-4 \le x \le 3, -2 \le y \le 5.$ Since the roots of the function are the solutions of $\left|x^2 + x - 1\right| = 1$, that is, $-2, -1, 0$ and 1, the graph is at or above the x-axis on the interval $[-\infty, -2) \cup [-1, 0] \cup [1, \infty).$

23. $\left|3x^2 - 8x + 2\right| < 2.$ Graph $\left|3x^2 - 8x + 2\right| - 2$ using the window $-2 \le x \le 4, -3 \le y \le 8.$ Since the roots of the function are the solutions of $\left|3x^2 - 8x + 2\right| = 2$, that is, $0, 2/3, 2$ and $8/3$, the graph is below the x-axis on the interval $(0, 2/3) \cup (2, 8/3).$

25. $\left|x^5 - x^3 + 1\right| < 2.$ Graph $\left|x^5 - x^3 + 1\right| - 2$ using the window $-2 \le x \le 2, -3 \le y \le 3.$ Since the roots of the function are approximately -1.426 and 1.237, the graph is below the x-axis on the interval $(-1.426, 1.237).$

27. $\left|x^4 - x^3 + x^2 - x + 1\right| > 4.$ Graph $\left|x^4 - x^3 + x^2 - x + 1\right| - 4$ using the window $-2 \le x \le 2, -5 \le y \le 5.$ Since the roots of the function are approximately -0.888 and 1.560, the graph is above the x-axis on the interval $(-\infty, -0.888) \cup (1.560, \infty).$

29. $\dfrac{x+2}{|x-3|} \le 4$. Graph $\dfrac{x+2}{|x-3|} - 4$ using the window $-5 \le x \le 10, -5 \le y \le 5$. Since the roots of the function are 2 and 14/3, the graph is at or below the x-axis on the interval $(-\infty, 2] \cup (14/3, \infty]$.

31. Graph $\left| \dfrac{2x^2 + 2x - 12}{x^3 - x^2 + x - 2} \right| - 2$ using the window $-3 \le x \le 3, -3 \le y \le 5$.

Since the roots of the function are approximately -1.130 and 1.670, the graph is undefined at the real root of $x^3 - x^2 + x - 2 = 0$, approximately 1.353. Thus the graph is above the x-axis on the interval $(-1.130, 1.353) \cup (1.353, 1.670)$.

33. If $|x - 3| < \dfrac{E}{5}$

$$-\frac{E}{5} < x - 3 < \frac{E}{5}$$

$$-E < 5x - 15 < E$$

$$-E < (5x - 11) - 4 < E$$

$$|(5x - 11) - 4| < E$$

4.7 Complex Numbers

1. $(2 + 3i) + (6 - i) = 2 + 3i + 6 - i = 8 + 2i$

3. $(2 - 8i) - (4 + 2i) = 2 - 8i - 4 - 2i = -2 - 10i$

5. $\dfrac{5}{4} - \left(\dfrac{7}{4} + 2i \right) = \dfrac{5}{4} - \dfrac{7}{4} - 2i = -\dfrac{1}{2} - 2i$

7. $\left(\dfrac{\sqrt{2}}{2} + i \right) - \left(\dfrac{\sqrt{3}}{2} - i \right) = \dfrac{\sqrt{2}}{2} + i - \dfrac{\sqrt{3}}{2} + i = \dfrac{\sqrt{2} - \sqrt{3}}{2} + 2i$

9. $(2 + i)(3 + 5i) = 6 + 10i + 3i + 5i^2 = 6 + 13i + 5(-1) = 1 + 13i$

11. $(-3 + 2i)(4 - i) = -12 + 3i + 8i - 2i^2 = -12 + 11i - 2(-1) = -10 + 11i$

13. $(2 - 5i)^2 = 2^2 - 2 \cdot 2(5i) + (5i)^2 = 4 - 20i + 25i^2 = 4 - 20i + 25(-1) = -21 - 20i$

15. $(\sqrt{3} + i)(\sqrt{3} - i) = (\sqrt{3})^2 - i^2 = 3 - (-1) = 4$

17. $i^{15} = (i^4)^3 i^3 = 1^3(-i) = -i$

19. $i^{33} = \left(i^4\right)^8 i = 1^8 i = i$

21. $(-i)^{107} = \left[(-i)^4\right]^{26}(-i)^3 = 1^{26}\left(-i^3\right) = -(-i) = i$

23. $\dfrac{1}{5-2i} = \dfrac{1}{(5-2i)}\dfrac{(5+2i)}{(5+2i)} = \dfrac{5+2i}{25-4i^2} = \dfrac{5+2i}{29} = \dfrac{5}{29} + \dfrac{2}{29}i$

25. $\dfrac{1}{3i} = \dfrac{1}{3i}\dfrac{-i}{-i} = \dfrac{-i}{3\left(-i^2\right)} = \dfrac{-i}{3\cdot 1} = -\dfrac{1}{3}i$

27. $\dfrac{3}{4+5i} = \dfrac{3}{(4+5i)}\dfrac{(4-5i)}{(4-5i)} = \dfrac{12-15i}{16-25i^2} = \dfrac{12-15i}{16-(-25)} = \dfrac{12-15i}{41} = \dfrac{12}{41} - \dfrac{15}{41}i$

29. $\dfrac{1}{i(4+5i)} = \dfrac{1}{4i+5i^2} = \dfrac{1}{-5+4i} = \dfrac{1}{(-5+4i)}\dfrac{(-5-4i)}{(-5-4i)} = \dfrac{-5-4i}{25-16i^2} = \dfrac{-5-4i}{41} = -\dfrac{5}{41} - \dfrac{4i}{41}$

31. $\dfrac{2+3i}{i(4+i)} = \dfrac{2+3i}{4i+i^2} = \dfrac{2+3i}{-1+4i} = \dfrac{(2+3i)}{(-1+4i)}\dfrac{(-1-4i)}{(-1-4i)} = \dfrac{-2-8i-3i-12i^2}{1-16i^2}$

$$= \dfrac{-2-11i+12}{1+16} = \dfrac{10-11i}{17} = \dfrac{10}{17} - \dfrac{11}{17}i$$

33. $\dfrac{2+i}{1-i} + \dfrac{1}{1+2i} = \dfrac{(2+i)(1+2i)}{(1-i)(1+2i)} + \dfrac{1-i}{(1-i)(1+2i)} = \dfrac{2+4i+i+2i^2+1-i}{1+2i-i-2i^2}$

$$= \dfrac{3+4i+2i^2}{1+i-2i^2} = \dfrac{1+4i}{3+i} = \dfrac{(1+4i)}{(3+i)}\dfrac{(3-i)}{(3-i)} = \dfrac{3+12i-i-4i^2}{9-i^2}$$

$$= \dfrac{7+11i}{10} = \dfrac{7}{10} + \dfrac{11}{10}i$$

35. $\dfrac{i}{3+i} - \dfrac{3+i}{4+i} = \dfrac{i(4+i)-(3+i)(3+i)}{(3+i)(4+i)} = \dfrac{4i+i^2-9-6i-i^2}{12+3i+4i+i^2} = \dfrac{-9-2i}{11+7i}$

$$= \dfrac{(-9-2i)}{(11+7i)}\dfrac{(11-7i)}{(11-7i)} = \dfrac{-99+63i-22i+14i^2}{121-49i^2} = \dfrac{-113+41i}{170}$$

$$= -\dfrac{113}{170} + \dfrac{41}{170}i$$

37. $6i$ **39.** $\sqrt{14}i$ **41.** $-4i$

43. $\sqrt{-16} + \sqrt{-49} = 4i + 7i = 11i$

45. $\sqrt{-15} - \sqrt{-18} = \sqrt{15}i - \sqrt{18}i = \left(\sqrt{15} - 3\sqrt{2}\right)i$

47. $\sqrt{-16}/\sqrt{-36} = \sqrt{16}i/\sqrt{36}i = 4i/6i = 2/3$

49. $\left(\sqrt{-25}+2\right)\left(\sqrt{-49}-3\right) = (5i+2)(7i-3) = 35i^2 -15i+14i-6 = -41-i$

51. $\left(2+\sqrt{-5}\right)\left(1-\sqrt{-10}\right) = \left(2+\sqrt{5}i\right)\left(1-\sqrt{10}i\right) = 2 - 2\sqrt{10}i + \sqrt{5}i - \sqrt{50}i^2$

$$= \left(2+\sqrt{50}\right) + \left(\sqrt{5}-2\sqrt{10}\right)i = \left(2+5\sqrt{2}\right) + \left(\sqrt{5}-2\sqrt{10}\right)i$$

53. $\dfrac{1}{1+\sqrt{-2}} = \dfrac{1}{1+i\sqrt{2}} = \dfrac{1}{\left(1+i\sqrt{2}\right)}\dfrac{\left(1-i\sqrt{2}\right)}{\left(1-i\sqrt{2}\right)} = \dfrac{1-i\sqrt{2}}{1-2i^2} = \dfrac{1-i\sqrt{2}}{3} = \dfrac{1}{3} - \dfrac{\sqrt{2}}{3}i$

55. $3x-4i = 6+2yi;$
$3x = 6$ and $-4 = 2y$
$\qquad x = 3 \qquad\quad y = -2$

57. $3+4xi = 2y-3i;$
$3 = 2y$ and $4x = -3$
$\qquad y = 3/2 \qquad x = -3/4$

59. $3x^2 - 2x + 5 = 0$

$$x = \frac{-(-2) \pm \sqrt{(-2)^2 - 4(3)(5)}}{2(3)}$$

$$x = \frac{2 \pm \sqrt{-56}}{6}$$

$$x = \frac{1 \pm i\sqrt{14}}{3}$$

$$\frac{1}{3} + \frac{\sqrt{14}}{3}i, \frac{1}{3} - \frac{\sqrt{14}}{3}i$$

61. $x^2 + x + 2 = 0$

$$x = \frac{-1 \pm \sqrt{1^2 - 4(1)(2)}}{2(1)}$$

$$x = \frac{-1 \pm \sqrt{-7}}{2}$$

$$x = \frac{-1 \pm i\sqrt{7}}{2}$$

$$-\frac{1}{2} + \frac{\sqrt{7}}{2}i, -\frac{1}{2} - \frac{\sqrt{7}}{2}i$$

63. $2x^2 - x = -4$
$2x^2 - x + 4 = 0$

$$x = \frac{-(-1) \pm \sqrt{(-1)^2 - 4(2)(4)}}{2(2)}$$

$$x = \frac{1 \pm \sqrt{-31}}{4}$$

$$x = \frac{1 \pm i\sqrt{31}}{4}$$

$$\frac{1}{4} + \frac{\sqrt{31}}{4}i, \frac{1}{4} - \frac{\sqrt{31}}{4}i$$

65. $2x^2 + 3 = 6x$
$2x^2 - 6x + 3 = 0$

$$x = \frac{-(-6) \pm \sqrt{(-6)^2 - 4(2)(3)}}{2(2)}$$

$$x = \frac{6 \pm \sqrt{12}}{4}$$

$$x = \frac{3 \pm \sqrt{3}}{2}$$

$$\frac{3+\sqrt{3}}{2}, \frac{3-\sqrt{3}}{2}$$

67. $x^3 - 8 = 0$

$(x-2)(x^2 + 2x + 4) = 0$

$x - 2 = 0$ or $x^2 + 2x + 4 = 0$

$x = 2 \qquad x = \dfrac{-2 \pm \sqrt{2^2 - 4(1)(4)}}{2(1)}$

$x = \dfrac{-2 \pm \sqrt{12}i}{2}$

$x = -1 \pm \sqrt{3}i$

$2, -1 + \sqrt{3}i, -1 - \sqrt{3}i$

69. $x^4 - 1 = 0$

$(x^2 - 1)(x^2 + 1) = 0$

$x^2 - 1 = 0$ or $x^2 + 1 = 0$

$x = \pm 1 \qquad x = \pm\sqrt{-1} = \pm i$

$1, -1, i, -i$

71. $i + i^2 + i^3 + \cdots + i^{15} = i + (-1) + (-i) + 1 +$

$\qquad i + (-1) + (-i) + 1 +$

$\qquad i + (-1) + (-i) + 1 +$

$\qquad i + (-1) + (-i)$

$\qquad = 4i - 4 - 4i + 3$

$\qquad = -1$

73. $\overline{z + w} = \overline{(a + bi) + (c + di)}$

$= \overline{(a + c) + (b + d)i}$

$= (a + c) - (b + d)i$

$= a - bi + c - di$

$= \overline{z} + \overline{w}$

75. $\overline{\left(\dfrac{z}{w}\right)} = \overline{\dfrac{a + bi}{c + di}} = \overline{\dfrac{a + bi}{c + di} \cdot \dfrac{c - di}{c - di}} = \overline{\dfrac{(ac + bd) + (bc - ad)i}{c^2 + d^2}} = \dfrac{(ac + bd) - (bc - ad)i}{c^2 + d^2}$

$\dfrac{\overline{z}}{\overline{w}} = \dfrac{\overline{a + bi}}{\overline{c + di}} = \dfrac{a - bi}{c - di} = \dfrac{a - bi}{c - di} \cdot \dfrac{c + di}{c + di} = \dfrac{(ac + bd) - (bc - ad)i}{c^2 + d^2}$

Thus $\overline{\left(\dfrac{z}{w}\right)} = \dfrac{\overline{z}}{\overline{w}}$

77. If z is a real number, $z = a = a = 0i, \overline{z} = \overline{a + 0i} = a = 0i = a = z$.

Conversely: If $\overline{z} = z, \overline{a + bi} = a + bi, a - bi = a + bi, -b = b, -b = 0, z = a, z$ is a real number.

79. $\dfrac{1}{z} = \dfrac{1}{a + bi} = \dfrac{1}{a + bi} \cdot \dfrac{a - bi}{a - bi} = \dfrac{a - bi}{a^2 + b^2} = \dfrac{a}{a^2 + b^2} - \dfrac{b}{a^2 + b^2}i$

4.8 Theory of Equations

1.   The remainder is $f(1) = 1^{10} + 1^8 = 2$.

3.   The remainder is $f(-1) = 3(-1)^4 - 6(-1)^3 + 2(-1) - 1 = 3 + 6 - 2 - 1 = 6$.

5.   The remainder is $f(-2) = (-2)^3 - 2(-2)^2 + 5(-2) - 4 = -8 - 8 - 10 - 4 = -30$.

7. 0 — multiplicity 54; $-\frac{4}{5}$ — multiplicity 1.

9. 0 — multiplicity 15; π — multiplicity 14; $\pi+1$ — multiplicity 13.

11. $x^2 - 2x + 5 = 0$

$\qquad (x-1)^2 = -4$

$\qquad\qquad x = 1 \pm 2i$

$\qquad [x - (1+2i)][x - (1-2i)]$

13. $3x^2 + 2x + 7 = 0$

$\qquad x = \dfrac{-2 \pm \sqrt{2^2 - 4(3)(7)}}{2(3)}$

$\qquad = \dfrac{-1 \pm 2\sqrt{5}i}{3}$

$\qquad 3\left(x - \dfrac{-1+2\sqrt{5}i}{3}\right)\left(x - \dfrac{-1-2\sqrt{5}i}{3}\right)$

15. $x^3 - 27 = 0$

$\qquad (x-3)(x^2 + 3x + 9) = 0$

$\qquad x = 3 \quad\text{or}\quad x = \dfrac{-3 \pm \sqrt{-27}}{2}$

$\qquad\qquad x = \dfrac{-3 \pm 3\sqrt{3}i}{2}$

$\qquad (x-3)\left(x - \dfrac{-3+3\sqrt{3}i}{2}\right)\left(x - \dfrac{-3-3\sqrt{3}i}{2}\right)$

17. $x^3 + 8 = 0$

$\qquad (x+2)(x^2 - 2x + 4) = 0$

$\qquad x = -2 \quad\text{or}\quad x = \dfrac{-(-2) \pm \sqrt{(-2)^2 - 4(1)(4)}}{2(1)}$

$\qquad\qquad x = \dfrac{2 \pm \sqrt{-12}}{2}$

$\qquad\qquad x = 1 \pm \sqrt{3}i$

$\qquad (x+2)\left[x - \left(1+\sqrt{3}i\right)\right]\left[x - \left(1-\sqrt{3}i\right)\right]$

19. $x^4 - 1 = 0$

$\qquad (x^2 - 1)(x^2 + 1) = 0$

$\qquad x^2 - 1 = 0 \ \text{ or } \ x^2 + 1 = 0$

$\qquad\quad x = \pm 1 \qquad\quad x = \pm i$

$\qquad (x-1)(x+1)(x-i)(x+i)$

21. $x^4 - 3x^2 - 10 = 0$

$\qquad (x^2 - 5)(x^2 + 2) = 0$

$\qquad x^2 - 5 = 0 \quad\text{or}\quad x^2 + 2 = 0$

$\qquad\quad x = \pm\sqrt{5} \qquad\quad x = \pm\sqrt{2}i$

$\qquad \left(x - \sqrt{5}\right)\left(x + \sqrt{5}\right)\left(x - \sqrt{2}i\right)\left(x + \sqrt{2}i\right)$

23. $(x-1)(x-7)(x+4)$

25. $(x-1)^2(x-2)^2(x-\pi)^2$

27. Let $f(x) = a(x+3)x(x-4)$
Since $f(5) = 80$,
$$80 = a(5+3)5(5-4)$$
$$a = 2$$
$$f(x) = 2x(x+3)(x-4)$$

29. $\left[x-(2+i)\right]\left[x-(2-i)\right] = (x-2)^2 - i^2 = x^2 - 4x + 5$

31. $2-i$ must also be a root:
$$(x-2)\left[x-(2-i)\right]\left[x-(2+i)\right] = (x-2)\left(x^2 - 4x + 5\right)$$
$$= x^3 - 6x^2 + 13x - 10$$

33. $1+i$ and $1-2i$ must also be roots:
$$(x+3)\left[x-(1-i)\right]\left[x-(1+i)\right]\left[x-(1-2i)\right]\left[x-(1+2i)\right]$$
$$= (x+3)\left[x^2 - 2x + 2\right]\left[x^2 - 2x + 5\right]$$
$$= (x+3)\left(x^4 - 4x^3 + 11x^2 - 14x + 10\right)$$
$$= x^5 - x^4 - x^3 + 19x^2 - 32x + 30$$

35. $\left[x-(1+2i)\right]\left[x-(1-2i)\right] = (x-1)^2 - 4i^2 = x^2 - 2x + 5$

37. $(x-4)^2\left[x-(3+i)\right]\left[x-(3-i)\right] = \left(x^2 - 8x + 16\right)\left(x^2 - 6x + 10\right)$
$$= x^4 - 14x^3 + 74x^2 - 176x + 160$$

39. $x^3(x-3)\left[x-(1+i)\right]\left[x-(1-i)\right] = \left(x^4 - 3x^3\right)\left(x^2 - 2x + 2\right) = x^6 - 5x^5 + 8x^4 - 6x^3$

41. $1-i$ must also be a root:
$$f(x) = a\left[x-(1+i)\right]\left[x-(1-i)\right]$$
Since $f(0) = 6$,
$$6 = a\left[-(1+i)\right]\left[-(1-i)\right]$$
$$6 = 2a$$
$$a = 3$$
$$f(x) = 3\left[x-(1+i)\right]\left[x-(1-i)\right]$$
$$= 3\left(x^2 - 2x + 2\right)$$
$$= 3x^2 - 6x + 6$$

43. $-i$ must also be a root:
$$f(x) = a(x-1)(x-i)(x+1)$$
$$= a(x-1)\left(x^2 + 1\right)$$
Since $f(-1) = 8$,
$$8 = a(-1-1)\left((-1)^2 + 1\right)$$
$$8 = -4a$$
$$a = -2$$
$$f(x) = -2(x-1)\left(x^2 + 1\right)$$
$$= -2\left(x^3 - x^2 + x - 1\right)$$
$$= -2x^3 + 2x^2 - 2x + 2$$

45. $(x-i)\left[x-(1-2i)\right]$

47. $(x-3)(x-i)\left[x-(2-i)\right]$

49. Since 3 is a root, $x - 3$ is a factor. Synthetic division by $x - 3$ yields $x^2 + x + 1$, which has roots $\dfrac{-1 \pm \sqrt{3}i}{2}$.

51. Since i is a root, $-i$ is also a root. Division by $(x - i)(x + i)$, or $x^2 + 1$, yields $x^2 + 3x + 2$, which has roots -1 and -2.

53. Since 1 is a root of multiplicity 2, $(x - 1)^2$ is a factor. Synthetic division by $x - 1$ twice yields $x^2 + 4$, which has roots $\pm 2i$.

55. Since $2 - i$ is a root, $2 + i$ is also a root. Division by $[x - (2 - i)][x - (2 + i)]$, or $x^2 - 4x + 5$, yields $x^2 + 1$, which has roots $\pm i$.

57. **a.** $\overline{z + w} = \overline{(a + bi) + (c + di)}$

$\qquad = \overline{(a + c) + (b + d)i}$

$\qquad = (a + c) - (b + d)i$

$\qquad = a - bi + c - di$

$\qquad = \bar{z} + \bar{w}$

b. $\overline{z \cdot w} = \overline{(a + bi) \cdot (c + di)}$

$\qquad = \overline{(ac - bd) + (bc + ad)i}$

$\qquad = (ac - bd) - (bc + ad)i$

$\qquad = (a - bi)(c - di)$

$\qquad = \bar{z} \cdot \bar{w}$

59. **a.** If z is a root of $f(x)$, $f(z) = 0$. Then $\overline{f(z)} = \bar{0} = 0$. Also

$\overline{f(z)} = \overline{az^3 + bz^2 + cz + d}$

$\qquad = \overline{az^3} + \overline{bz^2} + \overline{cz} + \overline{d}$

$\qquad = \overline{a}\,\overline{z^3} + \overline{b}\,\overline{z^2} + \overline{c}\,\overline{z} + \overline{d}$

$\qquad = a\bar{z}^3 + b\bar{z}^2 + c\bar{z} + d = f(\bar{z})$

b. Since $\overline{f(z)} = 0$, $a\bar{z}^3 + b\bar{z}^2 + c\bar{z} + d = 0$, therefore $\bar{z}$ is a root of $f(x)$.

61. For each non-real complex root z, there must be two factors of the polynomial: $(x - z)$ and $(x - \bar{z})$. This yields an even number of factors. There will remain at least one factor, hence at least one root, which must be real.

Chapter 4 Review Exercises

1. The graph of $f(x)$ is the graph of $g(x) = x^2$ shifted horizontally 2 units to the right and shifted 3 units upward. The vertex of the graph is $(2, 3)$.

3. We rewrite the rule of f as follows:

$$f(x) = x^2 - 8x + 12$$
$$= \left(x^2 - 8x + 16 - 16\right) + 12$$
$$= \left(x^2 - 8x + 16\right) - 16 + 12$$
$$= (x - 4)^2 - 4$$

The graph of $f(x)$ is the graph of $g(x) = x^2$ shifted horizontally 4 units to the right and shifted 4 units downward. The vertex of the graph is $(4, -4)$.

5. The graph is an upward-opening parabola (because f is a quadratic function and the coefficient of x^2 is positive). The x-coordinate of its vertex is

$$-\frac{b}{2a} = -\frac{-9}{2(3)} = \frac{9}{6} = 1.5$$

The y-coordinate of its vertex is $f(1.5) = 3(1.5)^2 - 9(1.5) + 1 = -5.75$

The vertex of the graph is $(1.5, -5.75)$.

7. a. The fence consists of four pieces, of lengths x, $x - 70$, y, and $y - 50$, respectively. Therefore, the total amount of fencing is given by:

$$x + (x - 70) + y + (y - 50) = 400$$

Solving this equation for y in terms of x, we obtain:

$$2x + 2y - 120 = 400$$
$$2y = 520 - 2x$$
$$y = 260 - x$$

b. Since the area of a rectangle is the product of the length and the width, the area of the playground is the area of the rectangle with dimensions x and y, minus the area of the school itself. Thus,

$$A(x) = xy - (70)(50) = x(260 - x) - 3500 = -x^2 + 260x - 3500$$

c. Since the function in part **b** is quadratic, and the coefficient of x^2 is negative, the function attains its maximum value at its vertex. The x-coordinate of its vertex is

$$-\frac{b}{2a} = -\frac{260}{2(-1)} = 130$$

Therefore, the dimensions of the playground with the largest possible area are $x = 130$ feet and $y = 130$ feet.

9. Sketch a figure.

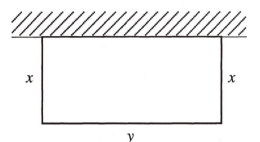

Let x be the width and y be the length of the garden. From the figure it is clear that:

$$2x + y = 120;\ y = 120 - 2x$$

$$A = xy = x(120 - 2x)$$

Rewriting this as $A(x) = 120x - 2x^2 = -2x^2 + 120x$ we have a quadratic function where the coefficient of x^2 is negative. The function attains its maximum value at the vertex of the graph of $A(x)$, when

$$x = -\frac{b}{2a} = -\frac{120}{2(-2)} = 30$$

Thus, the lengths of the other three sides are given by $x = 30$ feet, $y = 60$ feet, and $x = 30$ feet.

11. a, c, e, and **f** are polynomials.

13. To find the remainder when $f(x) = x^{112} - 2x^8 + 9x^5 - 4x^4 + x - 5$ is divided by $x - 1$, we apply the Remainder Theorem with $c = 1$. The remainder is
$$f(1) = 1^{112} - 2 \cdot 1^8 + 9 \cdot 1^5 - 4 \cdot 1^4 + 1 - 5 = 1 - 2 + 9 - 4 + 1 - 5 = 0.$$

15. The synthetic division is performed as follows:

```
2⌋1  −5   8   1  −17   16  −4
     2  −6   4   10  −14    4
  ─────────────────────────────
  1  −3   2   5   −7    2    0
```

The last row shows that the quotient is $x^5 - 3x^4 + 2x^3 + 5x^2 - 7x + 2$. This is the other factor; since the remainder is zero, $x - 2$ is a factor.

17. A general rule for a polynomial function of degree 3 is $f(x) = ax^3 + bx^2 + cx + d$. In this case, since $f(0) = 5$, substitution gives $5 = d$; thus $f(x) = ax^3 + bx^2 + cx + 5$. Furthermore, since $f(1) = 0$ and $f(-1) = 0$, substitution gives $0 = a + b + c + 5$ and $0 = -a + b - c + 5$. Therefore $2b + 10 = 0$ and $b = -5$, and $c = -a$. Thus any polynomial of the form $f(x) = ax^3 - 5x^2 - ax + 5$ meets the requirements of the problem, for example, choosing $a = 1$, $f(x) = x^3 - 5x^2 - x + 5$.

19. The roots of the polynomial $3x^2 - 2x - 5$ are the solutions of the quadratic equation $3x^2 - 2x - 5 = 0$. Solving this equation yields:

$$3x^2 - 2x - 5 = 0$$

$$(3x - 5)(x + 1) = 0$$

$$3x - 5 = 0 \quad \text{or} \quad x + 1 = 0$$

$$x = \tfrac{5}{3} \qquad\qquad x = -1$$

21. The roots of $x^6 - 4x^3 + 4$ are the solutions of $x^6 - 4x^3 + 4 = 0$. Solving this equation yields:

$$x^6 - 4x^3 + 4 = 0$$

$$\left(x^3 - 2\right)\left(x^3 - 2\right) = 0$$

$$x^3 - 2 = 0$$

The only real solution of this equation is $x = \sqrt[3]{2}$, and this is the only real root of the polynomial.

23. Clearly, 0 is a root of multiplicity 3 for $3y^3\left(y^4 - y^2 - 5\right)$. To find the other real roots, solve the equation in quadratic form $y^4 - y^2 - 5 = 0$, first for y^2. The quadratic formula applied to this equation yields:

$$y^2 = \frac{1 \pm \sqrt{21}}{2}$$

Since only the choice of the positive sign yields a positive value for y^2, the only real roots are the solutions of

$$y^2 = \frac{1 + \sqrt{21}}{2},$$

that is,

$$y = \pm\sqrt{\frac{1 + \sqrt{21}}{2}} \quad \text{or} \qquad y = \pm\frac{\sqrt{2 + 2\sqrt{21}}}{2}$$

25. a. If the polynomial has a rational root $r\,/\,s$, then by the Rational Root Test r must be a factor of the constant term 3. Therefore r must be one of ± 1 or ± 3. Similarly s must be a factor of the leading coefficient 2, so s must be one of ± 1 or ± 2. The only possible rational roots are therefore
$1, -1, 3, -3, \tfrac{1}{2}, -\tfrac{1}{2}, \tfrac{3}{2}, -\tfrac{3}{2}$

b. Graph the polynomial in a viewing window that includes all of these numbers on the x-axis, say $-4 \le x \le 4$ and $-5 \le y \le 5$.

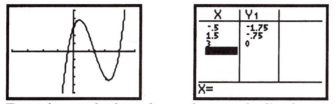

From the graph, the only numbers on the list that could possibly be roots are $-\frac{1}{2}$, $\frac{3}{2}$, and 3. The table shows that only 3 is a root. The other roots must be irrational numbers.

c. Synthetic division by $x - 3$ yields

$$
\begin{array}{r|rrrr}
3 & 2 & -8 & 5 & 3 \\
  &   & 6 & -6 & -3 \\
\hline
  & 2 & -2 & -1 & 0
\end{array}
$$

The other roots are therefore the solutions of the equation $2x^2 - 2x - 1 = 0$.

Solving this equation by the quadratic formula yields $x = \dfrac{1 \pm \sqrt{3}}{2}$ in addition to the root $x = 3$.

27. Since $x^3 + 4x = x(x^2 + 4)$, and $x^2 + 4$ has no real factors, the only real root of the polynomial is 0.

29. The equation $x^4 - 11x^2 + 18 = 0$ can be solved by factoring as follows:

$$x^4 - 11x^2 + 18 = 0$$

$$(x^2 - 9)(x^2 - 2) = 0$$

$$(x - 3)(x + 3)(x^2 - 2) = 0$$

$$x - 3 = 0 \quad \text{or} \quad x + 3 = 0 \quad \text{or} \quad x^2 - 2 = 0$$

$$x = 3 \qquad\qquad x = -3 \qquad\qquad x = \pm\sqrt{2}$$

Thus, the roots of the polynomial are 3, –3, $\sqrt{2}$, and $-\sqrt{2}$.

31. Synthetic division by $x - 5$ yields

$$
\begin{array}{r|rrrrr}
5 & 1 & -4 & 0 & 16 & -16 \\
  &   & 5 & 5 & 25 & 205 \\
\hline
  & 1 & 1 & 5 & 41 & 189
\end{array}
$$

Since every number in the last row in the synthetic division is nonnegative, 5 is an upper bound for the real roots of the polynomial.

33. If the polynomial has a rational root, by the Rational Root Test it can only be ± 1. Graph the polynomial in a viewing window that includes all of these numbers on the x-axis, say $-2 \le x \le 2$ and $-5 \le y \le 5$. The graph shows, and a table confirms, that -1 and 1 are both roots.

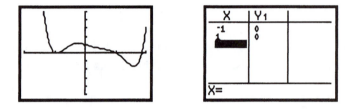

Synthetic division by $x-1$ yields a quotient of $x^5 - x^4 - 2x^3 + x^2 - 1$. Synthetic division of this reduced polynomial by $x+1$ yields a quotient of $x^4 - 2x^3 + x - 1$. This polynomial has no further rational roots; use the zero (root) routine to find two further real roots of -0.86676 and 1.86676.

35. The difference quotient is computed as follows:

$$\frac{f(x+h) - f(x)}{h} = \frac{\left[(x+h)^2 + (x+h)\right] - \left(x^2 + x\right)}{h}$$

$$= \frac{x^2 + 2xh + h^2 + x + h - x^2 - x}{h}$$

$$= \frac{2xh + h^2 + h}{h}$$

$$= \frac{h(2x + h + 1)}{h}$$

$$= 2x + h + 1$$

37. The graph below could not possibly be the graph of a polynomial function, because it has a sharp corner.

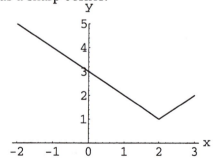

(Many answers are possible.)

39. Statement **c** is false. $(f \circ f)(0) = f(f(0)) = f(2)$ is negative.

41. Use the window $-2 \le x \le 9, -35 \le y \le 10$.

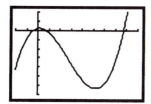

43. The window $-2 \le x \le 18, -500 \le y \le 1200$ provides a complete graph of the function.

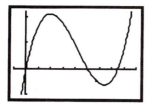

45. a. Entering the data into the TI-83 yields the following:

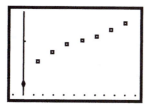

b. Applying the cubic regression routine yields:

c. The cost of making the seventy-first statue is $C(71) - C(70)$. Using the calculator values for $C(70)$ and $C(71)$,

$$C(71) - C(70) = 3466.53 - 3439.74 = \$26.79.$$

d. Average Cost $= \dfrac{C(x)}{x}$

$$\frac{C(35)}{35} = \frac{2992.3125}{35} = \$85.49$$

$$\frac{C(75)}{75} = \frac{3588.3168}{75} = \$47.84$$

47. x-intercepts: $-3, 0, 3$
local maximum: $(-1.732, 10.392)$
local minimum: $(1.732, -10.392)$

49. x-intercepts: $-1.908, -.376$
local maximum: $(.469, 2.941)$
local minima: $(-1.326, -4.914)$,
$\qquad (1.607, .617)$

51. x-intercepts: none
vertical asymptote: $x = -4$
horizontal asymptote: $y = 0$
holes: none

53. x-intercepts: -2.5
vertical asymptote: $x = 3$
horizontal asymptote: $y = \frac{4}{3}$
holes: none

55. $f(x) = \dfrac{(x+1)(x-1)}{(x-1)(x+2)(x-3)} = \dfrac{x+1}{(x+2)(x-3)}$

vertical asymptotes: $x = -2, x = 3$

horizontal asymptote: $y = 0$

57. Use windows $-4.7 \le x \le 4.7, -5 \le y \le 5$ and $2 \le x \le 20, -.2 \le y \le .1$

59. Use window $-18.8 \le x \le 18.8, -8 \le y \le 8$

61. Use $C(x) = .5x + 500,\ R(x) = xp = x\left(1.95 - \dfrac{x}{2000}\right)$.

Then $P(x) = R(x) - C(x) = 1.45x - 500 - \dfrac{x^2}{2000}$

Graph using the window $0 \le x \le 3000, 0 \le y \le 600$ and use the zero (root) routine to find that $P(x)$ is positive if x is between 400 and 2500 bags of Munchies. If $x = 400, p = \$1.75$; if $x = 2500, p = \$.70$.

63. a. Time $= \dfrac{\text{Distance driven}}{\text{Driving speed}} + \dfrac{\text{Train distance}}{\text{Train speed}}$

$t(x) = \dfrac{40}{x} + \dfrac{110}{x+25}$

The speed of the car should obey $0 \le x \le 55$.

b. Graph the function and the line $y = 2.5$ using the window $0 \le x \le 55, 0 < y < 3$ and use the intersection routine to find $x = 44$ miles per hour.

65. $\dfrac{f(x+h) - f(x)}{h} = \dfrac{\dfrac{x+h}{x+h+1} - \dfrac{x}{x+1}}{h}$

$= \dfrac{(x+h)(x+1) - x(x+h+1)}{h(x+1)(x+h+1)}$

$= \dfrac{x^2 + x + xh + h - x^2 - xh - x}{h(x+1)(x+h+1)}$

$= \dfrac{h}{h(x+1)(x+h+1)}$

$= \dfrac{1}{(x+1)(x+h+1)}$

67. (d) only

69. $\left|\dfrac{y+2}{3}\right| \geq 5$

$\dfrac{y+2}{3} \geq 5$ or $\dfrac{y+2}{3} \leq -5$

$y + 2 \geq 15 \qquad y + 2 \leq -15$

$\qquad y \geq 13 \qquad\qquad y \leq -17$

71. To solve $\dfrac{2x-1}{3x+1} < 1$, graph $\dfrac{2x-1}{3x+1} - 1$ using the window $-5 \leq x \leq 5, -5 \leq y \leq 5$ (disregard the erroneous vertical line near the asymptote $x = -\frac{1}{3}$). The graph is below the x-axis on the interval $(-\infty, -2) \cup \left(-\frac{1}{3}, \infty\right)$.

73. $\left|\dfrac{1}{1-x^2}\right| \geq \dfrac{1}{2}$. Graph $\left|\dfrac{1}{1-x^2}\right| - \dfrac{1}{2}$ using the window $-4.7 \leq x \leq 4.7, -3 \leq y \leq 3$. The graph is at or above the x-axis and the function is defined on the interval $\left[-\sqrt{3}, -1\right) \cup (-1, 1) \cup \left(1, \sqrt{3}\right]$.

75. $(x-1)^2(x^2-1)x \leq 0$

$(x-1)^3(x+1)x \leq 0$

Interval	$x < -1$	$-1 < x < 0$	$0 < x < 1$	$1 < x$
Test number	-2	$-.5$	$.5$	2
Sign	$-$	$+$	$-$	$+$
Graph	Below x-axis	Above x-axis	Below x-axis	Above x-axis

Solution: $(-\infty, -1] \cup [0, 1]$

77. The solution set of the original is $\left(-\frac{3}{2}, 0\right)$. If x is in this set then it cannot lie in any of the solution sets in **(a)**, **(b)**, **(c)**, or **(d)**, hence the answer is **(e)**.

79. $|3x+2| \geq 2$

$3x + 2 \geq 2 \qquad$ or $\qquad 3x + 2 \leq -2$

$\quad 3x \geq 0 \qquad\qquad\qquad 3x \leq -4$

$\qquad x \geq 0 \qquad\qquad\qquad x \leq -4/3$

81. $\dfrac{x-2}{x+4} \leq 3$ is equivalent to $\dfrac{x-2-3(x+4)}{x+4} \leq 0$, that is, to $\dfrac{-2x-14}{x+4} \leq 0$.

Interval	$x < -7$	$-7 < x < -4$	$-4 < x$
Test number	-8	-5	0
Sign	$-$	$+$	$-$
Graph	Below x-axis	Above x-axis	Below x-axis

Solution: $(-\infty, -7] \cup (-4, \infty)$

83.
$$\frac{x^2 + x - 9}{x + 3} < 1$$

$$\frac{x^2 + x - 9}{x + 3} - 1 < 0$$

$$\frac{x^2 + x - 9}{x + 3} - \frac{x + 3}{x + 3} < 0$$

$$\frac{x^2 - 12}{x + 3} < 0$$

Graph the function whose rule is given by the fractional expression using the window $-5 \le x \le 5, -10 \le y \le 10$. The graph is below the x-axis on the interval $\left(-\infty, -2\sqrt{3}\right) \cup \left(-3, 2\sqrt{3}\right)$.

85.
$$\frac{x^2 - x - 5}{x^2 + 2} > -2$$

$$\frac{x^2 - x - 5}{x^2 + 2} + 2 > 0$$

$$\frac{x^2 - x - 5}{x^2 + 2} + \frac{2x^2 + 4}{x^2 + 2} > 0$$

$$\frac{3x^2 - x - 1}{x^2 + 2} > 0$$

Graph the function whose rule is given by the fractional expression using the window $-5 \le x \le 5, -1 \le y \le 4$. The graph is above the x-axis on the interval $\left(-\infty, \frac{1 - \sqrt{13}}{6}\right) \cup \left(\frac{1 + \sqrt{13}}{6}, \infty\right)$ (the radical expressions being the zeros of the numerator).

87. $x^2 + 3x + 10 = 0$

$$x = \frac{-3 \pm \sqrt{3^2 - 4 \cdot 1 \cdot 10}}{2 \cdot 1}$$

$$x = \frac{-3 \pm \sqrt{-31}}{2 \cdot 1}$$

$$x = -\frac{3}{2} \pm i\frac{\sqrt{31}}{2}$$

89. $5x^2 + 2 = 3x$

$$5x^2 - 3x + 2 = 0$$

$$x = \frac{-(-3) \pm \sqrt{(-3)^2 - 4 \cdot 5 \cdot 2}}{2 \cdot 5}$$

$$x = \frac{3 \pm \sqrt{-31}}{10}$$

$$x = \frac{3}{10} \pm i\frac{\sqrt{31}}{10}$$

91. $3x^4 + x^2 - 2 = 0$

$$\left(3x^2 - 2\right)\left(x^2 + 1\right) = 0$$

$$3x^2 - 2 = 0 \quad \text{or} \quad x^2 + 1 = 0$$

$$x = \pm\sqrt{\frac{2}{3}} \qquad x = \pm\sqrt{-1}$$

$$x = \pm\frac{\sqrt{6}}{3} \qquad x = \pm i$$

93. $x^3 + 8 = 0$

$(x+2)(x^2 - 2x + 4) = 0$

$x + 2 = 0 \quad x^2 - 2x + 4 = 0$

$x = -2$

$\qquad x = \dfrac{-(-2) \pm \sqrt{(-2)^2 - 4 \cdot 1 \cdot 4}}{2 \cdot 1}$

$\qquad x = \dfrac{2 \pm \sqrt{-12}}{2}$

$\qquad x = 1 \pm i\sqrt{3}$

95. Since the polynomial has real coefficients, $-i$ must also be a root. Divide by $(x - i)(x + i) = x^2 + 1$ to obtain $x^2 - x - 2 = (x - 2)(x + 1)$. Thus the roots are i, $-i$, 2 and -1.

97. $x^2 [x - (1 + i)][x - (1 - i)] = x^2(x^2 - 2x + 2) = x^4 - 2x^3 + 2x^2$ is one possibility.

Chapter 5
Exponential and Logarithmic Functions
5.1 Radical and Rational Exponents

1. $\left(25k^2\right)^{\frac{3}{2}}\left(16k^{\frac{1}{3}}\right)^{\frac{3}{4}} = \left(25^{\frac{3}{2}}k^3\right)\left(16^{\frac{3}{4}}k^{\frac{1}{4}}\right) = 125k^3 \cdot 8 \cdot k^{\frac{1}{4}} = 1000k^{\frac{13}{4}}$

3. $\left(c^{\frac{2}{5}}d^{-\frac{2}{3}}\right)\left(c^6d^3\right)^{\frac{4}{3}} = c^{\frac{2}{5}}d^{-\frac{2}{3}}c^8d^4 = c^{\frac{42}{5}}d^{\frac{10}{3}}$

5. $\dfrac{\left(x^2\right)^{\frac{1}{3}}\left(y^2\right)^{\frac{2}{3}}}{3x^{\frac{2}{3}}y^2} = \dfrac{x^{\frac{2}{3}}y^{\frac{4}{3}}}{3x^{\frac{2}{3}}y^2} = \dfrac{1}{3y^{\frac{2}{3}}}$

7. $\dfrac{(7a)^2(5b)^{\frac{3}{2}}}{(5a)^{\frac{3}{2}}(7b)^4} = \dfrac{7^2a^2 5^{\frac{3}{2}}b^{\frac{3}{2}}}{5^{\frac{3}{2}}a^{\frac{3}{2}}7^4 b^4} = \dfrac{a^{\frac{1}{2}}}{49b^{\frac{5}{2}}}$

9. $\left(a^{x^2}\right)^{\frac{1}{x}} = a^x$

11. $x^{\frac{1}{2}}\left(x^{\frac{2}{3}} - x^{\frac{2}{3}}\right) = x^{\frac{1}{2}}x^{\frac{2}{3}} - x^{\frac{1}{2}}x^{\frac{2}{3}} = x^{\frac{7}{6}} - x^{\frac{11}{6}}$

13. $\left(x^{\frac{1}{2}} + y^{\frac{1}{2}}\right)\left(x^{\frac{1}{2}} - y^{\frac{1}{2}}\right) = x^{\frac{1}{2}}x^{\frac{1}{2}} - y^{\frac{1}{2}}y^{\frac{1}{2}} = x - y$

15. $(x+y)^{\frac{1}{2}}\left[(x+y)^{\frac{1}{2}} - (x+y)\right] = (x+y)^{\frac{1}{2}}(x+y)^{\frac{1}{2}} - (x+y)^{\frac{1}{2}}(x+y)^1 = (x+y) - (x+y)^{\frac{3}{2}}$

17. $\left(x^{\frac{1}{3}} + 3\right)\left(x^{\frac{1}{3}} - 2\right)$

19. $\left(x^{\frac{1}{2}} + 1\right)\left(x^{\frac{1}{2}} + 3\right)$

21. $x^{\frac{4}{3}} - 81 = \left(x^{\frac{2}{3}} - 9\right)\left(x^{\frac{2}{3}} + 9\right) = \left(x^{\frac{1}{3}} - 3\right)\left(x^{\frac{1}{3}} + 3\right)\left(x^{\frac{2}{3}} + 9\right)$

23. $\left(a^2 + b^2\right)^{\frac{1}{3}}$

25. $\left(a^{\frac{3}{4}}\right)^{\frac{1}{4}} = a^{\frac{3}{16}}$

27. $t^{\frac{1}{5}}16^{\frac{1}{2}}t^{\frac{1}{2}} = 4t^{\frac{7}{10}}$

29. $\sqrt{80} = \sqrt{16 \cdot 5} = \sqrt{16}\sqrt{5} = 4\sqrt{5}$

31. $\sqrt{6}\sqrt{12} = \sqrt{6}\sqrt{6}\sqrt{2} = 6\sqrt{2}$

33. $\dfrac{-6 + \sqrt{99}}{15} = \dfrac{-6 + 3\sqrt{11}}{15} = \dfrac{3\left(-2 + \sqrt{11}\right)}{15} = \dfrac{-2 + \sqrt{11}}{5}$

35. $\sqrt{50} - \sqrt{72} = \sqrt{25 \cdot 2} - \sqrt{36 \cdot 2} = 5\sqrt{2} - 6\sqrt{2} = -\sqrt{2}$

37. $5\sqrt{20} - \sqrt{45} + 2\sqrt{80} = 5\sqrt{4 \cdot 5} - \sqrt{9 \cdot 5} + 2\sqrt{16 \cdot 5} = 10\sqrt{5} - 3\sqrt{5} + 8\sqrt{5} = 15\sqrt{5}$

39. $\sqrt{16a^8b^{-2}} = 4a^4b^{-1} = \dfrac{4a^4}{b}$

41. $\dfrac{\sqrt{c^2d^6}}{\sqrt{4c^3d^{-4}}} = \sqrt{\dfrac{c^2d^6}{4c^3d^{-4}}} = \sqrt{\dfrac{cd^{10}}{4c^2}} = \dfrac{\sqrt{c}\sqrt{d^{10}}}{\sqrt{4c^2}} = \dfrac{d^5\sqrt{c}}{2c}$

43. $\dfrac{3}{\sqrt{8}} = \dfrac{3}{\sqrt{8}}\dfrac{\sqrt{2}}{\sqrt{2}} = \dfrac{3\sqrt{2}}{\sqrt{16}} = \dfrac{3\sqrt{2}}{4}$

45. $\dfrac{3}{2+\sqrt{12}} = \dfrac{3}{\left(2+\sqrt{12}\right)}\dfrac{\left(2-\sqrt{12}\right)}{\left(2-\sqrt{12}\right)} = \dfrac{6-3\sqrt{12}}{4-12} = \dfrac{6-6\sqrt{3}}{-8} = \dfrac{3\sqrt{3}-3}{4}$

47. $\dfrac{2}{\sqrt{x}+2} = \dfrac{2}{\left(\sqrt{x}+2\right)}\dfrac{\left(\sqrt{x}-2\right)}{\left(\sqrt{x}-2\right)} = \dfrac{2\left(\sqrt{x}-2\right)}{x-4}$

49. $\begin{aligned}\dfrac{f(x+h)-f(x)}{h} &= \dfrac{\sqrt{x+h+1}-\sqrt{x+1}}{h} \\[2mm] &= \dfrac{\sqrt{x+h+1}-\sqrt{x+1}}{h}\dfrac{\sqrt{x+h+1}+\sqrt{x+1}}{\sqrt{x+h+1}+\sqrt{x+1}} \\[2mm] &= \dfrac{(x+h+1)-(x+1)}{h\left(\sqrt{x+h+1}+\sqrt{x+1}\right)} \\[2mm] &= \dfrac{1}{\sqrt{x+h+1}+\sqrt{x+1}}\end{aligned}$

51. $\begin{aligned}\dfrac{f(x+h)-f(x)}{h} &= \dfrac{\sqrt{(x+h)^2+1}-\sqrt{x^2+1}}{h} \\[2mm] &= \dfrac{\sqrt{(x+h)^2+1}-\sqrt{x^2+1}}{h}\dfrac{\sqrt{(x+h)^2+1}+\sqrt{x^2+1}}{\sqrt{(x+h)^2+1}+\sqrt{x^2+1}} \\[2mm] &= \dfrac{(x+h)^2+1-\left(x^2+1\right)}{h\left(\sqrt{(x+h)^2+1}+\sqrt{x^2+1}\right)} \\[2mm] &= \dfrac{2xh+h^2}{h\left(\sqrt{(x+h)^2+1}+\sqrt{x^2+1}\right)} \\[2mm] &= \dfrac{2x+h}{\sqrt{(x+h)^2+1}+\sqrt{x^2+1}}\end{aligned}$

53. $92.8935(.24)^{.6669} = 35.8 \text{ (million mi)}$ **55.** $92.8935(29.46)^{.6669} = 886.8 \text{ (mil mi)}$

57. In 1800, $x = 10$. $3.9572(1.0299)^{10} = 5.3 \text{ (million)}$

59. In 1845, $x = 55$. $3.9572(1.0299)^{55} = 20.0 \text{ (million)}$

61. a. Since even powers of non-zero numbers, positive or negative, are positive, x^{2n} is positive and cannot equal -4.

   b. $(-8)^{\frac{1}{3}} = -2$ but $(-8)^{\frac{2}{6}} = \sqrt[6]{(-8)^2} = \sqrt[6]{64} = 2$.

63. a.

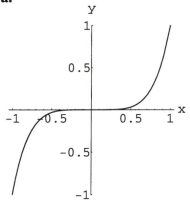

The function is one-to-one, as indicated by the fact that the graph passes the horizontal line test.

b. Set $y = x^5$

$$x = y^{1/5}$$

$$f^{-1}(y) = y^{1/5}$$

$$f^{-1}(x) = x^{1/5}$$

65. The graph of $g(x) = \sqrt{x+3}$ is the graph of $f(x) = \sqrt{x}$ shifted horizontally 3 units to the left.

67. The graph of $k(x) = \sqrt{x+4} - 4$ is the graph of $f(x) = \sqrt{x}$ shifted horizontally 4 units to the left, then vertically 4 units down.

69. a.

L	C	$Q = L^{1/4}C^{3/4}$
10	7	7.65
20	14	15.31
30	21	22.96
40	28	30.61
60	42	45.92

b. If both labor and capital are doubled, output is doubled. If both are tripled, output is tripled.

71. a.

L	C	$Q = L^{1/2}C^{3/4}$
10	7	13.61
20	14	32.37
30	21	53.73
40	28	76.98
60	42	127.79

b. If both labor and capital are doubled, output is multiplied by $2^{5/4}$. If both are tripled, output is multiplied by $3^{5/4}$.

5.1A Radical Equations

1. $\sqrt{x+2} = 3$

$\left(\sqrt{x+2}\right)^2 = 3^2$

$x + 2 = 9$

$x = 7$

The solution 7 checks.

3. $\sqrt{4x+9} = 5$

$\left(\sqrt{4x+9}\right)^2 = 5^2$

$4x + 9 = 25$

$x = 4$

The solution 4 checks.

5. $\sqrt[3]{5-11x} = 3$

$\left(\sqrt[3]{5-11x}\right)^3 = 3^3$

$5 - 11x = 27$

$-11x = 22$

$x = -2$

The solution −2 checks.

7. $\sqrt[3]{x^2-1} = 2$

$\left(\sqrt[3]{x^2-1}\right)^3 = 2^3$

$x^2 - 1 = 8$

$x^2 = 9$

$x = \pm 3$

The solutions 3 and −3 check.

9. $\sqrt{x^2-x-1} = 1$

$\left(\sqrt{x^2-x-1}\right)^2 = 1^2$

$x^2 - x - 1 = 1$

$x^2 - x - 2 = 0$

$x = 2 \text{ or } x = -1$

The solutions 2 and −1 check.

11. $\sqrt{x+7} = x - 5$

$\left(\sqrt{x+7}\right)^2 = (x-5)^2$

$x + 7 = x^2 - 10x + 25$

$0 = x^2 - 11x + 18$

$x = 2 \text{ or } x = 9$

The solution 9 checks.
(2 is extraneous.)

13. $\sqrt{3x^2+7x-2} = x + 1$

$\left(\sqrt{3x^2+7x-2}\right)^2 = (x+1)^2$

$3x^2 + 7x - 2 = x^2 + 2x + 1$

$2x^2 + 5x - 3 = 0$

$(2x-1)(x+3) = 0$

$x = \dfrac{1}{2} \text{ or } x = -3$

The solution $\frac{1}{2}$ checks.
(−3 is extraneous.)

15. $\sqrt[3]{x^3+x^2-4x+5} = x + 1$

$\left(\sqrt[3]{x^3+x^2-4x+5}\right)^3 = (x+1)^3$

$x^3 + x^2 - 4x + 5 = x^3 + 3x^2 + 3x + 1$

$0 = 2x^2 + 7x - 4$

$0 = (2x-1)(x+4)$

$x = \dfrac{1}{2} \text{ or } x = -4$

The solutions $\frac{1}{2}$ and −4 check.

17. Graph $y = \sqrt[5]{9-x^2}$ and $y = x^2 + 1$ using the window $-4 \le x \le 4, -2 \le y \le 5$ and apply the intersection routine to find $x = .730$ and, by symmetry, $x = -.730$.

19. Graph $y = \sqrt[3]{x^5 - x^3 - x}$ and $y = x + 2$ using the window $-2 \le x \le 4, -2 \le y \le 6$ and apply the intersection routine to find $x = -1.173, -1,$ and 2.591.

21. Graph $y = \sqrt{x^2 + 3x - 6}$ and $y = x^4 - 3x^2 + 2$ using the window $1 \le x \le 3, -1 \le y \le 2$ and apply the intersection routine to find $x = 1.658$.

23.
$$\sqrt{5x + 6} = 3 + \sqrt{x + 3}$$
$$5x + 6 = 9 + 6\sqrt{x + 3} + x + 3$$
$$4x - 6 = 6\sqrt{x + 3}$$
$$2x - 3 = 3\sqrt{x + 3}$$
$$4x^2 - 12x + 9 = 9x + 27$$
$$4x^2 - 21x - 18 = 0$$
$$(4x + 3)(x - 6) = 0$$
$$x = -\frac{3}{4} \text{ or } x = 6$$
The solution 6 checks.
$(-\frac{3}{4}$ is extraneous.)

25.
$$\sqrt{2x - 5} = 1 + \sqrt{x - 3}$$
$$2x - 5 = 1 + 2\sqrt{x - 3} + x - 3$$
$$x - 3 = 2\sqrt{x - 3}$$
$$x^2 - 6x + 9 = 4x - 12$$
$$x^2 - 10x + 21 = 0$$
$$(x - 3)(x - 7) = 0$$
$$x = 3 \text{ or } x = 7$$
The solutions 3 and 7 check.

27. Use the given formula with $h = 5, S = 100$
$$100 = \pi r\sqrt{r^2 + 25}$$
A calculator graph in the window $0 \le x \le 10, 0 \le y \le 150$:

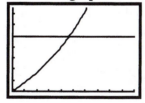

shows the root at 4.658 inches.

29. Use the given formula with $h = \frac{1}{3}r$, volume $= 180$.
$$180 = \pi r^2 \left(\frac{1}{3}r\right) / 3$$
$$180 = \frac{\pi r^3}{9}$$
$$r^3 = \frac{1620}{\pi}$$
$$r = 8.02 \text{ inches}$$

31.
$$A = \sqrt{1 + \frac{a^2}{b^2}}$$
$$A^2 = 1 + \frac{a^2}{b^2}$$
$$A^2 - 1 = \frac{a^2}{b^2}$$
$$b^2(A^2 - 1) = a^2$$
$$b^2 = \frac{a^2}{A^2 - 1}$$
$$b = \sqrt{\frac{a^2}{A^2 - 1}} = \frac{a}{\sqrt{A^2 - 1}}$$

33.
$$K = \sqrt{1 - \frac{x^2}{u^2}}$$
$$K^2 = 1 - \frac{x^2}{u^2}$$
$$1 - K^2 = \frac{x^2}{u^2}$$
$$u^2(1 - K^2) = x^2$$
$$u^2 = \frac{x^2}{1 - K^2}$$
$$u = \sqrt{\frac{x^2}{1 - K^2}} = \frac{x}{\sqrt{1 - K^2}}$$

35. $x - 4x^{1/2} + 4 = 0$
Let $u = x^{1/2}$, $u^2 = x$
$$u^2 - 4u + 4 = 0$$
$$(u - 2)^2 = 0$$
$$u = 2$$
$$x^{1/2} = 2$$
$$x = 4$$
The solution 4 checks.

37. $2x - \sqrt{x} - 6 = 0$
Let $u = \sqrt{x}$, $u^2 = x$
$$2u^2 - u - 6 = 0$$
$$(2u + 3)(u - 2) = 0$$
$$u = -\frac{3}{2} \quad \text{or} \quad u = 2$$
$$\sqrt{x} = -3/2 \qquad \sqrt{x} = 2$$
impossible $x = 4$
The solution 4 checks.

39. $x^{2/3} + 3x^{1/3} + 2 = 0$
Let $u = x^{1/3}$, $u^2 = x^{2/3}$
$$u^2 + 3u + 2 = 0$$
$$(u + 1)(u + 2) = 0$$
$$u = -1 \quad \text{or} \quad u = -2$$
$$x^{1/3} = -1 \qquad x^{1/3} = -2$$
$$x = -1 \qquad x = -8$$
The solutions −1 and −8 check.

41. $x^{1/2} - x^{1/4} - 2 = 0$
Let $u = x^{1/4}$, $u^2 = x^{1/2}$
$$u^2 - u - 2 = 0$$
$$(u + 1)(u - 2) = 0$$
$$u = -1 \quad \text{or} \quad u = 2$$
$$x^{1/4} = -1 \qquad x^{1/4} = 2$$
impossible $x = 16$
The solution 16 checks.

43. Let $u = x^{1/5}$. Then the equation becomes $u^3 - 2u^2 + u - 6 = 0$. Graph using the standard window and apply the zero (root) routine to find $u = 2.537656$. Then
$$x^{1/5} = 2.537656$$
$$x = (2.537656)^5 = 105.236$$

45. Since x cannot equal zero, multiply both sides by x^3 to obtain $1 + 2x - 4x^2 + 5x^3 = 0$. Graph using the window $-2 \le x \le 2, -10 \le y \le 10$ and apply the zero (root) routine to find $x = -.283$.

47. Label the text figure:

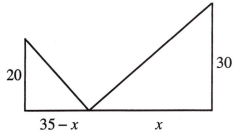

$35 - x$ x

From the Pythagorean Theorem, the length of the rope is given by

$$L = \sqrt{20^2 + (35 - x)^2} + \sqrt{30^2 + x^2}$$

a. Solve $63 = \sqrt{20^2 + (35 - x)^2} + \sqrt{30^2 + x^2}$. Graph the left and right sides of the equation in the window $0 \le x \le 35$, $50 \le y \le 70$ and apply the intersection routine to obtain $x = 11.47$ feet and $x = 29.91$ feet.

b. Graph the equation for L in the same window and apply the minimum routine to obtain $x = 21$ feet.

49. Redraw the text figure:

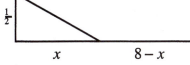

x $8 - x$

Cost (in thousands) = Underwater Cost (in thousands) + Land Cost (in thousands)

$$72 = 12\sqrt{x^2 + \left(\tfrac{1}{2}\right)^2} + 8(8 - x)$$

Solve by graphing the left and right sides of the equation in the window $0 \le x \le 8$, $0 \le y \le 100$ and apply the intersection routine to obtain $x = 1.795$ miles.

5.2 Exponential Functions

1.

3.

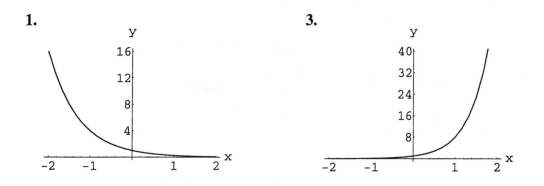

5.

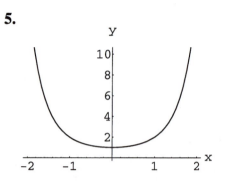

7. Shift vertically 5 units down.

9. Stretch by a factor of 3 with respect to the y-axis.

11. Shift horizontally 2 units to the left, then 5 units down.

13. $f(x) - C$; $g(x) - A$; $h(x) - B$

15. $f(-x) = 10^{-x}$. $f(x) \neq f(-x)$, so the function is not even.
$-f(x) = -10^x$. $-f(x) \neq f(-x)$, so the function is not odd.
Neither

17. $f(-x) = \dfrac{e^{-x} + e^{-(-x)}}{2} = \dfrac{e^{-x} + e^x}{2} = f(x)$, so the function is even.

19. $f(-x) = e^{-(-x)^2} = e^{-x^2} = f(x)$, so the function is even.

21. average rate of change $= \dfrac{f(3) - f(1)}{3 - 1} = \dfrac{3 \cdot 2^3 - 1 \cdot 2^1}{2} = \dfrac{24 - 2}{2} = 11$

23. average rate of change $= \dfrac{f(0) - f(-1)}{0 - (-1)} = \dfrac{5^{-0^2} - 5^{-(-1)^2}}{1} = 1 - \dfrac{1}{5} = \dfrac{4}{5}$

25. $\dfrac{f(x+h) - f(x)}{h} = \dfrac{10^{x+h} - 10^x}{h} = 10^x \dfrac{10^h - 1}{h}$

27. $\dfrac{f(x+h) - f(x)}{h} = \dfrac{\left(2^{x+h} + 2^{-(x+h)}\right) - \left(2^x + 2^{-x}\right)}{h} = \dfrac{2^x\left(2^h - 1\right) + 2^{-x}\left(2^{-h} - 1\right)}{h}$

29. $-3 \leq x \leq 3, 0 \leq y \leq 12$

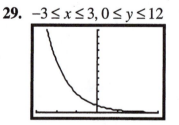

31. $-4 \leq x \leq 4, 0 \leq y \leq 10$

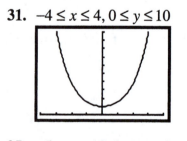

33. $-10 \leq x \leq 10, 0 \leq y \leq 20$

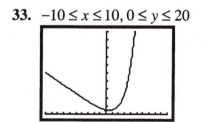

35. $-5 \leq x \leq 10, 0 \leq y \leq 6$

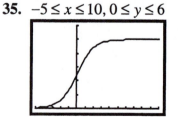

37. The negative x-axis is an asymptote. There is no vertical asymptote.
minimum: $(-1.44, -.53)$

39. There are no asymptotes.
minimum: $(0, 1)$

41. The x-axis is an asymptote.
There is no vertical asymptote.
maximum: $(0, 1)$

43. a. 15 days: $p(15) = 100 \cdot 3^{15/10} = 520$ flies

25 days: $p(25) = 100 \cdot 3^{25/10} = 1559$ flies

b. Solve $2500 = 100 \cdot 3^{t/10}$, or $25 = 3^{t/10}$, to obtain $t = 29.3$ days.

45. a. Since $p(0) = 15$, $15 = ke^{-.0000425(0)}$, $15 = k$.

b. Use $p(x) = 15e^{-.0000425x}$ to obtain:

$p(5000) = 15e^{-.0000425(5000)} = 12.1$ pounds per square inch

c. $p(160,000) = 15e^{-.0000425(160,000)} = .0167$ pounds per square inch

47. a. $D(1980) = \dfrac{79.257}{1 + 9.7135 \times 10^{24} e^{-.0304(1980)}} = 74$ years

$D(2000) = \dfrac{79.257}{1 + 9.7135 \times 10^{24} e^{-.0304(2000)}} = 76$ years

b. Solve: $60 = \dfrac{79.257}{1 + 9.7135 \times 10^{24} e^{-.0304t}}$ to obtain $t = 1930$

49. a. At $t = 0$, $p(0) = \dfrac{2000}{1 + 199e^{-.5544(0)}} = \dfrac{2000}{1 + 199} = 10$ beavers

At $t = 5$, $p(5) = \dfrac{2000}{1 + 199e^{-.5544(5)}} \approx 149$ beavers

b. Solve $1000 = \dfrac{2000}{1 + 199e^{-.5544t}}$ to obtain $t \approx 9.5$ years

51. a.

Time	Number of Cells
0	1
.25	2
.5	4
.75	8
1	16

b. $C(t) = 2^{4t}$ or $C(t) = 16^t$

53. a. Since $f(1) = 18$, $Pa = 18$. Since $f(2) = 54$, $Pa^2 = 54$.
Therefore, $a = 3$, $P = 6$. $f(x) = 6 \cdot 3^x$

b. 3

c. The park can accomodate $6.2 \times 5280^2 \div 10$, or $17,284,608$ frogs. Since
$f(12) = 3,188,646$, and $f(14) = 28,697,814$, there will be enough space after
12 weeks, but not after 14 weeks.

55. a.

Fold	Thickness
1	.004
2	.008
3	.016
4	.032

b. $f(x) = (.002)2^x$

c. $f(20) = (.002)2^{20} = 2097.152$ inches or 174.8 feet

d. Solve $243,000 \times 5280 \times 12 = (.002)2^x$ to obtain $x = 43$ folds.

57. a. $f(x) = 1200(1.04)^x$

b. After 3 years, $f(3) = 1200(1.04)^3 = \$1349.84$

After 5 years, $f(5) = 1200(1.04)^5 = \$1459.98$

After 5 years 9 months, $f(5.75) = 1200(1.04)^{5.75} = \$1503.57$

c. Solve $1850 = 1200(1.04)^x$ to obtain approximately $x = 11$ years and one month.

59. a. $f(x) = 100.4(1.014)^x$

b. In 2010, $x = 10$. $f(10) = 100.4(1.014)^{10} = 115.4$ million

c. Solve $125 = 100.4(1.014)^x$ to obtain $x = 15.8$. This corresponds to 2016.

61. a. Set $f(x) = 32.44a^x$. Since $f(50) = 98.23$, solve $98.23 = 32.44a^{50}$ to obtain

$a = 1.0224$. $f(x) = 32.44(1.0224)^x$.

b. In 2010, $x = 10$. $f(10) = 32.44(1.0224)^{10} = 40.49$ million

In 2025, $x = 25$. $f(25) = 32.44(1.0224)^{25} = 56.45$ million

c. Solve $55 = 32.44(1.0224)^x$ to obtain $x = 23.8$. This corresponds to 2024.

63. Since the population grows by a factor of $205/200 = 1.025$ every hour, we can write
$P(x) = 200(1.025)^x$. After 10 hours $P(10) = 256$. After 2 days, or 48 hours,
$P(48) = 200(1.025)^{48} = 654$.

65. a. $f(x) = (1 - .25)^x = .75^x$

b. Solve $1 - .90 = .75^x$ to obtain $x = 8$ feet.

67. a. Use the formula $M(x) = c\left(.5^{x/h}\right)$ with $c = 5$ and $h = 5730$.

$M(x) = 5\left(.5^{x/5730}\right)$

b. After 4000 years, $M(4000) = 5\left(.5^{4000/5730}\right) = 3.08$ grams.

After 8000 years, $M(8000) = 5\left(.5^{8000/5730}\right) = 1.90$ grams.

c. Solve $1 = 5\left(.5^{x/5730}\right)$ to obtain $x = 13,305$ years.

69. a. $f(x) = a^x$.

71. a.

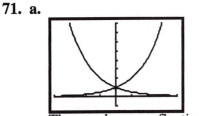

The graphs are reflections of each other in the y-axis.

b. As above, the graphs are reflections of each other in the y-axis.

73. a. The graphs appear to coincide near $x = 0$, but diverge further away.

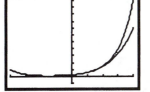

b. $f_8(x)$ is a good fit in this window on the TI-83.

c. This polynomial will not be a good fit, but, eventually, $f_{12}(x)$ will be.

5.2A Compound Interest and the Number e

1. a. Compounded annually for 5 years represents 5 periods.
$$A = 1000(1+.08)^5 = \$1469.32$$
b. Compounded quarterly for 5 years represents $4 \times 5 = 20$ periods.
$$A = 1000(1+.08/4)^{20} = \$1485.95$$
c. Compounded monthly for 5 years represents $12 \times 5 = 60$ periods.
$$A = 1000(1+.08/12)^{60} = \$1489.85$$
d. Compounded weekly for 5 years represents $\dfrac{365}{7} \times 5 = \dfrac{1825}{7}$ periods.
$$A = 1000\left(1+.08 \div \tfrac{365}{7}\right)^{1825/7} = \$1491.37$$

3. $A = 500(1+.02)^8 = \$585.83$

5. $A = 500(1+.02/4)^{10 \times 4} = \$610.40$

7. $A = 500(1+.029/365.25)^{8.5 \times 365.25} = \$639.76$

9. $A = 500e^{.03(4)} = \$563.75$

11. $A = 500e^{.0245(6.2)} = \$582.02$

13. Use $A = P(1+r)^t$ with $A = 5000$, $r = .06$, $t = 7$.

$$5000 = P(1+.06)^7$$

$$P = \frac{5000}{1.06^7}$$

$$P = \$3325.29$$

15. Use $A = P(1+r)^t$ with $A = 4800$, $r = .072 \div 4$ or $.018$, $t = 5 \times 4$ or 20.

$$4800 = P(1+.018)^{20}$$

$$P = \frac{4800}{1.018^{20}}$$

$$P = \$3359.59$$

17. Fund A: $10,000(1+.132)^2 = 10,000(1.132)^2 = \$12,814.24$

Fund B: $10,000(1+.127/4)^{4 \times 2} = 10,000(1.03175)^8 = \$12,840.91$

Fund C: $10,000(1+.126/12)^{12 \times 2} = 10,000(1.0105)^{24} = \$12,849.07$

Fund C will return the most money.

19. First calculate the amount of money paid:

$$A = 1200(1+.14/12)^{12 \times 2} = \$1585.18$$

Since $1200 of this is principal, $1585.18 – \$1200, or \$385.18, is interest.

21. Use $A = P(1+r)^t$ with $A = 1.5$ million, $r = .064 \div 12$, $t = 12 \times 4$.

$$1,500,000 = P(1+.064/12)^{12 \times 4}$$

$$P = \frac{1,500,000}{(1+.064/12)^{48}}$$

$$P = \$1,162,003.14$$

23. Use $A = P(1+r)^t$ with $A = 1407.1$, $P = 1000$, $t = 7$.

$$1407.1 = 1000(1+r)^7$$

$$(1+r)^7 = 1.4071$$

$$r = 1.4071^{1/7} - 1$$

$$r = .05 \text{ or } 5\%$$

25. Use $A = P(1+r)^t$ with $A = 4000$, $P = 3000$, $t = 5$.

$$4000 = 3000(1+r)^5$$

$$(1+r)^5 = 4/3$$

$$r = \left(\frac{4}{3}\right)^{1/5} - 1$$

$$r = .0592 \text{ or } 5.92\%$$

27. Use $A = P(1+r)^t$ with $r = .08$.

    a. $A = 200$, $P = 100$: Solve $200 = 100(1.08)^t$. $t = 9.006$ years.

    b. $A = 1000$, $P = 500$: Solve $1000 = 500(1.08)^t$. $t = 9.006$ years.

    c. $A = 2400$, $P = 1200$: Solve $2400 = 1200(1.08)^t$. $t = 9.006$ years.

    d. The doubling time is independent of the investment.

29. Use $A = Pe^{rt}$ with $A = 1000$, $P = 500$, $r = .07$.

    Solve $1000 = 500e^{.07t}$ to obtain $t = 9.9$ years.

5.3 Common and Natural Logarithmic Functions

1. To what power must 10 be raised to obtain 10,000? 4. $\log 10{,}000 = 4$

3. To what power must 10 be raised to obtain $\sqrt{10}/1000$? $-5/2$, since

$$\sqrt{10}/1000 = \sqrt{10}/\left(\sqrt{10}\right)^6 = 1/\left(\sqrt{10}\right)^5 = 10^{-5/2}. \quad \log \frac{\sqrt{10}}{1000} = -\frac{5}{2}$$

5. $10^3 = 1000$ **7.** $10^{2.88} = 750$ **9.** $e^{1.0986} = 3$

11. $e^{-4.6052} = .01$ **13.** $e^{z+w} = x^2 + 2y$ **15.** $\log .01 = -2$

17. $\log 3 = .4771$ **19.** $\ln 25.79 = 3.25$ **21.** $\ln 5.5527 = 12/7$

23. $\ln w = 2/r$ **25.** $\sqrt{43}$ **27.** 15

29. $\ln\sqrt{e} = \ln e^{1/2} = 1/2$ **31.** 931   **33.** $x + y$

35. x^2

37. $4(25)^x = 4\left(e^{\ln 25}\right)^x = 4e^{(\ln 25)x}$ **39.** $-16(30.5)^x = -16\left(e^{\ln 30.5}\right)^x = -16e^{(\ln 30.5)x}$

41. $x + 1 > 0$ hence the domain is $x > -1$.   **43.** $-x > 0$ hence the domain is $x < 0$.

45. a. The graphs are identical when $x > 0$, but not everywhere.

    b. The functions $f(x) = e^x$ and $g(x) = \ln x$ are inverse functions of one another. This means that $(f \circ g)(x) = e^{\ln x} = x$ everywhere *on the domain of g*. The domain of $g(x) = \ln x$ is precisely $x > 0$.

47. The graphs are identical when $x > 0$, but not everywhere.

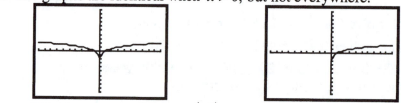

The domain of $f(x) = \log(x^2)$ is all real numbers except 0. The domain of $g(x) = 2\log x$ is only $x > 0$.

49. Stretch by a factor of 2 with respect to the y-axis.

51. Shift horizontally 4 units to the right.

53. Shift horizontally 3 units to the left, then 4 units vertically down.

55.   **57.**

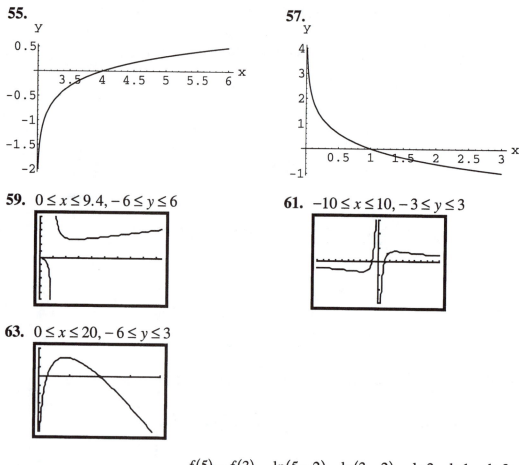

59. $0 \le x \le 9.4, -6 \le y \le 6$ **61.** $-10 \le x \le 10, -3 \le y \le 3$

63. $0 \le x \le 20, -6 \le y \le 3$

65. average rate of change $= \dfrac{f(5) - f(3)}{5 - 3} = \dfrac{\ln(5 - 2) - \ln(3 - 2)}{2} = \dfrac{\ln 3 - \ln 1}{2} = \dfrac{\ln 3}{2}$

67. average rate of change $= \dfrac{f(-3)-f(-5)}{(-3)-(-5)} = \dfrac{\ln\left((-3)^2+(-3)+1\right)-\ln\left((-5)^2+(-5)+1\right)}{2}$

$$= \dfrac{\ln 7 - \ln 21}{2}$$

69. a. average rate of change $= \dfrac{f(3+h)-f(3)}{(3+h)-3} = \dfrac{\ln(3+h)-\ln(3)}{h}$

b. Solve $.25 = \dfrac{\ln(3+h)-\ln(3)}{h}$ to obtain $h = 2.201$

71. Set $y = \dfrac{1}{1+e^{-x}}$

$y = \dfrac{e^x}{e^x+1}$

$ye^x + y = e^x$

$y = e^x - ye^x$

$\dfrac{y}{1-y} = e^x$

$x = \ln\dfrac{y}{1-y} = f^{-1}(y)$

$f^{-1}(x) = \ln\dfrac{x}{1-x}$

73. Let $f(x) = A\ln x + B$. Since $f(1) = 10$,

$10 = A\ln 1 + B$

$10 = B$

Thus $f(x) = A\ln x + 10$. Since $f(e) = 1$,

$1 = A\ln e + 10$

$1 = A + 10$

$A = -9$

Thus $f(x) = -9\ln x + 10$.

75. Use the given formula with $t = 7$, $c = 75.126$, $p = 44$ to obtain

$h = (30 \cdot 7 + 8000)\ln(75.126/44) = 4392\,\text{meters}$

77. a. 1999: $x = 99$ 2002: $x = 102$

$f(99) = -154.41 + 39.38\ln 99$ $f(102) = -154.41 + 39.38\ln 102$

$f(99) = 26.55$ billion pounds $f(102) = 27.72$ billion pounds

b. Solve $30 = -154.41 + 39.38 \ln x$

$$184.41 = 39.38 \ln x$$

$$\frac{184.41}{39.38} = \ln x$$

$$x = e^{184.41/39.38}$$

$$x = 108.1$$

Approximately, 2008.

79. a. Use the given formula with $x = 6000$.

$$T = -.93 \ln \left[\frac{7000 - 6000}{6999(6000)} \right] = 9.9$$

10 days.

b. Solve $14 = -.93 \ln \left[\frac{7000 - x}{6999x} \right]$ to obtain $x = 6986$

81. a. $d = 0$:

$N = 51 + 100 \ln(0/100 + 2)$

$N = 120$

$d = 1000$:

$N = 51 + 100 \ln(1000/100 + 2)$

$N = 300$

$d = 10,000$:

$N = 51 + 100 \ln(10,000/100 + 2)$

$N = 514$

b. Spending $1000 on advertising, selling 300 bikes at $25 each, generates a profit of $7500. This is worthwhile. Spending $9000 more to generate a profit of $25 \times (514 - 300)$, or $5350 more, might not be worthwhile.

c. Spending $1000 on advertising selling 300 bikes at $35 each, generates a profit of $10,500. This is worthwhile. Spending $9000 more to generate a profit of $35 \times (514 - 300)$, or $7490 more, might not be worthwhile.

83. It seems that $n = 30$ gives an approximation with an error of $\le .00001$.

5.4 Properties of Logarithms

1. $\ln x^2 + 3\ln y = \ln x^2 + \ln y^3 = \ln\left(x^2 y^3\right)$

3. $\log\left(x^2 - 9\right) - \log(x+3) = \log\left(\dfrac{x^2 - 9}{x+3}\right) = \log(x-3)$

5. $2(\ln x) - 3\left(\ln x^2 + \ln x\right) = 2\ln x - 3\ln x^2 - 3\ln x$
$$= -\ln x - 3\ln x^2$$
$$= -\ln x - \ln x^6$$
$$= \ln\frac{1}{x \cdot x^6}$$
$$= \ln\frac{1}{x^7}$$

7. $3\ln\left(e^2 - e\right) - 3 = 3\ln\left(e^2 - e\right) - 3\ln e$
$$= 3\ln\frac{e^2 - e}{e}$$
$$= 3\ln(e-1)$$
$$= \ln(e-1)^3$$

9. $\log(10x) + \log(20y) - 1 = \log(200xy) - \log 10$
$$= \log\frac{200xy}{10}$$
$$= \log 20xy$$

11. $\ln\left(x^2 y^5\right) = \ln x^2 + \ln y^5 = 2\ln x + 5\ln y = 2u + 5v$

13. $\ln\left(\sqrt{x} \cdot y^2\right) = \ln x^{\frac{1}{2}} + \ln y^2 = \frac{1}{2}\ln x + 2\ln y = \frac{1}{2}u + 2v$

15. $\ln\sqrt[3]{x^2\sqrt{y}} = \ln\left(x^2 y^{\frac{1}{2}}\right)^{\frac{1}{3}} = \ln\left(x^{\frac{2}{3}} y^{\frac{1}{6}}\right) = \frac{2}{3}\ln x + \frac{1}{6}\ln y = \frac{2}{3}u + \frac{1}{6}v$

17. Since $\ln\left|\frac{1}{e}\right| = -1$ and $\left|\ln\frac{1}{e}\right| = 1$, the statement is false.

19. The statement is true by the Power Law for Logarithms.

21. Since $\ln e^3 = 3$ and $(\ln e)^3 = 1$, the statement is false.

23. Let $a = 1, b = 10$. Then

$$\frac{\log a}{\log b} = \frac{\log 1}{\log 10} = \frac{0}{1} = 0$$

but

$$\log\left(\frac{a}{b}\right) = \log\frac{1}{10} = -1$$

25. e

27. $100 = 10^2$, hence $R = 2$

29. $350 = 10^{\log 350} = 10^{2.544}$, hence $R = 2.544$
$R = 2.544$

31. $L(100 i_0) = 10 \cdot \log\left(\frac{100 i_0}{i_0}\right) = 10\log 100 = 10 \cdot 2 = 20$ decibels

33. $L(4 \times 10^6 i_0) = 10 \cdot \log\left(\frac{4 \times 10^6 i_0}{i_0}\right) = 10\log(4 \times 10^6) = 10(6.6021) \approx 66$ decibels

35. 40 decibels – 20 decibels = 20 decibels louder.

37. a. Both equal 1.255…
b. Both equal 2.658…
c. Both equal 3.9518…

d. They suggest that $\log c = \dfrac{\ln c}{\ln 10}$ as long as the logarithms are defined.

39. a. .9421569…
b. 1.9421569…
c. 2.9421569…
d. 3.9421569…
e. 4.9421569…
f. Each number is 10 times the previous. Each logarithm is 1 more than the previous.
This suggests that $\log(c \cdot 10^k) = \log c + k$

5.4A Logarithmic Functions to Other Bases

1.

x	0	1	2	4
$f(x) = \log_4 x$	undefined	0	.5	1

3.

x	1/36	1/6	1	216
$h(x) = \log_6 x$	-2	-1	0	3

5.

x	0	1/7	$\sqrt{7}$	49
$f(x) = 2\log_7 x$	undefined	-2	1	4

7.

x	-2.75	-1	1	29
$h(x) = 3\log_2(x+3)$	-6	3	6	15

9. $\log_{10} .01 = -2$ **11.** $\log_{10} \sqrt[3]{10} = 1/3$ **13.** $\log_{10} r = 7k$

15. $\log_7 5,764,801 = 8$ **17.** $\log_3 1/9 = -2$ **19.** $10^4 = 10,000$

21. $10^{2.88} \approx 750$ **23.** $5^3 = 125$ **25.** $2^{-2} = 1/4$

27. $10^{z+w} = x^2 + 2y$ **29.** $\sqrt{97}$ **31.** $x^2 + y^2$

33. $\log_{16} 4 = \log_{16} 16^{1/2} = 1/2$ **35.** $\log_{\sqrt{3}} 27 = \log_{\sqrt{3}}\left(\sqrt{3}\right)^6 = 6$

37. Since $\log_b 3 = 1, b = 3$ **39.** Since $\log_b .05 = -1, b^{-1} = .05$, hence $b = 20$.

41. $\log_3 243 = \log_3 3^5 = 5$

43. $\log_{27} x = 1/3$
$x = 27^{1/3}$
$x = 3$

45. $\log_x 64 = 3$
$x^3 = 64$
$x = 4$

47. $2\log x + 3\log y - 6\log z = \log x^2 + \log y^3 - \log z^6 = \log \dfrac{x^2 y^3}{z^6}$

49. $\log x - \log(x+3) + \log(x^2 - 9) = \log \dfrac{x(x^2 - 9)}{x+3} = \log x(x-3)$

51. $\dfrac{1}{2}\log_2(25c^2) = \log_2(25c^2)^{1/2} = \log_2 5|c|$

53. $-2\log_4(7c) = \log_4(7c)^{-2} = \log_4 \dfrac{1}{49c^2}$

55. $2\ln(x+1) - \ln(x+2) = \ln(x+1)^2 - \ln(x+2) = \ln\dfrac{(x+1)^2}{x+2}$

57. $\log_2(2x) - 1 = \log_2(2x) - \log_2 2 = \log_2 \dfrac{2x}{2} = \log_2 x$

59. $2\ln\left(e^2 - e\right) - 2 = 2\ln\left(e^2 - e\right) - 2\ln e = 2\ln\dfrac{e^2 - e}{e} = 2\ln(e - 1) = \ln(e - 1)^2$

61. $\log_2 10 = \dfrac{\ln 10}{\ln 2} = 3.3219$

63. $\log_7 5 = \dfrac{\ln 5}{\ln 7} = .8271$

65. $\log_{500} 1000 = \dfrac{\ln 1000}{\ln 500} = 1.1115$

67. $\log_{12} 56 = \dfrac{\ln 56}{\ln 12} = 1.6199$

69. True by the quotient law for logarithms.

71. True by the power law for logarithms.

73. False. for example, let $x = 1$, $\log_5 5 \cdot 1 = 1$ but $5\log_5 1 = 0$.

75. Since $398\log 397 \approx 1034.3$ and $397\log 398 \approx 1032.1$,
$$398\log 397 > 397\log 398$$
$$10^{398\log 397} > 10^{397\log 398} \quad \left(\text{since } 10^x \text{ is an increasing function}\right)$$
$$397^{398} > 398^{397}$$

77. $\log_b u = \dfrac{\log_a u}{\log_a b}$ by the change-of-base formula.

79. $\log_{100} u = \dfrac{\log_{10} u}{\log_{10} 100} = \dfrac{\log_{10} u}{2}$

81. $\log_b x = \dfrac{1}{2}\log_b v + 3$

$\log_b x = \log_b v^{\frac{1}{2}} + \log_b b^3$

$\log_b x = \log_b b^3 \sqrt{v}$

$\quad x = b^3 \sqrt{v} \quad (\text{since } \log_b x \text{ is a one-to-one function})$

83. The graphs are not identical, and the statement is false. The window $0 \le x \le 10$, $-3 \le y \le 3$ is shown.

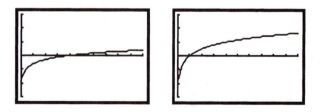

85. Graph using the window $-5 \le x \le 5, -5 \le y \le 5$.
Local maximum: $x = -.3679$
Local minimum: $x = .3679$
Hole: $(0,0)$

5.5 Algebraic Solutions of Exponential and Logarithmic Equations

1. $3^x = 81$
$3^x = 3^4$
$x = 4$

3. $3^{x+1} = 9^{5x}$
$3^{x+1} = 3^{2(5x)}$
$x + 1 = 2(5x)$
$1 = 9x$
$x = 1/9$

5. $3^{5x}9^{x^2} = 27$
$3^{5x} \cdot 3^{2x^2} = 3^3$
$5x + 2x^2 = 3$
$2x^2 + 5x - 3 = 0$
$(2x - 1)(x + 3) = 0$
$x = 1/2$ or $x = -3$

7. $9^{x^2} = 3^{-5x-2}$
$3^{2x^2} = 3^{-5x-2}$
$2x^2 = -5x - 2$
$2x^2 + 5x + 2 = 0$
$(2x + 1)(x + 2) = 0$
$x = -1/2$ or $x = -2$

9. $3^x = 5$
$\ln 3^x = \ln 5$
$x \ln 3 = \ln 5$
$x = \dfrac{\ln 5}{\ln 3}$
$x = 1.465$

11. $2^x = 3^{x-1}$
$\ln 2^x = \ln 3^{x-1}$
$x \ln 2 = (x - 1)\ln 3$
$x \ln 2 = x \ln 3 - \ln 3$
$x(\ln 2 - \ln 3) = -\ln 3$
$x = \dfrac{-\ln 3}{\ln 2 - \ln 3}$
$x = 2.7095$

13.
$$3^{1-2x} = 5^{x+5}$$
$$\ln 3^{1-2x} = \ln 5^{x+5}$$
$$(1-2x)\ln 3 = (x+5)\ln 5$$
$$\ln 3 - 2x\ln 3 = x\ln 5 + 5\ln 5$$
$$\ln 3 - 5\ln 5 = x(2\ln 3 + \ln 5)$$
$$x = \frac{\ln 3 - 5\ln 5}{2\ln 3 + \ln 5}$$
$$x = -1.825$$

15.
$$2^{1-3x} = 3^{x+1}$$
$$\ln 2^{1-3x} = \ln 3^{x+1}$$
$$(1-3x)\ln 2 = (x+1)\ln 3$$
$$\ln 2 - 3x\ln 2 = x\ln 3 + \ln 3$$
$$\ln 2 - \ln 3 = x(3\ln 2 + \ln 3)$$
$$x = \frac{\ln 2 - \ln 3}{3\ln 2 + \ln 3}$$
$$x = -.1276$$

17.
$$e^{2x} = 5$$
$$2x = \ln 5$$
$$x = \frac{\ln 5}{2}$$
$$x = .805$$

19.
$$6e^{-1.4x} = 21$$
$$e^{-1.4x} = 3.5$$
$$-1.4x = \ln 3.5$$
$$x = \frac{\ln 3.5}{-1.4}$$
$$x = -.895$$

21.
$$2.1e^{(x/2)\ln 3} = 5$$
$$e^{(x/2)\ln 3} = \frac{5}{2.1}$$
$$\frac{x}{2}\ln 3 = \ln\frac{5}{2.1}$$
$$x = \frac{2\ln\frac{5}{2.1}}{\ln 3}$$
$$x = 1.579$$

23.
$$9^x - 4\cdot 3^x + 3 = 0$$
Let $u = 3^x$, $u^2 = 9^x$
$$u^2 - 4u + 3 = 0$$
$$(u-1)(u-3) = 0$$
$$u = 1 \quad \text{or} \quad u = 3$$
$$3^x = 1 \qquad 3^x = 3$$
$$x = 0 \qquad x = 1$$

25.
$$e^{2x} - 5e^x + 6 = 0$$
Let $e^x = u$, $e^{2x} = u^2$
$$u^2 - 5u + 6 = 0$$
$$(u-3)(u-2) = 0$$
$$u = 3 \quad \text{or} \quad u = 2$$
$$e^x = 3 \qquad e^x = 2$$
$$x = \ln 3 \qquad x = \ln 2$$
$$x = 1.099 \qquad x = .693$$

27.
$$6e^{2x} - 16e^x = 6$$
Let $e^x = u$, $e^{2x} = u^2$
$$6u^2 - 16u = 6$$
$$3u^2 - 8u - 3 = 0$$
$$(u-3)(3u+1) = 0$$
$$u = 3 \qquad\qquad \text{or} \quad u = -1/3$$
$$e^x = 3 \qquad\qquad\qquad e^x = -1/3$$
$$x = \ln 3 = 1.099 \qquad \text{impossible}$$

29. $4^x + 6 \cdot 4^{-x} = 5$

$\left(4^x\right)^2 + 6 = 5 \cdot 4^x$

Let $4^x = u$, $\left(4^x\right)^2 = u^2$

$u^2 + 6 = 5u$

$u^2 - 5u + 6 = 0$

$(u-3)(u-2) = 0$

$u = 3$ or $u = 2$

$4^x = 3$ $4^x = 2$

$x = \ln 3 / \ln 4$ $x = 1/2$

$x = .792$

31. $\dfrac{e^x - e^{-x}}{2} = t$

Let $e^x = u$

$\dfrac{u - u^{-1}}{2} = t$

$\dfrac{u^2 - 1}{2u} = t$

$u^2 - 1 = 2ut$

$u^2 - 2ut - 1 = 0$

$u = \dfrac{2t \pm \sqrt{4t^2 + 4}}{2}$

$u = t \pm \sqrt{t^2 + 1}$

Since $t - \sqrt{t^2 + 1}$ is negative, the only

solution is $e^x = t + \sqrt{t^2 + 1}$

$x = \ln\left(t + \sqrt{t^2 + 1}\right)$

33. If $\ln u = \ln v$ we can write

$\ln u - \ln v = 0$

$\ln \dfrac{u}{v} = 0$

$\dfrac{u}{v} = 1$

$u = v$

35. $\ln(3x - 5) = \ln 11 + \ln 2$

$\ln(3x - 5) = \ln 22$

$3x - 5 = 22$

$3x = 27$

$x = 9$

37. $\log(3x - 1) + \log 2 = \log 4 + \log(x + 2)$

$\log 2(3x - 1) = \log 4(x + 2)$

$6x - 2 = 4x + 8$

$2x = 10$

$x = 5$

39. $2 \ln x = \ln 36$

$\ln x^2 = \ln 36$

$x^2 = 36$

$x = \pm 6$

Since x must be positive, the only
solution is 6.

41. $\ln x + \ln(x + 1) = \ln 3 + \ln 4$

$\ln x(x + 1) = \ln 12$

$x(x + 1) = 12$

$x^2 + x - 12 = 0$

$(x + 4)(x - 3) = 0$

$x = -4$ or $x = 3$

Since x must be positive, the only
solution is 3.

43. $\ln x = \ln 3 - \ln(x+5)$

$\ln x = \ln \dfrac{3}{x+5}$

$x = \dfrac{3}{x+5}$

$x^2 + 5x = 3$

$x^2 + 5x - 3 = 0$

$x = \dfrac{-5 \pm \sqrt{37}}{2}$

Since x must be positive, the only

solution is $\dfrac{-5 + \sqrt{37}}{2}$.

45. $\ln(x+9) - \ln x = 1$

$\ln \dfrac{x+9}{x} = 1$

$\dfrac{x+9}{x} = e$

$x + 9 = ex$

$9 = (e-1)x$

$x = \dfrac{9}{e-1}$

47. $\log x + \log(x-3) = 1$

$\log x(x-3) = 1$

$x(x-3) = 10$

$x^2 - 3x - 10 = 0$

$(x-5)(x+2) = 0$

$x = 5 \quad \text{or} \quad x = -2$

Since x must be greater than 3, the only solution is 5.

49. $\log \sqrt{x^2 - 1} = 2$

$\sqrt{x^2 - 1} = 10^2$

$x^2 - 1 = 10,000$

$x^2 = 10,001$

$x = \pm\sqrt{10,001}$

51. $\ln(x^2 + 1) - \ln(x-1) = 1 + \ln(x+1)$

$\ln \dfrac{x^2 + 1}{x-1} = \ln e + \ln(x+1)$

$\ln \dfrac{x^2 + 1}{x-1} = \ln e(x+1)$

$\dfrac{x^2 + 1}{x-1} = e(x+1)$

$x^2 + 1 = e(x^2 - 1)$

$x^2 + 1 = ex^2 - e$

$e + 1 = x^2(e-1)$

$x^2 = \dfrac{e+1}{e-1}$

$x = \pm\sqrt{\dfrac{e+1}{e-1}}$

Since x must be greater than 1, the only solution is $\sqrt{\dfrac{e+1}{e-1}}$.

53. Use $M(x) = c(.5)^{x/h}$ with $c = 300$, $x = .26$, $M(.26) = 200$.

$$200 = 300(.5)^{.26/h}$$

$$\frac{2}{3} = (.5)^{.26/h}$$

$$\ln\frac{2}{3} = \frac{.26}{h}\ln.5$$

$$h = \frac{.26\ln.5}{\ln 2/3}$$

$h = .444$ billion years, or $444,000,000$ years

55. Use $M(x) = c(.5)^{x/h}$ with $c = 3$, $x = 23.7$ and $M(23.7) = 1$.

$$1 = 3(.5)^{23.7/h}$$

$$\frac{1}{3} = (.5)^{23.7/h}$$

$$\ln\frac{1}{3} = \frac{23.7}{h}\ln.5$$

$$h = \frac{23.7\ln.5}{\ln 1/3}$$

$$h = 14.95 \text{ days}$$

57. Use $M(x) = c(.5)^{x/h}$ with $x = 6$, and $M(6) = .336c$.

$$.336c = c(.5)^{6/h}$$

$$.336 = (.5)^{6/h}$$

$$\ln.336 = \frac{6}{h}\ln.5$$

$$h = \frac{6\ln.5}{\ln.336}$$

$$h = 3.813 \text{ days}$$

59. Use $M(x) = c(.5)^{x/h}$ with $h = 5730$ (the half-life of carbon-14) and $M(c) = c - .36c = .64c$.

$$.64c = c(.5)^{x/5730}$$

$$.64 = (.5)^{x/5730}$$

$$\ln.64 = \frac{x}{5730}\ln.5$$

$$x = \frac{5730\ln.64}{\ln.5} = 3690 \text{ years}$$

61. Use $M(x) = c(.5)^{x/h}$ with $h = 5730$ (the half-life of carbon-14) and $M(c) = c - .264c = .736c$.

$$.736c = c(.5)^{x/5730}$$

$$.736 = (.5)^{x/5730}$$

$$\ln.736 = \frac{x}{5730}\ln.5$$

$$x = \frac{5730\ln.736}{\ln.5} = 2530 \text{ years}$$

63. Use $A = P(1+r)^t$ with $P = 1000$, $A = 2000$, $t = 10 \times 4$, and $r = i/4$.

$$2000 = 1000\left(1 + \frac{i}{4}\right)^{40}$$

$$2 = \left(1 + \frac{i}{4}\right)^{40}$$

$$1 + \frac{i}{4} = 2^{1/40}$$

$$i = 4\left(2^{1/40} - 1\right)$$

$$i = .0699 \text{ or } 6.99\%$$

65. a. Use $A = P(1+r)^t$ with $P = 500$, $A = 1500$, and $r = .05$.

$$1500 = 500(1 + .05)^t$$

$$3 = 1.05^t$$

$$\ln 3 = t \ln 1.05$$

$$t = \frac{\ln 3}{\ln 1.05} = 22.5 \text{ years}$$

b. Use $A = P(1+r)^t$ with $P = 500$, $A = 1500$, and $r = .05/4$, $t = 4n$.

$$1500 = 500(1 + .05/4)^{4n}$$

$$3 = 1.0125^{4n}$$

$$\ln 3 = 4n \ln 1.0125$$

$$n = \frac{\ln 3}{4 \ln 1.0125} = 22.1 \text{ years}$$

67. Use $A = P(1+r)^t$ with $A = 5000$, $t = 4 \times 9 = 36$, and $r = .05/4$.

$$5000 = P(1 + .05/4)^{36}$$

$$P = \frac{5000}{1.0125^{36}}$$

$$P = \$3197.05$$

69. Use $f(x) = Pe^{kx}$ with $P = 364$, $k = .004$, $f(x) = 500$.

$$500 = 364e^{.004x}$$

$$\frac{500}{364} = e^{.004x}$$

$$\ln \frac{500}{364} = .004x$$

$$x = \frac{1}{.004} \ln \frac{500}{364}$$

$$x = 79.4 \text{ years}$$

71. a. Use $f(x) = Pe^{kx}$ with $P = 151$,
$x = 10$, $f(10) = 173$.
$$173 = 151e^{k(10)}$$
$$\frac{173}{151} = e^{10k}$$
$$\ln\frac{173}{151} = .10k$$
$$k = \frac{1}{10}\ln\frac{173}{151}$$
$$k = .0136 \text{ or } 1.36\% \text{ per year}$$

b. Use $f(x) = Pe^{kx}$ with $P = 151$,
$f(x) = 250$ and this value of k.
$$250 = 151e^{.0136x}$$
$$\frac{250}{151} = e^{.0136x}$$
$$\ln\frac{250}{151} = .0136x$$
$$x = \frac{1}{.0136}\ln\frac{250}{151}$$
$$x = 37 \text{ years}$$

73. a. Use $P = e^{kt}$ with $P = 25$, $t = .15$
$$25 = e^{k(.15)}$$
$$\ln 25 = .15k$$
$$k = \frac{\ln 25}{.15} = 21.46$$

b. Use $P = e^{kt}$ with $P = 50$
and this value of k.
$$50 = e^{21.46t}$$
$$\ln 50 = 21.46t$$
$$t = \frac{\ln 50}{21.46} = .182$$

75. a. When $x = 1$, $f(1) = 100$. When
$x = 2$, $f(2) = 500$. Since the number
of bacteria is multiplied by 5 every
hour, $f(0) = 100/5 = 20$. When
$x = 3$,
$f(3) = 5 \times 500 = 2500$ bacteria.

b. Use $f(x) = 20 \cdot 5^x$ with $f(x) = 40$.
$$40 = 20 \cdot 5^x$$
$$2 = 5^x$$
$$\ln 2 = x\ln 5$$
$$x = \frac{\ln 2}{\ln 5} = .43 \text{ hours}$$

77. a. At $t = 0$,
$$f(0) = \frac{45,000}{1 + 224e^{-.899(0)}} = 200$$
At $t = 3$,
$$f(3) = \frac{45,000}{1 + 224e^{-.899(3)}} = 2795$$

b. Use $f(t) = 22,500$.
$$22,500 = \frac{45,000}{1 + 224e^{-.899t}}$$
$$1 + 224e^{-.899t} = 2$$
$$e^{-.899t} = \frac{1}{224}$$
$$t = \frac{1}{-.899}\ln\left(\frac{1}{224}\right)$$
$$t = 6.02 \text{ weeks}$$

79. a. $50 = c\left(1 - e^{-4k}\right)$ and $70 = c\left(1 - e^{-8k}\right)$ **b.** $50 = c\left(1 - e^{-4k}\right)$ and $90 = c\left(1 - e^{-8k}\right)$

Thus $\dfrac{70}{50} = \dfrac{c\left(1 - e^{-8k}\right)}{c\left(1 - e^{-4k}\right)}$

$\dfrac{70}{50} = \dfrac{c\left(1 - e^{-4k}\right)\left(1 + e^{-4k}\right)}{c\left(1 - e^{-4k}\right)}$

$\dfrac{7}{5} = 1 + e^{-4k}$

$\dfrac{2}{5} = e^{-4k}$

$k = -\dfrac{1}{4}\ln\dfrac{2}{5} = .229$

Then $50 = c\left(1 - e^{-4(.229)}\right)$

$c = \dfrac{50}{1 - e^{-4(.229)}} = 83.3$

Thus $\dfrac{90}{50} = \dfrac{c\left(1 - e^{-8k}\right)}{c\left(1 - e^{-4k}\right)}$

Solving as in part **a**,

$k = -\dfrac{1}{4}\ln\dfrac{4}{5} = .0558$

Then $50 = c\left(1 - e^{-4(.0558)}\right)$

$c = \dfrac{50}{1 - e^{-4(.0558)}} = 250$

Hence

$N = 250\left(1 - e^{-.0558t}\right)$. When $N = 125$

$125 = 250\left(1 - e^{-.0558t}\right)$

$.5 = 1 - e^{-.0558t}$

$.5 = e^{-.0558t}$

$t = \dfrac{\ln .5}{.0558} = 12.4$ weeks

5.6 Exponential, Logarithmic, and Other Models

1. The data appears to grow slowly, then more quickly, then level off. A logistic model seems best, although a cubic model might be possible.

3. The data appears to grow exponentially, and an exponential model seems best. The isolated point is called an 'outlier'.

5. The data appears to be decaying exponentially, and an exponential model (expect a negative exponent) seems best, although a power model might be possible.

7. The data appears to group itself into a parabola, and a quadratic model seems best. A cubic model might be possible.

9. The data appears to group itself into half of a parabola, and a quadratic model seems best. A cubic model might be possible.

11. Successive ratios: $\dfrac{15.2}{3} = 5.07$, $\dfrac{76.9}{15.2} = 5.06$, $\dfrac{389.2}{76.9} = 5.06$, $\dfrac{1975.5}{389.2} = 5.08$, and

$\dfrac{9975.8}{1975.5} = 5.05$. Since these ratios are very nearly the same, an exponential model would be appropriate.

13. The logistic model in Example 2 is $y = \dfrac{442.1}{1 + 56.33e^{-.0216x}}$. As x increases, $e^{-.0216x}$ decreases from a maximum of 1 toward 0. The denominator therefore decreases toward 1, and thus y increases toward 442.1 (in millions) but can never reach or exceed this value.

15. a. $0 \le y \le 5$

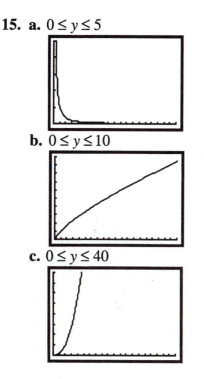

b. $0 \le y \le 10$

c. $0 \le y \le 40$

17. Shown is the graph of x vs. $\ln y$ using the window $0 \le x \le 12, 0 \le y \le 8$.

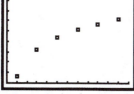

Shown next is the graph of $\ln x$ vs. $\ln y$ using the window $0 \le x \le 3, 0 \le y \le 8$.

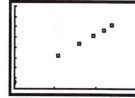

Shown next is the graph of $\ln x$ vs. y using the window $0 \le x \le 3, 0 \le y \le 500$.

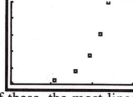

Of these, the most linear is the graph of $\ln x$ vs. $\ln y$ indicating that a power model would be most appropriate.

19. Shown is the graph of x vs. $\ln y$ using the window $0 \le x \le 40, 0 \le y \le 4$.

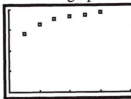

Shown next is the graph of $\ln x$ vs. $\ln y$ using the window $0 \le x \le 4, 0 \le y \le 4$.

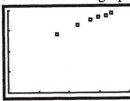

Shown next is the graph of $\ln x$ vs. y using the window $0 \le x \le 4, 0 \le y \le 50$.

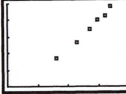

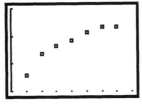

Of these, the most linear is the graph of $\ln x$ vs. $\ln y$ indicating that a power model would be most appropriate.

21. a. Use the window $0 \le x \le 8, 90,000 \le y \le 105,000$.

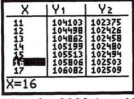

b. The logarithmic model is shown on the left, the logistic model on the right.

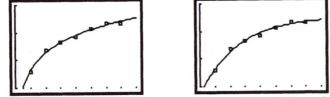

c. In the table below, the logarithmic model is Y_1 and the logistic model is Y_2

X	Y₁	Y₂
11	104103	102375
12	104498	102426
13	104862	102458
14	105199	102480
15	105513	102494
16	105806	102503
17	106082	102509

X=16

Thus for 2000 ($x = 12$) the logarithmic model predicts 104,498 while the logistic model predicts 102,426. 2005 is shown as $x = 17$.

d. As a general rule, logistic models are better for ordinary population growth data, but in this case the prediction of slow steadily decreasing growth from the logarithmic model might be better than the prediction of levelling off at 102,519 from the logistic model.

23. a. Use the window $12 \le x \le 108, 0 \le y \le 90$.

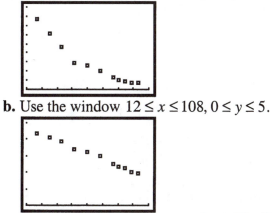

b. Use the window $12 \le x \le 108, 0 \le y \le 5$.

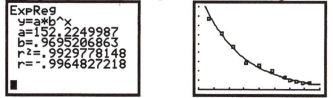

c. An exponential model seems best. The TI-83 result is shown, as well as the result of graphing it and the original data in the original window.

```
ExpReg
 y=a*b^x
 a=152.2249987
 b=.9695206863
 r²=.9929778148
 r=-.9964827218
```

25. a. Use the window $0 \le x \le 25, 0 \le y \le 36$

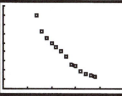

b. Entering the data into the Ti-83 and applying the ExpReg feature yields the following:

```
ExpReg
 y=a*b^x
 a=91.16295745
 b=.8591926825
 r²=.9869547611
 r=-.9934559684
```

c. In the year 2003, the model yields 2.8 students/computer.
d. The model indicates 1 student/computer when $x = 29.74$, that is, in late 2009.
e. Any projection of future trends becomes tenuous with passing time. The availability of new technology could have unforeseen effects.

27. a. Use the window $0 \le x \le 8, 95 \le y \le 110$.

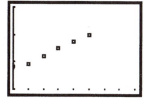

b. Successive ratios: $\dfrac{99.6}{99} = 1.0061$, $\dfrac{101}{99.6} = 1.0141$, $\dfrac{102.5}{101} = 1.0149$, $\dfrac{103.6}{102.5} = 1.0107$,

and $\dfrac{104.8}{103.6} = 1.0116$. Since these ratios are very nearly the same, an exponential

model would be appropriate.

c.   **d.** With $x = 0$ corresponding to 1995:

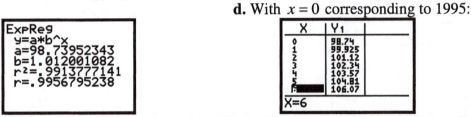

e. The table shown in part **d** predicts 108.6 million households in 2003 ($x = 8$).

29. a. Entering the data into the TI-83 and applying the LnReg feature yields:

b. The model predicts 77.4 years ($x = 86$).
c. The model predicts 80.9 years in 2012 and 81.05 years in 2023.

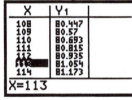

Chapter 5 Review Exercises

1. $\sqrt{\sqrt[3]{c^{12}}} = \left(\left(c^{12} \right)^{1/3} \right)^{1/2} = c^2$

3. $\left(a^{-2/3} b^{2/5} \right)\left(a^3 b^6 \right)^{4/3} = a^{-2/3} b^{2/5} a^4 b^8 = a^{10/3} b^{42/5}$

5. $\left(u^{1/4} - v^{1/4} \right)\left(u^{1/4} + v^{1/4} \right) = \left(u^{1/4} \right)^2 - \left(v^{1/4} \right)^2 = u^{1/2} - v^{1/2}$

7. $\dfrac{\sqrt[3]{6c^4 d^{14}}}{\sqrt[3]{48 c^{-2} d^2}} = \sqrt[3]{\dfrac{6 c^4 d^{14}}{48 c^{-2} d^2}} = \sqrt[3]{\dfrac{c^6 d^{12}}{8}} = \dfrac{c^2 d^4}{2}$

9. $\dfrac{\sqrt{2x+2h+1}-\sqrt{2x+1}}{h}\cdot\dfrac{\sqrt{2x+2h+1}+\sqrt{2x+1}}{\sqrt{2x+2h+1}+\sqrt{2x+1}}=\dfrac{(2x+2h+1)-(2x+1)}{h\left(\sqrt{2x+2h+1}+\sqrt{2x+1}\right)}$

$$=\dfrac{2h}{h\left(\sqrt{2x+2h+1}+\sqrt{2x+1}\right)}=\dfrac{2}{\sqrt{2x+2h+1}+\sqrt{2x+1}}$$

11. $\sqrt{x-1}=2-x$

$\left(\sqrt{x-1}\right)^2=(2-x)^2$

$x-1=x^2-4x+4$

$0=x^2-5x+5$

$x=\dfrac{5\pm\sqrt{5}}{2}$ The solution $\dfrac{5-\sqrt{5}}{2}$ checks. $\left(\dfrac{5+\sqrt{5}}{2}\text{ is extraneous.}\right)$

13. $\sqrt{x+1}+\sqrt{x-1}=1$

$\sqrt{x+1}=1-\sqrt{x-1}$

$x+1=1-2\sqrt{x-1}+x-1$

$1=-2\sqrt{x-1}$

Since the left side is positive and the right side is negative, the equation has no solution.

15. Graph $y=\sqrt[3]{x^4-2x^3+6x-7}$ and $y=x+3$ using the window $-10\le x\le10$, $-5\le y\le15$ and apply the intersection routine to find $x=-1.733$ and $x=5.521$.

17. $-3\le x\le3, 0\le y\le2$

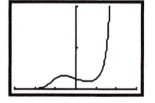

19. Horizontal asymptote: $y=-1$

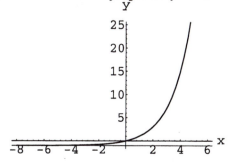

21. Vertical asymptote: $x = -4$

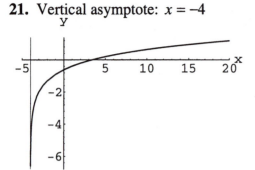

23. Graph the function using the window $0 \le x \le 20, 0 \le y \le 100$.

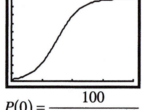

a. $P(0) = \dfrac{100}{1 + 48.2e^{-.52(0)}} = 2.03\%$ **b.** $P(6) = \dfrac{100}{1 + 48.2e^{-.52(6)}} = 31.97\%$

c. The graph seems steepest between $t = 6$ and $t = 10$.

d. Since $P = 100\%$ is a horizontal asymptote for the graph, the program will never be completely mastered.

25. $\ln 756 = 6.628$ **27.** $\ln(u + v) = r^2 - 1$ **29.** $\log 756 = 2.8785$

31. $e^{7.118} = 1234$ **33.** $e^t = rs$ **35.** $5^u = cd - k$

37. 3   **39.** 3/4

41. $3\ln\sqrt{x} + (1/2)\ln x = \ln\sqrt{x}^3 + \ln\sqrt{x}$
$$= \ln\sqrt{x}^4 = \ln x^2$$

43. $\ln 3x - 3\ln x + \ln 3y = \ln 3x - \ln x^3 + \ln 3y = \ln\dfrac{3x \cdot 3y}{x^3} = \ln\dfrac{9y}{x^2}$

45. $4\ln x - 2(\ln x^3 + 4\ln x) = 4\ln x - 2(3\ln x + 4\ln x) = -10\ln x = \ln x^{-10}$

47. $\log_{20} 400 = \log_{20} 20^2 = 2$

49. (c)   **51.** (c)

53. $\log_3 9^{x^2} = 4$
$x^2 \log_3 9 = 4$
$2x^2 = 4$
$x = \pm\sqrt{2}$

55. $8^x = 4^{x^2-3}$

$2^{3x} = 2^{2(x^2-3)}$

$3x = 2x^2 - 6$

$0 = 2x^2 - 3x - 6$

$x = \dfrac{3 \pm \sqrt{57}}{4}$

57. $2 \cdot 4^x - 5 = -4$

$2 \cdot 4^x = 1$

$4^x = \dfrac{1}{2}$

$4^x = 4^{-\frac{1}{2}}$

$x = -\dfrac{1}{2}$

59. $u = c + d \ln x$

$\dfrac{u-c}{d} = \ln x$

$x = e^{\frac{u-c}{d}}$

61. $\ln x + \ln(3x-5) = \ln 2$

$\ln x(3x-5) = \ln 2$

$3x^2 - 5x = 2$

$3x^2 - 5x - 2 = 0$

$(3x+1)(x-2) = 0$

$x = -\dfrac{1}{3}$ or $x = 2$

Since x must be positive, 2 is the only solution.

63. $\log(x^2 - 1) = 2 + \log(x+1)$

$\log(x^2 - 1) = \log 100 + \log(x+1)$

$x^2 - 1 = 100(x+1)$

$x^2 - 100x - 101 = 0$

$(x - 101)(x + 1) = 0$

$x = 101$ or $x = -1$

Since x cannot equal -1, 101 is the only solution.

65. Use $M(x) = c(.5)^{x/h}$ with $c = 10$, $h = 1590$, $x = 365$.

$M(365) = 10(.5)^{365/1590} = 1.64$ milligrams

67. If P is the initial amount, then after 1 year $P - .0559P = .9441P$ remains. Use
$M(x) = P(.5)^{x/h}$ with $x = 1$, $M(1) = .9441P$.

$.9441P = P(.5)^{1/h}$

$.9441 = (.5)^{1/h}$

$\dfrac{1}{h} = \dfrac{\ln .9441}{\ln .5}$

$h = \dfrac{\ln .5}{\ln .9441}$

$h = 12.05$ years

69. Use $A = P(1+r)^t$ with $A = 1000, r = .08/4, t = 4 \times 10$.

$$1000 = P(1+.08/4)^{40}$$

$$P = \frac{1000}{1.02^{40}}$$

$$P = \$452.89$$

71. Increasing the intensity by a factor of 10^3 increases the Richter magnitude by 3 units; the second earthquake measures 7.6.

73. a. The table shows that a 20-mph wind makes 25°F feel like 11°F. **b.**

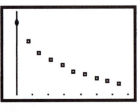

c. The plotted data strongly suggest exponential decay. One might argue that the ratios of successive values are very roughly equal. **d.**

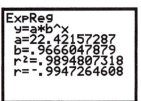

e. Using the Table feature of the TI-83, the model yields 10.3°F.

Chapter 6
Trigonometric Functions

6.1 Angles and Their Measurement

1. $\dfrac{1}{9}(2\pi) = \dfrac{2\pi}{9}$ 　　　　**3.** $\dfrac{1}{18}(2\pi) = \dfrac{\pi}{9}$ 　　　　**5.** $\dfrac{1}{36}(2\pi) = \dfrac{\pi}{18}$

7. Four of many possibilities would be $\dfrac{\pi}{4} + 2\pi = \dfrac{9\pi}{4}$, $\dfrac{\pi}{4} + 4\pi = \dfrac{17\pi}{4}$, $\dfrac{\pi}{4} - 2\pi = -\dfrac{7\pi}{4}$, $\dfrac{\pi}{4} - 4\pi = -\dfrac{15\pi}{4}$.

9. Four of many possibilities would be $-\dfrac{\pi}{6} + 2\pi = \dfrac{11\pi}{6}$, $-\dfrac{\pi}{6} + 4\pi = \dfrac{23\pi}{6}$, $-\dfrac{\pi}{6} - 4\pi = -\dfrac{13\pi}{6}$, $-\dfrac{\pi}{6} - 4\pi = -\dfrac{25\pi}{6}$.

11. $-\dfrac{\pi}{3} + 2\pi = \dfrac{5\pi}{3}$ 　　　　**13.** $\dfrac{19\pi}{4} - 4\pi = \dfrac{3\pi}{4}$

15. $-\dfrac{7\pi}{5} + 2\pi = \dfrac{3\pi}{5}$ 　　　　**17.** $7 - 2\pi$

19. $6° \cdot \dfrac{\pi}{180°} = \dfrac{\pi}{30}$ 　　　　**21.** $-12° \cdot \dfrac{\pi}{180°} = -\dfrac{\pi}{15}$

23. $75° \cdot \dfrac{\pi}{180°} = \dfrac{5\pi}{12}$ 　　　　**25.** $135° \cdot \dfrac{\pi}{180°} = \dfrac{3\pi}{4}$

27. $-225° \cdot \dfrac{\pi}{180°} = -\dfrac{5\pi}{4}$ 　　　　**29.** $930° \cdot \dfrac{\pi}{180°} = \dfrac{31\pi}{6}$

31. $\dfrac{\pi}{5} \cdot \dfrac{180°}{\pi} = 36°$ 　　　　**33.** $-\dfrac{\pi}{10} \cdot \dfrac{180°}{\pi} = -18°$

35. $\dfrac{3\pi}{4} \cdot \dfrac{180°}{\pi} = 135°$ 　　　　**37.** $\dfrac{\pi}{45} \cdot \dfrac{180°}{\pi} = 4°$

39. $-\dfrac{5\pi}{12} \cdot \dfrac{180°}{\pi} = -75°$ 　　　　**41.** $\dfrac{27\pi}{5} \cdot \dfrac{180°}{\pi} = 972°$

43. $\dfrac{40}{60} \cdot 2\pi = \dfrac{4\pi}{3}$ 　　　　**45.** $\dfrac{35}{60} \cdot 2\pi = \dfrac{7\pi}{6}$

47. $\left(3 + \dfrac{25}{60}\right) \cdot 2\pi = \dfrac{41\pi}{3}$

49. Use $s = \theta r$ with $\theta = \dfrac{4\pi}{3}$ (from Exercise **43**) and $r = 6$.

$$s = \dfrac{4\pi}{3} \cdot 6 = 8\pi \approx 25.13 \text{ cm}$$

51. Use $s = \theta r$ with $s = 85$ and $r = 20$.
$$85 = \theta \cdot 20$$
$$\theta = 4.25 \text{ radians}$$

53. $\quad s = r\theta$
$$200 = 36\theta$$
$$\theta = 5.56 \text{ radians}$$

55. $\quad s = r\theta$
$$72000 = 36\theta$$
$$\theta = 2000 \text{ radians}$$

57. $s = r\theta$
$$s = 5 \cdot 1$$
$$s = 5$$

59. $s = r\theta$
$$s = 5 \cdot 1.75$$
$$s = 8.75$$

61. $s = r\theta$
$$s = 4000\left(90° - 40°\right) \cdot \dfrac{\pi}{180°}$$
$$s = 3491 \text{ miles}$$

63. $s = r\theta$
$$s = 4000\left(41.5° - 28°\right) \cdot \dfrac{\pi}{180°}$$
$$s = 942.5 \text{ miles}$$

65. 3.5 minutes at 1 rpm equals 3.5 revolutions, or $3.5(2\pi) = 7\pi$ radians.

67. 1 minute at 2 rpm equals 2 revolutions, or $2(2\pi) = 4\pi$ radians.

69. 4.25 minutes at 5 rpm equals $4.25 \times 5 = 21.25$ revolutions, or $21.25(2\pi) = 42.5\pi$ radians.

71. 1 minute at k rpm equals k revolutions, or $2\pi k$ radians.

73. Use $s = r\theta$ with $s = 6$ and $r = 2$.
$$6 = 2\theta$$
$$\theta = 3 \text{ radians}$$

75. a. 200 revolutions per minute equals $200(2\pi) = 400\pi$ radians per minute.
   b. linear speed $= r(\text{angular speed})$
$$= 2(400\pi)$$
$$= 800\pi$$
$$= 2513.3 \text{ inches per minute}$$
$$= 209.4 \text{ feet per minute}$$

77. a. 2.5 revolutions per second equals $2.5(2\pi) = 5\pi$ radians per second.

b. linear speed $= r(\text{angular speed})$

$$= 7.5(5\pi)$$

$$= 37.5\pi \text{ inches per second}$$

$$= 37.5\pi \frac{\text{inches}}{\text{second}} \times \frac{1}{12} \frac{\text{foot}}{\text{inches}} \times \frac{1}{5280} \frac{\text{miles}}{\text{foot}} \times 3600 \frac{\text{seconds}}{\text{hour}}$$

$$= 6.69 \text{ miles per hour}$$

79. 6 revolutions per minute equals $6(2\pi) \div 60$ or $\pi/5$ radians per second.

Since linear speed $= r(\text{angular speed})$

$$10 = r\left(\frac{\pi}{5}\right)$$

$$r = \frac{50}{\pi} = 15.9 \text{ feet}$$

6.2 The Sine, Cosine and Tangent Functions

1. $\sin\left(\dfrac{3\pi}{2}\right) = -1$ **3.** $\cos\left(\dfrac{3\pi}{2}\right) = 0$ **5.** $\tan 4\pi = \dfrac{\sin 4\pi}{\cos 4\pi} = \dfrac{0}{1} = 0$

7. $\cos\left(-\dfrac{3\pi}{2}\right) = 0$ **9.** $\cos\left(-\dfrac{11\pi}{2}\right) = 0$

11. $\sin t = y = \dfrac{1}{\sqrt{5}}$, $\cos t = x = -\dfrac{2}{\sqrt{5}}$, $\tan t = \dfrac{y}{x} = \dfrac{1}{\sqrt{5}} \div -\dfrac{2}{\sqrt{5}} = -\dfrac{1}{2}$

13. $\sin t = y = -\dfrac{4}{5}$, $\cos t = x = -\dfrac{3}{5}$, $\tan t = \dfrac{y}{x} = -\dfrac{4}{5} \div -\dfrac{3}{5} = \dfrac{4}{3}$

15. $\sin\dfrac{\pi}{3} = \dfrac{\sqrt{3}}{2}$, $\cos\dfrac{\pi}{3} = \dfrac{1}{2}$, $\tan\dfrac{\pi}{3} = \sin\dfrac{\pi}{3} \div \cos\dfrac{\pi}{3} = \dfrac{\sqrt{3}}{2} \div \dfrac{1}{2} = \sqrt{3}$

17. $\sin\dfrac{\pi}{4} = \dfrac{\sqrt{2}}{2}$, $\cos\dfrac{\pi}{4} = \dfrac{\sqrt{2}}{2}$, $\tan\dfrac{\pi}{4} = \sin\dfrac{\pi}{4} \div \cos\dfrac{\pi}{4} = \dfrac{\sqrt{2}}{2} \div \dfrac{\sqrt{2}}{2} = 1$

19. $\sin\dfrac{3\pi}{4} = \dfrac{\sqrt{2}}{2}$, $\cos\dfrac{3\pi}{4} = -\dfrac{\sqrt{2}}{2}$, $\tan\dfrac{3\pi}{4} = \sin\dfrac{3\pi}{4} \div \cos\dfrac{3\pi}{4} = \dfrac{\sqrt{2}}{2} \div -\dfrac{\sqrt{2}}{2} = -1$

21. $\sin\dfrac{5\pi}{6} = \dfrac{1}{2}$, $\cos\dfrac{5\pi}{6} = -\dfrac{\sqrt{3}}{2}$, $\tan\dfrac{5\pi}{6} = \sin\dfrac{5\pi}{6} \div \cos\dfrac{5\pi}{6} = \dfrac{1}{2} \div -\dfrac{\sqrt{3}}{2} = -\dfrac{\sqrt{3}}{3}$

23. $\theta = -\dfrac{23\pi}{6}$ is coterminal with $-\dfrac{23\pi}{6} + 4\pi = \dfrac{\pi}{6}$.

$$\sin\left(-\dfrac{23\pi}{6}\right) = \sin\dfrac{\pi}{6} = \dfrac{1}{2}, \quad \cos\left(-\dfrac{23\pi}{6}\right) = \cos\dfrac{\pi}{6} = \dfrac{\sqrt{3}}{2}, \quad \tan\left(-\dfrac{23\pi}{6}\right) = \tan\dfrac{\pi}{6} = \dfrac{\sqrt{3}}{3}$$

25. $\theta = -\dfrac{19\pi}{3}$ is coterminal with $-\dfrac{19\pi}{3} + 8\pi = \dfrac{5\pi}{3}$. This angle has terminal side in

quadrant IV and the angle formed by the terminal side and the x-axis is $\pi/3$. Therefore,

$$\sin\left(-\dfrac{19\pi}{3}\right) = -\sin\dfrac{\pi}{3} = -\dfrac{\sqrt{3}}{2}, \quad \cos\left(-\dfrac{19\pi}{3}\right) = \cos\dfrac{\pi}{3} = \dfrac{1}{2}, \quad \tan\left(-\dfrac{19\pi}{3}\right) = -\tan\dfrac{\pi}{3} = -\sqrt{3}$$

27. $\theta = -\dfrac{15\pi}{4}$ is coterminal with $-\dfrac{15\pi}{4} + 4\pi = \dfrac{\pi}{4}$.

$$\sin\left(-\dfrac{15\pi}{4}\right) = \sin\dfrac{\pi}{4} = \dfrac{\sqrt{2}}{2}, \quad \cos\left(-\dfrac{15\pi}{4}\right) = \cos\dfrac{\pi}{4} = \dfrac{\sqrt{2}}{2}, \quad \tan\left(-\dfrac{15\pi}{4}\right) = \tan\dfrac{\pi}{4} = 1$$

29. The terminal side of $\theta = -17\pi/2$ in standard position is along the negative y-axis. Using the point $(0,-1)$ on this terminal side yields

$$\sin\left(-\dfrac{17\pi}{2}\right) = -1, \quad \cos\left(-\dfrac{17\pi}{2}\right) = 0, \quad \tan\left(-\dfrac{17\pi}{2}\right) \text{ is undefined}.$$

31. $\sin\dfrac{\pi}{3}\cos\pi + \sin\pi\cos\dfrac{\pi}{3} = \dfrac{\sqrt{3}}{2}(-1) + 0\left(\dfrac{1}{2}\right) = -\dfrac{\sqrt{3}}{2}$

33. $\cos\dfrac{\pi}{2}\cos\dfrac{\pi}{4} - \sin\dfrac{\pi}{2}\sin\dfrac{\pi}{4} = 0\left(\dfrac{\sqrt{2}}{2}\right) - 1\left(\dfrac{\sqrt{2}}{2}\right) = -\dfrac{\sqrt{2}}{2}$

35. $\sin\dfrac{3\pi}{4}\cos\dfrac{5\pi}{6} - \cos\dfrac{3\pi}{4}\sin\dfrac{5\pi}{6} = \dfrac{\sqrt{2}}{2}\left(-\dfrac{\sqrt{3}}{2}\right) - \left(-\dfrac{\sqrt{2}}{2}\right)\left(\dfrac{1}{2}\right) = \dfrac{-\sqrt{6} + \sqrt{2}}{4}$

37. $(x,y) = (2,7)$; $r = \sqrt{x^2 + y^2} = \sqrt{2^2 + 7^2} = \sqrt{53}$

$$\sin t = \dfrac{y}{r} = \dfrac{7}{\sqrt{53}}, \quad \cos t = \dfrac{x}{r} = \dfrac{2}{\sqrt{53}}, \quad \tan t = \dfrac{y}{x} = \dfrac{7}{2}$$

39. $(x,y) = (-5,-6)$; $r = \sqrt{x^2 + y^2} = \sqrt{(-5)^2 + (-6)^2} = \sqrt{61}$

$$\sin t = \dfrac{y}{r} = \dfrac{-6}{\sqrt{61}}, \quad \cos t = \dfrac{x}{r} = \dfrac{-5}{\sqrt{61}}, \quad \tan t = \dfrac{y}{x} = \dfrac{-6}{-5} = \dfrac{6}{5}$$

41. $(x,y) = (\sqrt{3},-10);$ $r = \sqrt{x^2 + y^2} = \sqrt{(\sqrt{3})^2 + (-10)^2} = \sqrt{103}$

$\sin t = \dfrac{y}{r} = \dfrac{-10}{\sqrt{103}},$ $\cos t = \dfrac{x}{r} = \dfrac{\sqrt{3}}{\sqrt{103}} = \dfrac{\sqrt{309}}{103},$ $\tan t = \dfrac{y}{x} = -\dfrac{10}{\sqrt{3}}$

43. Let $(x,y) = (1,-3);$ $r = \sqrt{x^2 + y^2} = \sqrt{1^2 + (-3)^2} = \sqrt{10}.$

$\sin t = \dfrac{y}{r} = \dfrac{-3}{\sqrt{10}},$ $\cos t = \dfrac{x}{r} = \dfrac{1}{\sqrt{10}},$ $\tan t = \dfrac{y}{x} = \dfrac{-3}{1} = -3$

45. Let $(x,y) = (-3,5);$ $r = \sqrt{x^2 + y^2} = \sqrt{(-3)^2 + 5^2} = \sqrt{34}.$

$\sin t = \dfrac{y}{r} = \dfrac{5}{\sqrt{34}},$ $\cos t = \dfrac{x}{r} = \dfrac{-3}{\sqrt{34}},$ $\tan t = \dfrac{y}{x} = -\dfrac{5}{3}$

47. The terminal side of the angle is on the line $2y + x = 0,$ with x negative.

Let $(x,y) = (-2,1);$ $r = \sqrt{x^2 + y^2} = \sqrt{(-2)^2 + 1^2} = \sqrt{5}.$

$\sin t = \dfrac{y}{r} = \dfrac{1}{\sqrt{5}},$ $\cos t = \dfrac{x}{r} = \dfrac{-2}{\sqrt{5}},$ $\tan t = \dfrac{y}{x} = \dfrac{1}{-2} = -\dfrac{1}{2}$

49. Quadrant I: $\sin t, \cos t, \tan t$ all $+$
Quadrant II: $\sin t\ +, \cos t\ -, \tan t\ -$
Quadrant III: $\sin t\ -, \cos t\ -, \tan t\ +$
Quadrant IV: $\sin t\ -, \cos t\ +, \tan t\ -$

51. Since $0 < 1 < \pi/2,$ the terminal side of an angle of 1 radian lies in the first quadrant, so $y = \sin 1$ is positive.

53. Since $\pi/2 < 3 < \pi,$ the terminal side of an angle of 3 radians lies in the second quadrant, so $y/x = \tan 3$ is negative.

55. Since $0 < 1/5 < \pi/2,$ the terminal side of an angle of 1.5 radians lies in the first quadrant, so $y/x = \tan 1.5$ is positive.

57. The terminal side of t must pass through $(0,1).$ Hence $t = \dfrac{\pi}{2} + 2\pi n,$ n any integer.

59. The terminal side of t must pass through $(1,0)$ or $(-1,0).$ Hence $t = \pi n,$ n any integer.

61. The terminal side of t must pass through $(0,-1)$ or $(0,1).$ Hence $t = \dfrac{\pi}{2} + \pi n,$ n any integer.

63. $\sin(\cos 0) = \sin 1$ $\cos(\sin 0) = \cos 0 = 1.$ Since $\sin 1 < 1,$ $\sin(\cos 0) < \cos(\sin 0).$

65. a. Horse A will reach the position occupied by horse B at a time t after rotation of $\pi/4$ radians, which is an eighth of a complete circle. This takes $1/8$ of a minute. Thus $B(t) = A(t + 1/8)$.

    b. Horse A will reach the position occupied by horse C at a time t after rotation of $1/3$ of a complete circle. This takes $1/3$ of a minute. Thus $C(t) = A(t + 1/3)$.

    c. $E(t) = D(t + 1/8)$; $F(t) = D(t + 1/3)$

    d. $D(t) / A(t) = 5/1; D(t) = 5A(t); D(t) = 5A(t)$

    e. $E(t) = D(t + 1/8) = 5A(t + 1/8); F(t) = D(t + 1/3) = 5A(t + 1/3)$

    f. In t minutes, the merry-go-round rotates $2\pi t$ radians. Therefore, $A(t) = \sin 2\pi t$.

    g. $B(t) = \sin 2\pi(t + 1/8) = \sin(2\pi t + \pi/4); C(t) = \sin 2\pi(t + 1/3) = \sin(2\pi t + 2\pi/3)$

    h. $D(t) = 5\sin 2\pi t; E(t) = 5\sin(2\pi t + \pi/4); F(t) = 5\sin(2\pi t + 2\pi/3)$

6.3 Algebra and Identities

1. $(fg)(t) = f(t)g(t) = 3\sin t(\sin t + 2\cos t) = 3\sin^2 t + 6\sin t \cos t$

3. $(fg)(t) = f(t)g(t) = 3\sin^2 t(\sin t + \tan t) = 3\sin^3 t + 3\sin^2 t \tan t$

5. $(\cos t + 2)(\cos t - 2)$ **7.** $(\sin t + \cos t)(\sin t - \cos t)$

9. $(\tan t + 3)^2$ **11.** $(3\sin t + 1)(2\sin t - 1)$

13. $\cos^4 t + 4\cos^2 t - 5 = (\cos^2 t + 5)(\cos^2 t - 1) = (\cos^2 t + 5)(\cos t + 1)(\cos t - 1)$

15. $(f \circ g)(t) = f(g(t)) = \cos(g(t)) = \cos(2t + 4)$
    $(g \circ f)(t) = g(f(t)) = 2f(t) + 4 = 2\cos t + 4$

17. $(f \circ g)(t) = f(g(t)) = \tan(g(t) + 3) = \tan(t^2 - 1 + 3) = \tan(t^2 + 2)$
    $(g \circ f)(t) = g(f(t)) = (f(t))^2 - 1 = \tan^2(t + 3) - 1$

19. By the Pythagorean identity, $\sin^2 t + \cos^2 t = 1$. Since $(5/13)^2 + (12/13)^2 = 1$, these values are possible.

21. By the Pythagorean identity, $\sin^2 t + \cos^2 t = 1$. Since $(-1)^2 + 1^2 = 2$, these values are not possible.

23. By the Pythagorean identity, $\sin^2 t + \cos^2 t = 1$. Also, $\tan t = \dfrac{\sin t}{\cos t}$, so $1 = \dfrac{1}{\cos t}$ and $\cos t = 1$ in this case. But since $1^2 + 1^2 = 2$, these values are not possible.

25. By the Pythagorean identity,

$\sin^2 t = 1 - \cos^2 t = 1 - (-.5)^2 = 3/4$. Hence $\sin t = -\sqrt{3/4}$ or $\sin t = \sqrt{3/4}$.

Since $\pi < t < 3\pi/2$, $\sin t$ is negative, hence $\sin t = -\sqrt{3/4} = -\sqrt{3}/2$.

27. By the Pythagorean identity,

$\sin^2 t = 1 - \cos^2 t = 1 - (-1/2)^2 = 3/4$. Hence $\sin t = -\sqrt{3/4}$ or $\sin t = \sqrt{3/4}$.

Since $0 < t < \pi/2$, $\sin t$ is positive, hence $\sin t = \sqrt{3/4} = \sqrt{3}/2$.

29. $\sin(-t) = -\sin t = -3/5$ **31.** $\sin(2\pi - t) = -\sin t = -3/5$

33. By the Pythagorean identity,

$\cos^2 t = 1 - \sin^2 t = 1 - (3/5)^2 = 16/25$.

Since $0 < t < \pi/2$, $\cos t$ is positive, hence $\cos t = \sqrt{16/25} = 4/5$.

$\tan t = \sin t \div \cos t = 3/5 \div 4/5 = 3/4$

35. $\tan(2\pi - t) = \tan(-t) = -\tan t = -3/4$ (using Exercise 33)

37. By the Pythagorean identity,

$\sin^2 t = 1 - \cos^2 t = 1 - (-2/5)^2 = 1/5$.

Since $\pi < t < 3\pi/2$, $\sin t$ is negative, hence $\sin t = -\sqrt{1/5} = -\sqrt{5}/5$.

39. $\cos(2\pi - t) = \cos(-t) = \cos t = -2/5$

41. $\sin(4\pi + t) = \sin t = \sqrt{5}/5$ (using Exercise 37)

43. By the Pythagorean identity,

$$\cos^2 \frac{\pi}{8} = 1 - \sin^2 \frac{\pi}{8} = 1 - \left(\frac{\sqrt{2-\sqrt{2}}}{2} \right)^2 = \frac{4-2+\sqrt{2}}{4} = \frac{2+\sqrt{2}}{4}.$$

Since $0 < \frac{\pi}{8} < \frac{\pi}{2}$, $\cos \frac{\pi}{8}$ is positive, hence $\cos \frac{\pi}{8} = \frac{\sqrt{2+\sqrt{2}}}{2}$.

45. $\sin \dfrac{17\pi}{8} = \sin \dfrac{\pi}{8} = \dfrac{\sqrt{2-\sqrt{2}}}{2}$

47. $\sin^2 t - \cos^2 t$

49. $\tan t \cos t = \dfrac{\sin t}{\cos t} \cos t = \sin t$

51. $\sqrt{\sin^3 t \cos t} \sqrt{\cos t} = \sqrt{\sin^3 t \cos^2 t} = \sin t \cos t \sqrt{\sin t}$

53. $\left(\dfrac{4\cos^2 t}{\sin^2 t}\right)\left(\dfrac{\sin t}{4\cos t}\right)^2 = \dfrac{4\cos^2 t}{\sin^2 t} \cdot \dfrac{\sin^2 t}{16\cos^2 t} = \dfrac{1}{4}$

55. $\dfrac{\cos^2 t + 4\cos t + 4}{\cos t + 2} = \dfrac{(\cos t + 2)^2}{\cos t + 2} = \cos t + 2$

57. $\dfrac{1}{\cos t} - \sin t \tan t = \dfrac{1}{\cos t} - \sin t \dfrac{\sin t}{\cos t} = \dfrac{1 - \sin^2 t}{\cos t} = \dfrac{\cos^2 t}{\cos t} = \cos t$

59. Use $k = \pi$. $f(t + k) = \sin 2(t + \pi) = \sin(2t + 2\pi) = \sin 2t = f(t)$ for all t.

61. Use $k = \pi$. $f(t + k) = \sin 4(t + \pi) = \sin(4t + 4\pi) = \sin 4t = f(t)$ for all t.

63. Use $k = \pi/2$. $f(t + k) = \tan 2(t + \pi/2) = \tan(2t + \pi) = \tan 2t = f(t)$ for all t in the domain.

65. a. If $0 < k < 2\pi$, there is no number such that $\cos k = 1$.
   b. If $\cos(t + k) = \cos t$ for every number t, then use $t = 0$.
   $\cos(0 + k) = \cos 0$
   $\cos k = 1$
   c. Therefore there is no number k with $0 < k < 2\pi$ such that $\cos(t + k) = \cos t$ for every number t.

6.4 Basic Graphs

1. $t = n\pi$ where n is any integer.   **3.** $t = \dfrac{\pi}{2} + 2n\pi$ where n is any integer.

5. $t = \pi + 2n\pi$ where n is any integer.

7. Since $\dfrac{y}{x} = 11$, $\tan t = 11$.   **9.** Since $\dfrac{y}{x} = 1.4$, $\tan t = 1.4$.

11. Shift 3 units vertically up.

13. Reflect the graph in the horizontal axis.   **15.** Shift 5 units vertically up.

17. Stretch by a factor of 3 away from the horizontal axis.

19. Stretch by a factor of 3 away from the horizontal axis, then shift 2 units vertically up.

21. Shift 2 units horizontally to the right.

23. Graph $y = \sin x$ and the horizontal line $y = \frac{3}{5}$ using the window $0 \le x \le 2\pi$, $-1 \le y \le 1$. Since there are two points of intersection, conclude that the equation has two solutions in the interval.

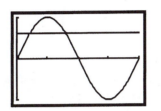

25. Graph $y = \tan x$ and the horizontal line $y = 4$ using the window $0 \le x \le 2\pi$, $-5 \le y \le 5$. Ignore the erroneous vertical lines near the asymptotes of $y = \tan x$. Since there are two points of intersection, conclude that the equation has two solutions in the interval.

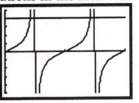

27. Graph $y = \sin x$ and the horizontal line $y = -\frac{1}{2}$ using the window $0 \le x \le 2\pi$, $-1 \le y \le 1$. Since there are two points of intersection, conclude that the equation has two solutions in the interval.

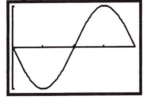

29. Graph $y = \cos x$ and any horizontal line, say $y = \frac{1}{2}$, using the window $0 \le x \le 2\pi$, $-1 \le y \le 1$. Since there are two points of intersection, conclude that any such equation has two solutions in the interval.

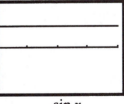

31. Graph $y = \sin(-x)$ and $y = -\sin x$ using the window $0 \le x \le 2\pi$, $-1 \le y \le 1$.
The graphs appear to coincide, hence the equation could be an identity.

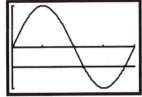

33. Graph $y = \sin^2 x + \cos^2 x$ and $y = 1$ using the window $0 \le x \le 2\pi$, $-1 \le y \le 1$. The graphs appear to coincide, hence the equation could be an identity (and is a well-known one).

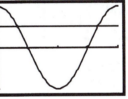

35. Graph $y = \sin x$ and $y = \cos\left(x - \frac{\pi}{2}\right)$ using the window $0 \le x \le 2\pi$, $-1 \le y \le 1$. The graphs appear to coincide, hence the equation could be an identity.

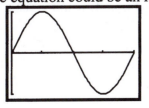

37. Graph $y = \dfrac{\sin x}{1 + \cos x}$ and $y = \tan x$ using the window $0 \le x \le 2\pi$, $-5 \le y \le 5$. The graphs do not coincide, hence the equation could not be an identity.

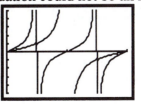

39. Graph $y = \cos\left(\frac{\pi}{2} + x\right)$ and $y = -\sin x$ using the window $0 \le x \le 2\pi,\ -1 \le y \le 1$. The graphs appear to coincide, hence the equation could be an identity.

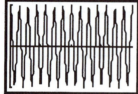

41. Graph $y = (1 + \tan x)^2$ and $y = 1/\cos x$ using the window $0 \le x \le 2\pi,\ -5 \le y \le 5$. The graphs do not coincide, hence the equation could not be an identity.

43. a. The graphs will appear identical, but the functions are not the same.
  b. The calculator plots 95 points, each of which has an x-coordinate that is a multiple of 2π. Since all y-coordinates are 1, the graph looks like the horizontal line $y = 1$.

45. a. There should be 80 full waves in an interval of length $80 \cdot 2\pi$.
  b. The calculator graph can't show 80 waves, since it plots only 95 points. As a signal that something is wrong, note that the waves shown are different.

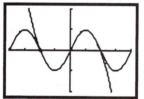

47. a. Shown is the window $-2\pi < x \le 2\pi,\ -2 \le y \le 2$. Agreement is good between $x = -\pi$ and $x = \pi$.

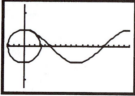

  b. Agreement is good over the entire interval for f_{15}. In fact:
  $\sin(2) = .9092974268$. $f_{15}(2) = .9092974265$

49. Since $\tan t = \dfrac{\sin t}{\cos t}$, a function of the form $\dfrac{f_m(t)}{g_n(t)}$, where the numerator and denominator functions are as defined in Exercises 47 and 48, is needed. The graph of $y = \dfrac{f_{15}(t)}{g_{16}(t)}$ is indistinguishable from the graph of $y = \tan t$ in the specified window.

51. The y-coordinate of the point on the graph of the cosine function is the same as the x-coordinate of the point on the unit circle, because of the definition of the cosine function as the x-coordinate of a point on the unit circle.

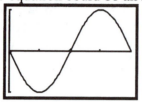

6.5 Periodic Graphs and Simple Harmonic Motion

1. Amplitude $= |3| = 3$

   Period $= \dfrac{2\pi}{2} = \pi$

   Phase shift $= -\dfrac{-\pi}{2} = \dfrac{\pi}{2}$

3. Amplitude $= |-7| = 7$

   Period $= \dfrac{2\pi}{7}$

   Phase shift $= -\dfrac{1}{7} \div 7 = -\dfrac{1}{49}$

5. Amplitude $= 1$

   Period $= \dfrac{2\pi}{2\pi} = 1$

   Phase shift $= 0$

7. Amplitude $= |6| = 6$

   Period $= \dfrac{2\pi}{3\pi} = \dfrac{2}{3}$

   Phase shift $= -\dfrac{1}{3\pi}$

9. Period $= \dfrac{2\pi}{b} = \dfrac{\pi}{4}$. Thus $b = 8$. Phase shift $= -\dfrac{c}{b} = \dfrac{\pi}{5}$. Thus $c = -\dfrac{\pi}{5}b = -\dfrac{8\pi}{5}$.

   $f(t) = 3\sin(8t - 8\pi/5)$ or $f(t) = 3\cos(8t - 8\pi/5)$

11. Period $= \dfrac{2\pi}{b} = 1$. Thus $b = 2\pi$. Phase shift $= -\dfrac{c}{b} = 0$. Thus $c = 0$.

   $f(t) = 2/3\sin 2\pi t$ or $f(t) = 2/3\cos 2\pi t$

13. Period $= \dfrac{2\pi}{b} = \dfrac{5}{3}$. Thus $b = \dfrac{6\pi}{5}$. Phase shift $= -\dfrac{c}{b} = -\dfrac{\pi}{2}$. Thus $c = \dfrac{\pi}{2}b = \dfrac{3\pi^2}{5}$.

   $f(t) = 7\sin\left(\dfrac{6\pi}{5}t + \dfrac{3\pi^2}{5}\right)$ or $f(t) = 7\cos\left(\dfrac{6\pi}{5}t + \dfrac{3\pi^2}{5}\right)$

15. Amplitude $= 2$. Period $= \dfrac{2\pi}{b} = \dfrac{\pi}{2}$. Thus $b = 4$. The shape is that of a sine function:

   $f(t) = 2\sin 4t$

17. Amplitude $= 1.5$. Period $= \dfrac{2\pi}{b} = 4\pi$. Thus $b = \dfrac{1}{2}$. The shape is that of a cosine

   function: $f(t) = 1.5\cos\left(\tfrac{1}{2}t\right)$

19. a. Period $= 2\pi / 300 = \pi / 150$.
   b. The graph should have 300 complete waves between 0 and 2π.
   c. Shown is the window $0 \le x \le 4\pi / 150, -2 \le y \le 2$.

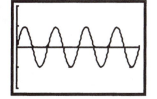

21. a. Period $= 2\pi / 900 = \pi / 450$.

b. The graph should have 900 complete waves between 0 and 2π.

c. Shown is the window $0 \le x \le 4\pi / 450, -2 \le y \le 2$.

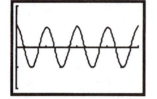

23. a. The amplitude is $A = 12$ and the period is $\pi / 5$, hence

$$\frac{2\pi}{b} = \frac{\pi}{5}$$

$$b = 10$$

The phase shift is 1/4 of a period to the right, hence

$$-\frac{c}{b} = \frac{\pi / 5}{4}$$

$$c = -\frac{\pi}{20}b = -\frac{\pi}{20} \cdot 10$$

$$c = -\frac{\pi}{2}$$

Thus, $f(t) = 12\sin\left(10t - \frac{\pi}{2}\right)$

b. This is a standard cosine curve reflected about the x-axis, with amplitude and period as above and no phase shift. Thus,

$$g(t) = -12\cos(10t)$$

25. a. This is a standard sine curve reflected about the x-axis. The amplitude is $A = 1$ and the period is π, hence

$$\frac{2\pi}{b} = \pi$$

$$b = 2$$

Thus, $f(t) = -\sin(2t)$

b. As in **a**, the amplitude is $A = 1$ and the period is π, but there is a phase shift of 1/4 of a period to the left, hence

$$-\frac{c}{b} = -\frac{\pi}{4}$$

$$c = \frac{\pi}{4}b = \frac{\pi}{4} \cdot 2$$

$$c = \frac{\pi}{2}$$

Thus, $g(t) = \cos\left(2t + \frac{\pi}{2}\right)$

27. Use the window $0 \le x \le 2\pi, -3 \le y \le 3$.   **29.** Use the window $0 \le x \le 2\pi, -1 \le y \le 1$.

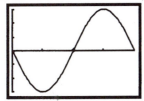   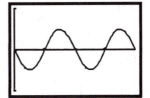

31. Use the window $0 \le x \le 2\pi, -3 \le y \le 3$.

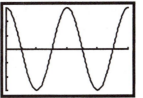

33. Use the window $0 \le x \le 2\pi, -1 \le y \le 1$.

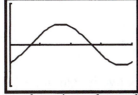

Since there is a phase shift of $\pi/3$, the local maximum and the local minimum must be at $x = \dfrac{\pi}{2} + \dfrac{\pi}{3} = \dfrac{5\pi}{6}$ and at $x = \dfrac{3\pi}{2} + \dfrac{\pi}{3} = \dfrac{11\pi}{6}$ respectively.

35. Use the window $0 \le x \le 2\pi, -3 \le y \le 3$.

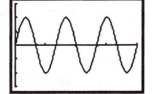

The maxima and minima must occur when

$$3x - \pi = -\frac{\pi}{2}, \frac{3\pi}{2}, \frac{7\pi}{2} \qquad 3x - \pi = \frac{\pi}{2}, \frac{5\pi}{2}, \frac{9\pi}{2}$$
$$\text{and}$$
$$x = \frac{\pi}{6}, \frac{5\pi}{6}, \frac{3\pi}{2} \qquad x = \frac{\pi}{2}, \frac{7\pi}{6}, \frac{11\pi}{6}$$

respectively.

37. Graph the function using the window $-2\pi \le x \le 2\pi, -6 \le y \le 6$

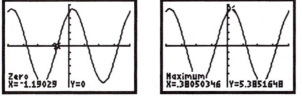

Note that the function has a maximum value of $A = 5.3852$. One complete period occurs between $x = -1.1903$ and $x = -1.1903 + 2\pi$, hence $b = 1$ and $-c/b = -1.1903$, thus $c = 1.1903$. $f(t) = 5.3852 \sin(t + 1.1903)$

39. Graph the function using the window $-2\pi \le x \le 2\pi, -6 \le y \le 6$

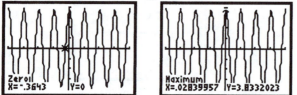

Note that the function has a maximum value of $A = 3.8332$. Four complete periods occur between $x = 0$ and $x = 2\pi$, hence $b = 4$. One complete period occurs between $x = -.3643$ and $x = -.3643 + \pi / 2$, hence $b = 1$ and $-c / b = -c / 4 = -.3643$, thus $c = 1.4572$.

$f(t) = 3.8332 \sin(4t + 1.4572)$

41. Graph the function using the window $0 \le x \le 2\pi, -4 \le y \le 4$.

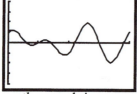

Since the graph is not a sine wave nor a transformed sine wave it cannot be represented as $g(t) = A\sin(bt + c)$.

43. The period of the function $f(t) = A\sin 1,960,000\pi t$ is

$$\frac{2\pi}{1,960,000\pi} = \frac{1}{980,000} \text{ seconds.}$$

The frequency is the reciprocal of the period, that is, 980,000 cycles per second (known as 980 KiloHertz).

45. The amplitude is $A = 125$. If time is measured in minutes, the period is 10, hence $b = 2\pi / 10 = \pi / 5$. At time $t = 0$ the car is at $y = 0$, hence the phase shift is 0.

$$f(t) = 125\sin\left(\frac{\pi}{5}t\right)$$

47. Since the wheel is rotating 10 times per second, the period is 1/10 second, we can write $2\pi / b = 1 / 10$, $b = 20\pi$, and the coordinates of W are given by $\left(\cos(20\pi t), \sin(20\pi t)\right)$. The coordinates of the shadow S of W on the x-axis are given by $\left(\cos(20\pi t), 0\right)$. From the Pythagorean Theorem the distance from S to P is given by $\sqrt{16 - \sin^2(20\pi t)}$ and hence the x-coordinate of P is $\cos(20\pi t) + \sqrt{16 - \sin^2(20\pi t)}$.

49. Since the weight starts at equilibrium, use the sine function. $A = 6$. $2\pi / b = 4$, $b = \pi / 2$. $h(t) = 6\sin(\pi t / 2)$.

51. Since the weight starts above equilibrium, use the cosine function with positive coefficient. $h(t) = 6\cos(\pi t / 2)$

53. $A = \frac{1}{2}(20) = 10.$ $2\pi / b = 4,$ $b = \pi / 2.$ $d(t) = 10\sin(\pi t / 2).$

55. a. For a very rough picture, plot the maximum, the minimum, and the two end points, four in all.
   b. There are 100 waves in this interval. After plotting the first end point, 3 are needed for each succeeding wave, hence 301 would be needed.
   c. Every calculator is different; the TI-83 plots 95 points.
   d. Obviously, 95 points is not enough when the absolute minimum is 301.

57. a.

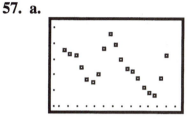

b. The SinReg feature of the TI-83 yields:

c. Graphing the model found in part **b** over the scatter plot yields:

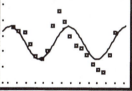

   Using the value of b from the TI-83 yields a period of
   $2\pi / b = 2\pi/.6176392135 = 10.2$ years.
   It is perfectly possible that the business cycle will continue to behave according to this model. It is also perfectly possible that other factors (elections, wars, who knows?)will make the model invalid after 2002.

59. a. The SinReg feature of the TI-83 yields:

   b. Using this value of b yields a period of $2\pi / b = 2\pi/.4 = 15.7$ months. This seems very unlikely for precipitation figures.

c. Plotting another year of the same data versus the model yields:

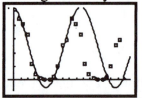

Clearly the 15.7 month period produces a very poor fit.

d. Now we obtain the following model and the following graph:

Using this value of b yields a period of $2\pi / b = 2\pi/.513333335 = 12.2$ months. This seems more reasonable, and the model fits the data better.

61. a. If the period is 2, then

$$\frac{2\pi}{\omega} = 2, \omega = \pi$$

Solve $\pi = \sqrt{\dfrac{9.8}{k}}$ to obtain $k = .9929$ meters.

b. If k is increased by .01% to $1.0001k$, then the period becomes approximately 2.000099998 seconds, hence the clock loses about .000099998 seconds every 2 seconds. Multiply this by 7,948,800/2 to obtain a loss of about 397.43 seconds.

6.5A Other Trigonometric Graphs

1. Graph the function using the window $-2\pi \le x \le 2\pi, -3 \le y \le 3$

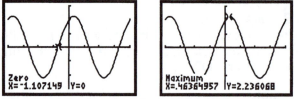

Note that the function has a maximum value of $A = 2.2361$. One complete period occurs between $x = 0$ and $x = 2\pi$, hence $b = 1$. The start of one period is at $x = -1.1071$, hence $-c / b = -c / 1 = -1.1071$, thus $c = 1.1071$.

$f(t) = 2.2361\sin(t + 1.1071)$

3. Graph the function using the window $-2\pi \le x \le 2\pi, -6 \le y \le 6$

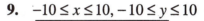

Note that the function has a maximum value of $A = 5.3852$. Four complete periods occur between $x = 0$ and $x = 2\pi$, hence $b = 4$. The start of one period is at $x = .29757$, hence $-c/b = -c/4 = .29757$, thus $c = -1.1903$.
$$f(t) = 5.3852\sin(4t - 1.1903)$$

5. Graph the function using the window $-2\pi \le x \le 2\pi, -6 \le y \le 6$

Note that the function has a maximum value of $A = 5.1164$. Three complete periods occur between $x = 0$ and $x = 2\pi$, hence $b = 3$. The start of one period is at $x = .24808$, hence $-c/b = -c/3 = .24808$, thus $c = -.7442$.
$$f(t) = 5.1164\sin(3t - .7442)$$

7. $0 \le x \le 2\pi, -5 \le y \le 5$

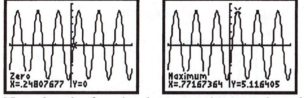

9. $-10 \le x \le 10, -10 \le y \le 10$

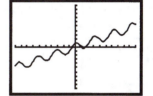

11. $0 \le x \le .02\pi, -2 \le y \le 2$

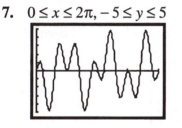

13. $0 \le x \le .04, -7 \le y \le 7$

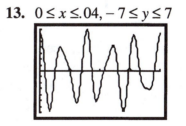

15. $0 \le x \le 10, -10 \le y \le 10$

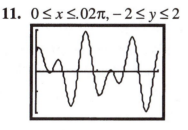

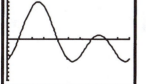

17. The graph has amplitude 1 and has one complete wave between $x = \ln(2\pi)$ and $x = \ln(4\pi)$, and then between $x = \ln(2n\pi)$ and $x = \ln((2n+2)\pi)$, where n is any positive integer. The negative x-axis is a horizontal asymptote. The window $-3 \le x \le \ln(8\pi), -1.5 \le y \le 1.5$ is shown.

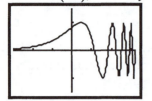

19. The function has one complete wave every 2π units. The graph lies between the graph of $y = \sqrt{|t|}$ and the graph of $y = -\sqrt{|t|}$ as shown in the right-hand graph. Note that this is an even function, hence the graph has y-axis symmetry. The window $-30 \le x \le 30, -6 \le y \le 6$ is shown.

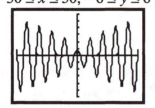 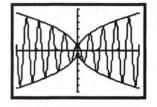

21. The function has one complete wave every 2π units. The graph lies between the graph of $y = 1/t$ and the graph of $y = -1/t$ as shown in the right-hand graph. There is a hole at the point $(0,1)$. The x-axis is a horizontal asymptote (even though the graph crosses this horizontal asymptote an infinite number of times). The window $-6\pi \le x \le 6\pi, -1 \le y \le 1$ is shown.

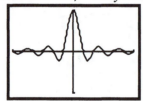 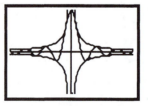

23. The function has one complete period every π units. There is a vertical asymptote at each point where $\cos t = 0$, that is, where $t = \frac{\pi}{2} + n\pi$, where n is an any integer. The window $-2\pi \le x \le 2\pi, -3 \le y \le 1$ is shown.

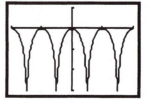

6.6 Other Trigonometric Functions

1. IV   **3.** II   **5.** IV

7. $r = \sqrt{x^2 + y^2} = \sqrt{3^2 + 4^2} = 5$

$\sin t = y/r = 4/5 \quad \cos t = x/r = 3/5 \quad \tan t = y/x = 4/3$

$\cot t = x/y = 3/4 \quad \sec t = r/x = 5/3 \quad \csc t = r/y = 5/4$

9. $r = \sqrt{x^2 + y^2} = \sqrt{(-5)^2 + 12^2} = 13$

$\sin t = y/r = 12/13 \quad \cos t = x/r = -5/13 \quad \tan t = y/x = -12/5$

$\cot t = x/y = -5/12 \quad \sec t = r/x = -13/5 \quad \csc t = r/y = 13/12$

11. $r = \sqrt{x^2 + y^2} = \sqrt{(-1/5)^2 + 1^2} = \sqrt{26/25} = \sqrt{26}/5$

$\sin t = \dfrac{y}{r} = 1 \div \dfrac{\sqrt{26}}{5} = \dfrac{5}{\sqrt{26}} \qquad \cot t = \dfrac{x}{y} = \left(-\dfrac{1}{5}\right) \div 1 = -\dfrac{1}{5}$

$\cos t = \dfrac{x}{r} = \left(-\dfrac{1}{5}\right) \div \dfrac{\sqrt{26}}{5} = -\dfrac{1}{\sqrt{26}} \qquad \sec t = \dfrac{r}{x} = \dfrac{\sqrt{26}}{5} \div \left(-\dfrac{1}{5}\right) = -\sqrt{26}$

$\tan t = \dfrac{y}{x} = 1 \div \left(-\dfrac{1}{5}\right) = -5 \qquad \csc t = \dfrac{r}{y} = \dfrac{\sqrt{26}}{5} \div 1 = \dfrac{\sqrt{26}}{5}$

13. $r = \sqrt{x^2 + y^2} = \sqrt{\left(\sqrt{2}\right)^2 + \left(\sqrt{3}\right)^2} = \sqrt{5}$

$\sin t = \dfrac{y}{r} = \dfrac{\sqrt{3}}{\sqrt{5}} = \dfrac{\sqrt{15}}{5} \qquad \cot t = \dfrac{x}{y} = \dfrac{\sqrt{2}}{\sqrt{3}} = \dfrac{\sqrt{6}}{3}$

$\cos t = \dfrac{x}{r} = \dfrac{\sqrt{2}}{\sqrt{5}} = \dfrac{\sqrt{10}}{5} \qquad \sec t = \dfrac{r}{x} = \dfrac{\sqrt{5}}{\sqrt{2}} = \dfrac{\sqrt{10}}{2}$

$\tan t = \dfrac{y}{x} = \dfrac{\sqrt{3}}{\sqrt{2}} = \dfrac{\sqrt{6}}{2} \qquad \csc t = \dfrac{r}{y} = \dfrac{\sqrt{5}}{\sqrt{3}} = \dfrac{\sqrt{15}}{3}$

15. $r = \sqrt{x^2 + y^2} = \sqrt{\left(1 + \sqrt{2}\right)^2 + 3^2} = \sqrt{12 + 2\sqrt{2}}$

$\sin t = \dfrac{y}{r} = \dfrac{3}{\sqrt{12 + 2\sqrt{2}}} \qquad \cot t = \dfrac{x}{y} = \dfrac{\sqrt{12 + 2\sqrt{2}}}{3}$

$\cos t = \dfrac{x}{r} = \dfrac{1 + \sqrt{2}}{\sqrt{12 + 2\sqrt{2}}} \qquad \sec t = \dfrac{r}{x} = \dfrac{\sqrt{12 + 2\sqrt{2}}}{1 + \sqrt{2}}$

$\tan t = \dfrac{y}{x} = \dfrac{3}{\sqrt{12 + 2\sqrt{2}}} \qquad \csc t = \dfrac{r}{y} = \dfrac{\sqrt{12 + 2\sqrt{2}}}{3}$

17. The terminal side of an angle of $\dfrac{4\pi}{3}$ in standard position lies in quadrant III.

$$\sin\frac{4\pi}{3} = -\sin\frac{\pi}{3} = -\frac{\sqrt{3}}{2} \qquad \cot\frac{4\pi}{3} = \cot\frac{\pi}{3} = \frac{1}{\sqrt{3}}$$

$$\cos\frac{4\pi}{3} = -\cos\frac{\pi}{3} = -\frac{1}{2} \qquad \sec\frac{4\pi}{3} = -\sec\frac{\pi}{3} = -2$$

$$\tan\frac{4\pi}{3} = \tan\frac{\pi}{3} = \sqrt{3} \qquad \csc\frac{4\pi}{3} = -\csc\frac{\pi}{3} = -\frac{2}{\sqrt{3}}$$

19. The terminal side of an angle of $\dfrac{7\pi}{4}$ in standard position lies in quadrant IV.

$$\sin\frac{7\pi}{4} = -\sin\frac{\pi}{4} = -\frac{\sqrt{2}}{2} \qquad \cot\frac{7\pi}{4} = -\cot\frac{\pi}{4} = -1$$

$$\cos\frac{7\pi}{4} = \cos\frac{\pi}{4} = \frac{\sqrt{2}}{2} \qquad \sec\frac{7\pi}{4} = \sec\frac{\pi}{4} = \sqrt{2}$$

$$\tan\frac{7\pi}{4} = -\tan\frac{\pi}{4} = -1 \qquad \csc\frac{7\pi}{4} = -\csc\frac{\pi}{4} = -\sqrt{2}$$

21. average rate of change $= \dfrac{\cot 3 - \cot 1}{3 - 1} = \dfrac{1}{2}\left(\dfrac{1}{\tan 3} - \dfrac{1}{\tan 1}\right) = -3.8287$

23. a. average rate of change $= \dfrac{\tan 2.01 - \tan 2}{h}$

$\quad h = .01 \qquad$ rate $= \dfrac{\tan 2.01 - \tan 2}{.01} = 5.6511$

$\quad h = .001 \qquad$ rate $= \dfrac{\tan 2.001 - \tan 2}{.001} = 5.7618$

$\quad h = .0001 \qquad$ rate $= \dfrac{\tan 2.0001 - \tan 2}{.0001} = 5.7731$

$\quad h = .00001 \qquad$ rate $= \dfrac{\tan 2.00001 - \tan 2}{..00001} = 5.7743$

$\quad$ **b.** $(\sec 2)^2 = 5.7744$. This seems to be the limit of the process in part **a.**

25. $\cos t \sin t (\csc t + \sec t) = \cos t \sin t \csc t + \cos t \sin t \sec t$

$$= \cos t \sin t \frac{1}{\sin t} + \cos t \sin t \frac{1}{\cos t}$$

$$= \cos t + \sin t$$

27. $(1 - \sec t)^2 = 1 - 2\sec t + \sec^2 t$

29. $(\cot t - \tan t)(\cot^2 t + 1 + \tan^2 t)$
$$= \cot^3 t - \tan^3 t$$

31. $\csc t (\sec t - \csc t)$

33. $\tan^4 t - \sec^4 t = \left(\tan^2 t - \sec^2 t\right)\left(\tan^2 t + \sec^2 t\right)$
$$= (-1)\left(\tan^2 t + \sec^2 t\right)$$
$$= -\tan^2 t - \sec^2 t$$

35. $\cos^3 t - \sec^3 t = \left(\cos t - \sec t\right)\left(\cos^2 t + \cos \sec t + \sec^2 t\right)$
$$= \left(\cos t - \sec t\right)\left(\cos^2 t + 1 + \sec^2 t\right)$$

37. $\dfrac{\cos^2 t \sin t}{\sin^2 t \cos t} = \dfrac{\cos t}{\sin t}$
$$= \cot t$$

39. $\dfrac{4\tan t \sec t + 2\sec t}{6\sin t \sec t + 2\sec t} = \dfrac{2\sec t(2\tan t + 1)}{2\sec t(3\sin t + 2)}$
$$= \dfrac{2\tan t + 1}{3\sin t + 2}$$

41. $4 - \tan t$

43. $\sin^2 t + \cos^2 t = 1$
$$\dfrac{\sin^2 t}{\sin^2 t} + \dfrac{\cos^2 t}{\sin^2 t} = \dfrac{1}{\sin^2 t}$$
$$1 + \cot^2 t = \csc^2 t$$

45. $\sec(-t) = \dfrac{1}{\cos(-t)}$
$$= \dfrac{1}{\cos t}$$
$$= \sec t$$

47. $\sin t = +\sqrt{1 - \cos^2 t} = \sqrt{1 - \left(-\dfrac{1}{2}\right)^2}$
$$= \dfrac{\sqrt{3}}{2}$$
$$\cos t = -\dfrac{1}{2}$$
$$\tan t = \dfrac{\sin t}{\cos t} = \dfrac{\sqrt{3}/2}{-1/2} = -\sqrt{3}$$
$$\cot t = \dfrac{1}{\tan t} = \dfrac{1}{-\sqrt{3}} = -\dfrac{1}{\sqrt{3}}$$
$$\sec t = \dfrac{1}{\cos t} = \dfrac{1}{-1/2} = -2$$
$$\csc t = \dfrac{1}{\sin t} = \dfrac{1}{\sqrt{3}/2} = \dfrac{2}{\sqrt{3}}$$

49. $\sin t = 1$
$$\cos t = 0$$
$$\tan t = \dfrac{\sin t}{\cos t} = \dfrac{1}{0} \text{ is undefined}$$
$$\cot t = \dfrac{1}{\tan t} = \dfrac{0}{1} = 0$$
$$\sec t = \dfrac{1}{\cos t} = \dfrac{1}{0} \text{ is undefined}$$
$$\csc t = \dfrac{1}{\sin t} = \dfrac{1}{1} = 1$$

51. $\tan t = -\sqrt{\sec^2 t - 1} = -\sqrt{\left(-\dfrac{13}{5}\right)^2 - 1} = -\dfrac{12}{5}$

$\cos t = \dfrac{1}{\sec t} = \dfrac{1}{-13/5} = -\dfrac{5}{13}$

$\sin t = \tan t \cos t = \left(-\dfrac{12}{5}\right)\left(-\dfrac{5}{13}\right) = \dfrac{12}{13}$

$\cot t = \dfrac{1}{\tan t} = \dfrac{1}{-12/5} = -\dfrac{5}{12}$

$\sec t = -\dfrac{13}{5}$

$\csc t = \dfrac{1}{\sin t} = \dfrac{1}{12/13} = \dfrac{13}{12}$

53. Graph $y = \tan x$ and $y = \cot\left(\frac{\pi}{2} - x\right)$ using the window $-2\pi \le x \le 2\pi$, $-5 \le y \le 5$. The graphs appear to coincide, hence the equation could be an identity.

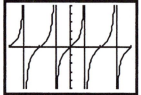

55. Graph $y = \dfrac{\sin x}{1 - \cos x}$ and $y = \cot x$ using the window $0 \le x \le 2\pi$, $-5 \le y \le 5$. The graphs do not coincide, hence the equation could not be an identity.

57. Graphing $y = \sec x$ and $y = x$ using the window $-2\pi \le x \le 2\pi$, $-5 \le y \le 5$ shows two, possibly three solutions, and indicates that there are an infinite number; however, none of them lie between $-\pi/2$ and $\pi/2$.

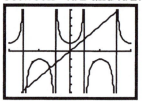

59. a. area $OCA = \frac{1}{2}AC \cdot OA = \frac{1}{2}\sin\theta\cos\theta$

b. area $ODB = \frac{1}{2}DB \cdot OB = \frac{1}{2}\tan\theta \cdot 1 = \frac{1}{2}\tan\theta$

c. area $OCB = \frac{1}{2}OB^2 \cdot \theta = \frac{1}{2}1^2 \cdot \theta = \frac{1}{2}\theta$

Chapter 6 Review Exercises

1. $-\dfrac{23\pi}{3} + 8\pi = \dfrac{\pi}{3}$

3. $\dfrac{9\pi}{5} \cdot \dfrac{180°}{\pi} = 324°$

5. $220° \cdot \dfrac{\pi}{180°} = \dfrac{11\pi}{9}$ radians

7. $-\dfrac{11\pi}{4} \cdot \dfrac{180°}{\pi} = -495°$

9. In the second quadrant, cosine is negative. Hence
$$\cos r = -\sqrt{1 - \sin^2 r} = -\sqrt{1 - \left(\sqrt{\tfrac{8}{9}}\right)^2} = -\sqrt{\tfrac{1}{9}} = -\tfrac{1}{3}$$

11. $\sin(-13\pi) = \sin \pi = 0$ **13.** $.809$ **15.** $-\sqrt{3}/2$ **17.** $\sqrt{3}/2$

19.

t	0	$\pi/6$	$\pi/4$	$\pi/3$	$\pi/2$
$\sin t$	0	$1/2$	$\sqrt{2}/2$	$\sqrt{3}/2$	1
$\cos t$	1	$\sqrt{3}/2$	$\sqrt{2}/2$	$1/2$	0

21. $\left(\sin\dfrac{\pi}{6} + 1\right)^2 = \left(\dfrac{1}{2} + 1\right)^2 = \left(\dfrac{3}{2}\right)^2 = \dfrac{9}{4}$

23. $(f \circ g)(\pi) = f(g(\pi)) = f(-\cos \pi) = f(1) = \log_{10} 1 = 0$

25. Since the terminal side lies in the second quadrant, cosine is negative.
$$\cos t = -\sqrt{1 - \sin^2 t} = -\sqrt{1 - \left(\dfrac{1}{\sqrt{3}}\right)^2} = -\sqrt{\dfrac{2}{3}} \text{ or } -\dfrac{\sqrt{6}}{3}$$

27. Since the terminal side lies in the third quadrant, cosine is negative.
$$\cos t = -\sqrt{1 - \sin^2 t} = -\sqrt{1 - \left(-\dfrac{4}{5}\right)^2} = -\sqrt{\dfrac{9}{25}} = -\dfrac{3}{5}$$

29. Since the terminal side lies in the second quadrant, cosine is negative.
$$\cos t = -\sqrt{1 - \sin^2 t} = -\sqrt{1 - \left(\dfrac{5}{13}\right)^2} = -\sqrt{\dfrac{144}{169}} = -\dfrac{12}{13}$$

31. $\cos\dfrac{2\pi}{3} = -\cos\dfrac{\pi}{3} = -\dfrac{1}{2}$ **33.** $\dfrac{\sqrt{3}}{2}$ **35.** (c)

37. $r = \sqrt{x^2 + y^2} = \sqrt{\left(\dfrac{-3}{\sqrt{50}}\right)^2 + \left(\dfrac{7}{\sqrt{50}}\right)^2} = \sqrt{\dfrac{9}{50} + \dfrac{49}{50}} = \sqrt{\dfrac{58}{50}} = \dfrac{\sqrt{58}}{\sqrt{50}}$

$\sin t = \dfrac{y}{r} = \dfrac{7}{\sqrt{50}} \div \dfrac{\sqrt{58}}{\sqrt{50}} = \dfrac{7}{\sqrt{58}}$

39. $\tan t = \dfrac{y}{x} = \dfrac{7}{\sqrt{50}} \div -\dfrac{3}{\sqrt{50}} = -\dfrac{7}{3}$

41. The slope of the line is $\tan\left(\frac{5\pi}{3}\right) = -\sqrt{3}$. The line passes through the origin. Hence $y = -\sqrt{3}x$ is the required equation.

43. $\tan w = \dfrac{\sin w}{\cos w} = -\dfrac{3}{\sqrt{13}} \div \dfrac{2}{\sqrt{13}} = -\dfrac{3}{2}$ **45.** $\cos(-w) = \cos w = \dfrac{2}{\sqrt{13}}$

47. The dashed line is the graph of $h(t) = \csc t$; the solid curve is the graph of $f(t) = \sin t$.

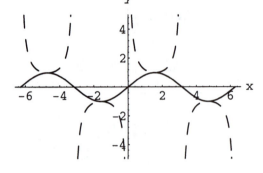

49. (d)   **51.** $\tan\theta = y/x = 3/(-2) = -3/2$ **53.** 0   **55. (d)**

57. $\cot\left(\dfrac{2\pi}{3}\right) = -\cot\left(\dfrac{\pi}{3}\right) = -\dfrac{1}{\sqrt{3}}$

59. a. 3/2
   b. $f(t) = 0$ if $5t = n\pi$. The smallest possible value is therefore $t = n\pi/5$.

61.

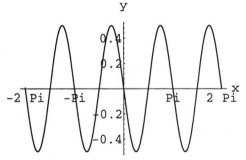

63. Graph $y = \cos x$ and $y = \sin\left(x - \frac{\pi}{2}\right)$ using the window $0 \le x \le 2\pi$, $-1 \le y \le 1$. The graphs do not coincide, hence the equation could not be an identity.

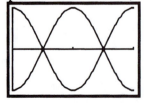

65. Graph $y = \dfrac{\sin x - \sin 3x}{\cos x + \cos 3x}$ and $y = \tan x$
using the window $0 \le x \le 2\pi$, $-5 \le y \le 5$.
The graphs appear to coincide, hence
the equation could be an identity.

67. The period of $\sin bt$ is $\dfrac{2\pi}{b}$, hence the period of $g(t)$ is $\dfrac{2\pi}{4\pi} = \dfrac{1}{2}$.

69. The graph has the general appearance of a cosine graph.
Amplitude = 2
Period $= \dfrac{4\pi}{5} = \dfrac{2\pi}{b}$, hence $b = \dfrac{5}{2}$.
The rule can be written $g(t) = 2\cos \frac{5}{2} t$.

71. $A = 8$. $\dfrac{2\pi}{b} = 5$, hence $b = \dfrac{2\pi}{5}$. $-\dfrac{c}{b} = 14$, hence $c = -14b = -\dfrac{28\pi}{5}$.

$f(t) = 8\sin\left(\dfrac{2\pi}{5} t - \dfrac{28\pi}{5} \right)$ or $f(t) = 8\cos\left(\dfrac{2\pi}{5} t - \dfrac{28\pi}{5} \right)$

73. Graph the function using the window $-2\pi \le x \le 2\pi$, $-11 \le y \le 11$.

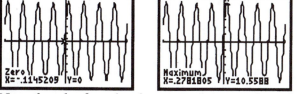

Note that the function has a maximum value of $A = 10.5588$. Four complete periods
occur between $x = 0$ and $x = 2\pi$, hence $b = 4$. The start of one period is at
$x = -.1145$, hence $-c / b = -c / 4 = -.1145$, thus $c = .4580$.
$f(t) = 10.5588 \sin(4t + .4580)$

75. $0 \le x \le \pi / 50$, $-5 \le y \le 5$

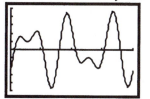

Chapter 7
Trigonometric Identities and Equations
7.1 Basic Identities and Proofs

1. If the functions $f(x) = \dfrac{\sec x - \cos x}{\sec x}$ and $g(x) = \sin^2 x$ are graphed in a window such as $0 \le x \le 2\pi$, $0 \le y \le 1$, they will appear identical, hence this equation could be an identity.

3. If the functions $f(x) = \dfrac{1 - \cos 2x}{2}$ and $g(x) = \sin^2 x$ are graphed in a window such as $0 \le x \le 2\pi$, $0 \le y \le 1$, they will appear identical, hence this equation could be an identity.

5. B

7. E

9. $\tan x \cos x = \dfrac{\sin x}{\cos x} \cos x$
$$= \sin x$$

11. $\cos x \sec x = \cos x \dfrac{1}{\cos x}$
$$= 1$$

13. $\tan x \csc x = \dfrac{\sin x}{\cos x} \dfrac{1}{\sin x}$
$$= \dfrac{1}{\cos x}$$
$$= \sec x$$

15. $\dfrac{\tan x}{\sec x} = \dfrac{\sin x}{\cos x} \div \dfrac{1}{\cos x}$
$$= \dfrac{\sin x}{\cos x} \cdot \dfrac{\cos x}{1}$$
$$= \sin x$$

17. $(1 + \cos x)(1 - \cos x) = 1 - \cos^2 x$
$$= \sin^2 x + \cos^2 x - \cos^2 x$$
$$= \sin^2 x$$

19. This is not an identity. For example, let $x = \dfrac{3\pi}{2}$.

$\sin \dfrac{3\pi}{2} = -1$ but $\sqrt{1 - \cos^2\left(\dfrac{3\pi}{2}\right)} = 1$

21. This is an identity.
$\dfrac{\sin(-x)}{\cos(-x)} = \dfrac{-\sin x}{\cos x}$
$$= -\tan x$$

23. This is an identity.
$\cot(-x) = \dfrac{\cos(-x)}{\sin(-x)} = \dfrac{\cos x}{-\sin x}$
$$= -\cot x$$

25. This is not an identity. For example, let $x = 0$.
$1 + \sec^2 0 = 2$ but $\tan^2 0 = 0$

27. This is an identity.

$$\sec^2 x - \csc^2 x = \left(\tan^2 x + 1\right) - \left(\cot^2 x + 1\right)$$

$$= \tan^2 x + 1 - \cot^2 x - 1$$

$$= \tan^2 x - \cot^2 x$$

29. This is an identity.

$$\sin^2 x(\cot x + 1)^2 = \sin^2 x\left(\frac{\cos x}{\sin x} + 1\right)^2$$

$$= \sin^2 x\left(\frac{\cos x + \sin x}{\sin x}\right)^2$$

$$= (\cos x + \sin x)^2$$

Also

$$\cos^2 x(\tan x + 1)^2 = \cos^2 x\left(\frac{\sin x}{\cos x} + 1\right)^2$$

$$= \cos^2 x\left(\frac{\sin x + \cos x}{\cos x}\right)^2$$

$$= (\sin x + \cos x)^2$$

Since left and right sides are equal to the same quantity, they are equal to each other.

31. This is an identity.

$$\sin^2 x - \tan^2 x = \sin^2 x - \frac{\sin^2 x}{\cos^2 x}$$

$$= \frac{\sin^2 x \cos^2 x - \sin^2 x}{\cos^2 x}$$

$$= \frac{\sin^2 x\left(\cos^2 x - 1\right)}{\cos^2 x}$$

$$= \tan^2 x\left(\cos^2 x - 1\right)$$

$$= \tan^2 x\left(-\sin^2 x\right)$$

$$= -\sin^2 x \tan^2 x$$

33. This is an identity.

$$\left(\cos^2 x - 1\right)\left(\tan^2 x + 1\right) = \left(\cos^2 x - 1\right)\sec^2 x$$

$$= \cos^2 x \sec^2 x - \sec^2 x$$

$$= \cos^2 x \frac{1}{\cos^2 x} - \sec^2 x$$

$$= 1 - \sec^2 x$$

$$= -\tan^2 x$$

35. This is an identity.
Start with the right side:
$$\frac{\sec x}{\csc x} = \frac{1/\cos x}{1/\sin x}$$
$$= \frac{1}{\cos x} \cdot \frac{\sin x}{1}$$
$$= \frac{\sin x}{\cos x} = \tan x$$

37. This is an identity.
$$\cos^4 x - \sin^4 x = \left(\cos^2 x\right)^2 - \left(\sin^2 x\right)^2$$
$$= \left(\cos^2 x + \sin^2 x\right)\left(\cos^2 x - \sin^2 x\right)$$
$$= 1\left(\cos^2 x - \sin^2 x\right)$$
$$= \cos^2 x - \sin^2 x$$

39. This is not an identity. For example, let $x = \pi/4$.
$$\left(\sin\frac{\pi}{4} + \cos\frac{\pi}{4}\right)^2 = \left(\frac{\sqrt{2}}{2} + \frac{\sqrt{2}}{2}\right)^2 = \left(\sqrt{2}\right)^2 = 2 \quad \text{but } \sin^2\frac{\pi}{4} + \cos^2\frac{\pi}{4} = 1$$

41. This is an identity. Using Exercise **35**:
$$\frac{\sec x}{\csc x} + \frac{\sin x}{\cos x} = \tan x + \frac{\sin x}{\cos x}$$
$$= \tan x + \tan x$$
$$= 2\tan x$$

43. This is an identity.
$$\frac{\sec x + \csc x}{1 + \tan x} = \frac{1/\cos x + 1/\sin x}{1 + \sin x/\cos x}$$
$$= \left(\frac{1}{\cos x} + \frac{1}{\sin x}\right) \div \left(1 + \frac{\sin x}{\cos x}\right)$$
$$= \left(\frac{\sin x + \cos x}{\sin x \cos x}\right) \div \left(\frac{\cos x + \sin x}{\cos x}\right)$$
$$= \frac{\sin x + \cos x}{\sin x \cos x} \cdot \frac{\cos x}{\cos x + \sin x} = \frac{1}{\sin x}$$
$$= \csc x$$

45. This is an identity.
$$\frac{1}{\csc x - \sin x} = \frac{1}{\dfrac{1}{\sin x} - \sin x}$$
$$= \frac{\sin x}{1 - \sin^2 x}$$
$$= \frac{\sin x}{\cos^2 x}$$
$$= \frac{1}{\cos x} \cdot \frac{\sin x}{\cos x}$$
$$= \sec x \tan x$$

47. This is not an identity. For example, let $x = \pi/4$.

$$\frac{\sin\frac{\pi}{4} - \cos\frac{\pi}{4}}{\tan\frac{\pi}{4}} = \frac{\sqrt{2}/2 - \sqrt{2}/2}{1} = 0 \text{ but } \frac{\tan\frac{\pi}{4}}{\sin\frac{\pi}{4} + \cos\frac{\pi}{4}} = \frac{1}{\sqrt{2}/2 + \sqrt{2}/2} = \frac{1}{\sqrt{2}}$$

49. The graph of the function is shown using the window $-2\pi \le x \le 2\pi, -1 \le y \le 1$.

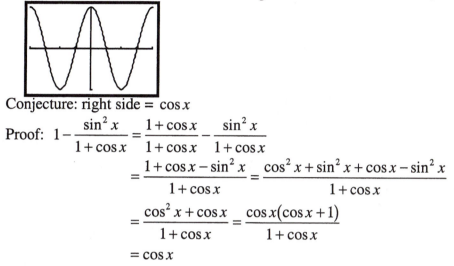

Conjecture: right side $= \cos x$

Proof:
$$1 - \frac{\sin^2 x}{1 + \cos x} = \frac{1 + \cos x}{1 + \cos x} - \frac{\sin^2 x}{1 + \cos x}$$
$$= \frac{1 + \cos x - \sin^2 x}{1 + \cos x} = \frac{\cos^2 x + \sin^2 x + \cos x - \sin^2 x}{1 + \cos x}$$
$$= \frac{\cos^2 x + \cos x}{1 + \cos x} = \frac{\cos x(\cos x + 1)}{1 + \cos x}$$
$$= \cos x$$

51. The graph of the function is shown using the window $-2\pi \le x \le 2\pi, -5 \le y \le 5$.

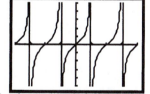

Conjecture: right side $= \tan x$

Proof: $(\sin x + \cos x)(\sec x + \csc x) - \cot x - 2 =$
$$= (\sin x + \cos x)\left(\frac{1}{\cos x} + \frac{1}{\sin x}\right) - \cot x - 2$$
$$= \tan x + 1 + 1 + \cot x - \cot x - 2 = \tan x$$

53.
$$\frac{\cos^3 x}{1 + \sin x} = \frac{\cos^3 x(1 - \sin x)}{(1 + \sin x)(1 - \sin x)}$$
$$= \frac{\cos^3 x(1 - \sin x)}{1 - \sin^2 x}$$
$$= \frac{\cos^3 x(1 - \sin x)}{\cos^2 x}$$
$$= \cos x(1 - \sin x)$$
$$= \frac{1 - \sin x}{\sec x}$$

55.
$$\frac{\cos x}{1 - \sin x} = \frac{\cos x(1 + \sin x)}{(1 - \sin x)(1 + \sin x)}$$
$$= \frac{\cos x(1 + \sin x)}{1 - \sin^2 x}$$
$$= \frac{\cos x(1 + \sin x)}{\cos^2 x}$$
$$= \frac{1 + \sin x}{\cos x}$$
$$= \frac{1}{\cos x} + \frac{\sin x}{\cos x}$$
$$= \sec x + \tan x$$

57. Use Strategy 4: Prove that $(\cos x \cot x)(\cos x \cot x) = (\cot x + \cos x)(\cot x - \cos x)$

$(\cot x + \cos x)(\cot x - \cos x) = \cot^2 x - \cos^2 x$

$$= \frac{\cos^2 x}{\sin^2 x} - \cos^2 x$$

$$= \frac{\cos^2 x - \cos^2 x \sin^2 x}{\sin^2 x}$$

$$= \frac{\cos^2 x(1 - \sin^2 x)}{\sin^2 x}$$

$$= \frac{\cos^2 x}{\sin^2 x}(1 - \sin^2 x)$$

$$= \cot^2 x \cos^2 x$$

$$= (\cos x \cot x)(\cos x \cot x)$$

Therefore $\dfrac{\cos x \cot x}{\cot x - \cos x} = \dfrac{\cot x + \cos x}{\cos x \cot x}$

59. $\log_{10}(\cot x) = \log_{10} \dfrac{1}{\tan x}$

$\qquad = \log_{10} 1 - \log_{10} \tan x$

$\qquad = 0 - \log_{10} \tan x = -\log_{10} \tan x$

61. $\log_{10}(\csc x + \cot x) = \log_{10}\left(\dfrac{(\csc x + \cot x)(\csc x - \cot x)}{\csc x - \cot x}\right)$

$\qquad = \log_{10} \dfrac{\csc^2 x - \cot^2 x}{\csc x - \cot x}$

$\qquad = \log_{10} \dfrac{1}{\csc x - \cot x}$

$\qquad = -\log_{10}(\csc x - \cot x)$

63. $-\tan x \tan y(\cot x - \cot y) = -\tan x \cot x \tan y + \tan x \tan y \cot y$

$\qquad\qquad\qquad\qquad\qquad = -\tan y + \tan x = \tan x - \tan y$

65. Use Strategy 4: Prove that $(\cos x - \sin y)(\cos x + \sin y) = (\cos y + \sin x)(\cos y - \sin x)$

$(\cos x - \sin y)(\cos x + \sin y) = \cos^2 x - \sin^2 y$

$$= 1 - \sin^2 x - (1 - \cos^2 y)$$

$$= \cos^2 y - \sin^2 x$$

$$= (\cos y + \sin x)(\cos y - \sin x)$$

Therefore $\dfrac{\cos x - \sin y}{\cos y - \sin x} = \dfrac{\cos y + \sin x}{\cos x + \sin y}$

7.2 Addition and Subtraction Identities

1. $\sin\dfrac{\pi}{12} = \sin\left(\dfrac{\pi}{3} - \dfrac{\pi}{4}\right)$

$= \sin\dfrac{\pi}{3}\cos\dfrac{\pi}{4} - \cos\dfrac{\pi}{3}\sin\dfrac{\pi}{4}$

$= \dfrac{\sqrt{3}}{2}\dfrac{\sqrt{2}}{2} - \dfrac{1}{2}\dfrac{\sqrt{2}}{2}$

$= \dfrac{\sqrt{6} - \sqrt{2}}{4}$

3. $\tan\dfrac{\pi}{12} = \tan\left(\dfrac{\pi}{3} - \dfrac{\pi}{4}\right)$

$= \dfrac{\tan\dfrac{\pi}{3} - \tan\dfrac{\pi}{4}}{1 + \tan\dfrac{\pi}{3}\tan\dfrac{\pi}{4}} = \dfrac{\sqrt{3} - 1}{1 + \sqrt{3}\cdot 1}$

$= \dfrac{\sqrt{3} - 1}{\sqrt{3} + 1}\cdot\left(\dfrac{\sqrt{3} - 1}{\sqrt{3} - 1}\right)$

$= \dfrac{3 - 2\sqrt{3} + 1}{3 - 1} = 2 - \sqrt{3}$

5. $\cot\dfrac{5\pi}{12} = \tan\left(\dfrac{\pi}{2} - \dfrac{5\pi}{12}\right)$

$= \tan\dfrac{\pi}{12}$

$= 2 - \sqrt{3}$ (from Exercise **3.**)

7. $\tan\dfrac{7\pi}{12} = \tan\left(\dfrac{\pi}{3} + \dfrac{\pi}{4}\right)$

$= \dfrac{\tan\dfrac{\pi}{3} + \tan\dfrac{\pi}{4}}{1 - \tan\dfrac{\pi}{3}\tan\dfrac{\pi}{4}}$

$= \dfrac{\sqrt{3} + 1}{1 - \sqrt{3}} = -2 - \sqrt{3}$

9. $\cot\dfrac{11\pi}{12} = \dfrac{\cos\dfrac{11\pi}{12}}{\sin\dfrac{11\pi}{12}} = \dfrac{\cos\left(\dfrac{\pi}{4} + \dfrac{2\pi}{3}\right)}{\sin\left(\dfrac{\pi}{4} + \dfrac{2\pi}{3}\right)}$

$= \dfrac{\cos\dfrac{\pi}{4}\cos\dfrac{2\pi}{3} - \sin\dfrac{\pi}{4}\sin\dfrac{2\pi}{3}}{\sin\dfrac{\pi}{4}\cos\dfrac{2\pi}{3} + \cos\dfrac{\pi}{4}\sin\dfrac{2\pi}{3}}$

$= \dfrac{\dfrac{\sqrt{2}}{2}\left(-\dfrac{1}{2}\right) - \dfrac{\sqrt{2}}{2}\dfrac{\sqrt{3}}{2}}{\dfrac{\sqrt{2}}{2}\left(-\dfrac{1}{2}\right) + \dfrac{\sqrt{2}}{2}\dfrac{\sqrt{3}}{2}} = \dfrac{-\sqrt{2} - \sqrt{6}}{-\sqrt{2} + \sqrt{6}} = -2 - \sqrt{3}$

11. $\sin 105° = \sin(60° + 45°)$

$= \sin 60°\cos 45° + \cos 60°\sin 45°$

$= \dfrac{\sqrt{3}}{2}\dfrac{\sqrt{2}}{2} + \dfrac{1}{2}\dfrac{\sqrt{2}}{2}$

$= \dfrac{\sqrt{6} + \sqrt{2}}{4}$

13. $\sin\left(\dfrac{\pi}{2}+x\right)=\sin\dfrac{\pi}{2}\cos x+\cos\dfrac{\pi}{2}\sin x$
$$=1\cdot\cos x+0\cdot\sin x$$
$$=\cos x$$

15. $\cos\left(x-\dfrac{3\pi}{2}\right)=\cos x\cos\dfrac{3\pi}{2}+\sin x\sin\dfrac{3\pi}{2}$
$$=\cos x\cdot 0+\sin x\cdot(-1)$$
$$=-\sin x$$

17. $\sec(x-\pi)=\dfrac{1}{\cos(x-\pi)}$

$$=\dfrac{1}{\cos x\cos\pi-\sin x\sin\pi}$$

$$=\dfrac{1}{\cos x(-1)-\sin x\cdot 0}$$

$$=-\dfrac{1}{\cos x}$$

$$=-\sec x$$

19. $\sin 3\cos 5-\cos 3\sin 5=\sin(3-5)=\sin(-2)=-\sin 2$

21. $\cos(x+y)\cos y+\sin(x+y)\sin y=\cos\big((x+y)-y\big)=\cos x$

23. $\cos(x+y)-\cos(x-y)=(\cos x\cos y-\sin x\sin y)-(\cos x\cos y+\sin x\sin y)$
$$=-2\sin x\sin y$$

25. First note that $\cos x=\sqrt{1-\sin^2 x}=\sqrt{1-\left(\dfrac{1}{3}\right)^2}=\dfrac{2\sqrt{2}}{3}$.

$$\sin\left(\dfrac{\pi}{4}+x\right)=\sin\dfrac{\pi}{4}\cos x+\cos\dfrac{\pi}{4}\sin x$$

$$=\dfrac{\sqrt{2}}{2}\dfrac{2\sqrt{2}}{3}+\dfrac{\sqrt{2}}{2}\dfrac{1}{3}$$

$$=\dfrac{4+\sqrt{2}}{6}$$

27. First note that $\sin x=-\sqrt{1-\cos^2 x}=-\sqrt{1-\left(\dfrac{1}{5}\right)^2}=-\dfrac{2\sqrt{6}}{5}$.

$$\sin\left(\dfrac{\pi}{3}-x\right)=\sin\dfrac{\pi}{3}\cos x-\cos\dfrac{\pi}{3}\sin x$$

$$=\dfrac{\sqrt{3}}{2}\left(-\dfrac{1}{5}\right)-\dfrac{1}{2}\left(-\dfrac{2\sqrt{6}}{5}\right)$$

$$=\dfrac{-\sqrt{3}+2\sqrt{6}}{10}$$

For Exercise **29 - 32**, note that $\cos x = \sqrt{1 - \sin^2 x} = \sqrt{1 - (.8)^2} = .6$ and $\cos y = \sqrt{1 - \sin^2 x} = \sqrt{1 - (\sqrt{.75})^2} = .5.$

29. $\begin{aligned} \sin(x + y) &= \sin x \cos y + \cos x \sin y \\ &= .8(.5) + (.6)\sqrt{.75} \\ &= -.9196 \end{aligned}$

31. $\begin{aligned} \sin(x - y) &= \sin x \cos y - \cos x \sin y \\ &= .8(.5) - (.6)\sqrt{.75} \\ &= -.1196 \end{aligned}$

33. From the figure we see that $\sin t = \dfrac{4}{5}$ and $\cos t = \dfrac{3}{5}$. Hence

$$\begin{aligned} 5\sin(x + t) &= 5(\sin x \cos t + \cos x \sin t) \\ &= \sin x \cdot 5\cos t + \cos x \cdot 5\sin t \\ &= \sin x \cdot 5\left(\frac{3}{5}\right) + \cos x \cdot 5\left(\frac{4}{5}\right) \\ &= 3\sin x + 4\cos x \end{aligned}$$

35.
$$\begin{aligned} \frac{f(x + h) - f(x)}{h} &= \frac{\cos(x + h) - \cos x}{h} \\ &= \frac{\cos x \cos h - \sin x \sin h - \cos x}{h} \\ &= \frac{\cos x(\cos h - 1) - \sin x \sin h}{h} \\ &= \cos x\left(\frac{\cos h - 1}{h}\right) - \sin x \frac{\sin h}{h} \end{aligned}$$

37. First note that $\cos x = \sqrt{1 - \sin^2 x} = \sqrt{1 - \left(\dfrac{24}{25}\right)^2} = \dfrac{7}{25}$ and

$$\cos y = -\sqrt{1 - \sin^2 y} = -\sqrt{1 - \left(\frac{4}{5}\right)^2} = -\frac{3}{5}. \text{ Then}$$

$$\tan x = \frac{\sin x}{\cos x} = \frac{24/25}{7/25} = \frac{24}{7} \text{ and } \tan y = \frac{\sin y}{\cos y} = \frac{4/5}{-3/5} = -\frac{4}{3}.$$

$$\begin{aligned} \sin(x + y) &= \sin x \cos y + \cos x \sin y \\ &= \frac{24}{25}\left(-\frac{3}{5}\right) + \frac{7}{25}\left(\frac{4}{5}\right) \\ &= \frac{-72 + 28}{125} \\ &= -\frac{44}{125} \end{aligned}$$

$$\begin{aligned} \tan(x + y) &= \frac{\tan x + \tan y}{1 - \tan x \tan y} \\ &= \frac{\dfrac{24}{7} + \left(-\dfrac{4}{3}\right)}{1 - \left(\dfrac{24}{7}\right)\left(-\dfrac{4}{3}\right)} \\ &= \frac{72 - 28}{21 + 96} = \frac{44}{117} \end{aligned}$$

Since $\sin(x + y)$ is negative and $\tan(x + y)$ is positive, $x + y$ lies in quadrant III.

39. First note that $\cos x = \sqrt{1 - \sin^2 x} = \sqrt{1 - \left(\dfrac{4}{5}\right)^2} = \dfrac{3}{5}$ and

$$\sin y = \sqrt{1 - \cos^2 y} = \sqrt{1 - \left(-\dfrac{12}{13}\right)^2} = \dfrac{5}{13}. \text{ Then}$$

$$\tan x = \dfrac{\sin x}{\cos x} = \dfrac{4/5}{3/5} = \dfrac{4}{3} \text{ and } \tan y = \dfrac{\sin y}{\cos y} = \dfrac{5/13}{-12/13} = -\dfrac{5}{12}.$$

$$\cos(x + y) = \cos x \cos y - \sin x \sin y \qquad\qquad \tan(x + y) = \dfrac{\tan x + \tan y}{1 - \tan x \tan y}$$

$$= \dfrac{3}{5}\left(-\dfrac{12}{13}\right) - \dfrac{4}{5}\left(\dfrac{5}{13}\right) \qquad\qquad = \dfrac{\dfrac{4}{3} + \left(-\dfrac{5}{12}\right)}{1 - \left(\dfrac{4}{3}\right)\left(-\dfrac{5}{12}\right)}$$

$$= \dfrac{-36 - 20}{65} \qquad\qquad\qquad\qquad = \dfrac{48 - 15}{36 + 20} = \dfrac{33}{56}$$

$$= -\dfrac{56}{65}$$

Since $\cos(x + y)$ is negative and $\tan(x + y)$ is positive, $x + y$ lies in quadrant III.

41. $\sin(u + v + w) = \sin((u + v) + w)$

$\qquad\qquad = \sin(u + v)\cos w + \cos(u + v)\sin w$

$\qquad\qquad = (\sin u \cos v + \cos u \sin v)\cos w + (\cos u \cos v - \sin u \sin v)\sin w$

$\qquad\qquad = \sin u \cos v \cos w + \cos u \sin v \cos w + \cos u \cos v \sin w - \sin u \sin v \sin w$

43. If $x + y = \dfrac{\pi}{2}$, then $y = \dfrac{\pi}{2} - x$ and $\sin y = \sin\left(\dfrac{\pi}{2} - x\right) = \cos x$. Thus

$\sin^2 x + \sin^2 y = \sin^2 x + \cos^2 x = 1$

45. $\sin(x - \pi) = \sin x \cos \pi - \cos x \sin \pi$ **47.** $\cos(\pi - x) = \cos \pi \cos x + \sin \pi \sin x$

$\qquad\quad = \sin x(-1) - \cos x(0) \qquad\qquad\qquad\quad = (-1)\cos x + (0)\sin x$

$\qquad\quad = -\sin x \qquad\qquad\qquad\qquad\qquad\qquad = -\cos x$

49. $\sin(x + \pi) = \sin x \cos \pi + \cos x \sin \pi$

$\qquad\quad = \sin x \cdot (-1) + \cos x \cdot (0)$ **51.** $\tan(x + \pi) = \dfrac{\tan x + \tan \pi}{1 + \tan x \tan \pi}$

$\qquad\quad = -\sin x \qquad\qquad\qquad\qquad\qquad\qquad = \dfrac{\tan x + 0}{1 + \tan x \cdot 0} = \tan x$

53. $\dfrac{1}{2}[\cos(x - y) - \cos(x + y)] = \dfrac{1}{2}[\cos x \cos y + \sin x \sin y - (\cos x \cos y - \sin x \sin y)]$

$\qquad\qquad\qquad\qquad\qquad\qquad = \dfrac{1}{2} \cdot 2 \sin x \sin y$

$\qquad\qquad\qquad\qquad\qquad\qquad = \sin x \sin y$

55. $\cos(x + y)\cos(x - y) = (\cos x \cos y - \sin x \sin y)(\cos x \cos y + \sin x \sin y)$

$\qquad\qquad\qquad\qquad\quad = \cos^2 x \cos^2 y - \sin^2 x \sin^2 y$

57. This is an identity.

$$\frac{\cos(x-y)}{\sin x \cos y} = \frac{\cos x \cos y + \sin x \sin y}{\sin x \cos y}$$

$$= \frac{\cos x}{\sin x} + \frac{\sin y}{\cos y}$$

$$= \cot x + \tan y$$

59. This is not an identity. For example, let $x = \pi$ and $y = \pi/2$, then
$\sin(x-y) = \sin(\pi - \pi/2) = 1$ but $\sin x - \sin y = \sin \pi - \sin(\pi/2) = -1$

61. This is an identity.

$$\frac{\tan x + \tan y}{\tan x - \tan y} = \frac{\dfrac{\sin x}{\cos x} + \dfrac{\sin y}{\cos y}}{\dfrac{\sin x}{\cos x} - \dfrac{\sin y}{\cos y}}$$

$$= \frac{\dfrac{\sin x}{\cos x} + \dfrac{\sin y}{\cos y}}{\dfrac{\sin x}{\cos x} - \dfrac{\sin y}{\cos y}} \cdot \frac{\cos x \cos y}{\cos x \cos y}$$

$$= \frac{\sin x \cos y + \sin y \cos x}{\sin x \cos y - \sin y \cos x}$$

$$= \frac{\sin(x+y)}{\sin(x-y)}$$

63. This is not an identity. For example, let $x = \pi/3$ and $y = -\pi/3$. Then

$$\frac{\cos(x+y)}{\cos(x-y)} = \frac{\cos\left(\dfrac{\pi}{3} + \left(-\dfrac{\pi}{3}\right)\right)}{\cos\left(\dfrac{\pi}{3} - \left(-\dfrac{\pi}{3}\right)\right)} = -2 \quad \text{but} \quad \frac{\cot x + \tan y}{\cot x - \tan y} = \frac{\cot\dfrac{\pi}{3} + \tan\left(-\dfrac{\pi}{3}\right)}{\cot\dfrac{\pi}{3} - \tan\left(-\dfrac{\pi}{3}\right)} = -\frac{1}{2}.$$

65. This is not an identity. For example, let $x = y = \dfrac{\pi}{3}$. Then

$$\tan(x+y) = \tan\left(\frac{\pi}{3} + \frac{\pi}{3}\right) = -\sqrt{3} \quad \text{but} \quad \tan x + \tan y = \tan\frac{\pi}{3} + \tan\frac{\pi}{3} = 2\sqrt{3}.$$

7.2A Lines and Angles

1. $\tan\theta = \text{slope} = \dfrac{y_2 - y_1}{x_2 - x_1} = \dfrac{5-2}{3-(-1)} = \dfrac{3}{4}$ **3.** $\tan\theta = \text{slope} = \dfrac{y_2 - y_1}{x_2 - x_1} = \dfrac{0-4}{6-1} = -\dfrac{4}{5}$

5. $\tan\theta = \text{slope} = \dfrac{y_2 - y_1}{x_2 - x_1} = \dfrac{5-(-7)}{3-3}$ undefined

7. $\tan\theta = \left|\dfrac{m-k}{1+mk}\right| = \left|\dfrac{3/2-(-1)}{1+(3/2)(-1)}\right| = 5$, hence $\theta = \tan^{-1}5 = 1.373$ radians

9. $\tan\theta = \left|\dfrac{m-k}{1+mk}\right| = \left|\dfrac{-1-0}{1+(-1)\cdot 0}\right| = 1$, hence $\theta = \tan^{-1}1 = \dfrac{\pi}{4}$ radians

11. Line L: $m = \dfrac{6-2}{5-3} = 2$ Line M: $k = \dfrac{0-3}{4-0} = -\dfrac{3}{4}$

$\tan\theta = \left|\dfrac{m-k}{1+mk}\right| = \left|\dfrac{2-(-3/4)}{1+2(-3/4)}\right| = \dfrac{11}{2}$, hence $\theta = \tan^{-1}\dfrac{11}{2} = 1.391$ radians.

13. Line L: $m = 3$ Line M: k satisfies $k(-.5) = -1$, hence $k = 2$.

$\tan\theta = \left|\dfrac{m-k}{1+mk}\right| = \left|\dfrac{3-2}{1+3\cdot 2}\right| = \dfrac{1}{7}$, hence $\theta = \tan^{-1}\dfrac{1}{7} = .142$ radians.

7.3 Other Identities

1. $\cos\dfrac{\pi}{8} = \cos\dfrac{1}{2}\left(\dfrac{\pi}{4}\right) = \sqrt{\dfrac{1+\cos(\pi/4)}{2}} = \sqrt{\dfrac{1+\sqrt{2}/2}{2}} = \dfrac{\sqrt{2+\sqrt{2}}}{2}$

3. $\sin\dfrac{3\pi}{8} = \sin\dfrac{1}{2}\left(\dfrac{3\pi}{4}\right) = \sqrt{\dfrac{1-\cos(3\pi/4)}{2}} = \sqrt{\dfrac{1-\left(-\sqrt{2}/2\right)}{2}} = \dfrac{\sqrt{2+\sqrt{2}}}{2}$

5. $\tan\dfrac{\pi}{12} = \tan\dfrac{1}{2}\left(\dfrac{\pi}{6}\right) = \dfrac{1-\cos(\pi/6)}{\sin(\pi/6)} = \dfrac{1-\sqrt{3}/2}{1/2} = 2-\sqrt{3}$

7. $\cos\dfrac{\pi}{12} = \cos\dfrac{1}{2}\left(\dfrac{\pi}{6}\right) = \sqrt{\dfrac{1+\cos(\pi/6)}{2}} = \sqrt{\dfrac{1+\sqrt{3}/2}{2}} = \dfrac{\sqrt{2+\sqrt{3}}}{2}$

9. $\sin\dfrac{7\pi}{8} = \sin\dfrac{1}{2}\left(\dfrac{7\pi}{4}\right) = \sqrt{\dfrac{1-\cos(7\pi/4)}{2}} = \sqrt{\dfrac{1-\sqrt{2}/2}{2}} = \dfrac{\sqrt{2-\sqrt{2}}}{2}$

11. $\tan\dfrac{7\pi}{8} = \tan\dfrac{1}{2}\left(\dfrac{7\pi}{4}\right) = \dfrac{1-\cos(7\pi/4)}{\sin(7\pi/4)} = \dfrac{1-\sqrt{2}/2}{-\sqrt{2}/2} = 1-\sqrt{2}$

13. $\begin{aligned}\sin 4x\cos 6x &= \tfrac{1}{2}\big[\sin(4x+6x)+\sin(4x-6x)\big]\\ &= \tfrac{1}{2}\big[\sin(10x)+\sin(-2x)\big]\\ &= \tfrac{1}{2}\sin 10x - \tfrac{1}{2}\sin 2x\end{aligned}$

15. $\begin{aligned}\cos 2x\cos 4x &= \tfrac{1}{2}\big[\cos(2x+4x)+\cos(2x-4x)\big]\\ &= \tfrac{1}{2}\big[\cos(6x)+\cos(-2x)\big]\\ &= \tfrac{1}{2}\cos 6x + \tfrac{1}{2}\cos 2x\end{aligned}$

17. $\begin{aligned}\sin 17x\sin(-3x) &= \tfrac{1}{2}\big[\cos(17x-(-3x))-\cos(17x+(-3x))\big]\\ &= \tfrac{1}{2}\big[\cos(20x)-\cos(14x)\big]\\ &= \tfrac{1}{2}\cos 20x - \tfrac{1}{2}\cos 14x\end{aligned}$

19. $\begin{aligned}\sin 3x + \sin 5x &= 2\sin\left(\dfrac{3x+5x}{2}\right)\cos\left(\dfrac{3x-5x}{2}\right)\\ &= 2\sin 4x\cos(-x)\\ &= 2\sin 4x\cos x\end{aligned}$

21. $\begin{aligned}\sin 9x - \sin 5x &= 2\cos\left(\dfrac{9x+5x}{2}\right)\sin\left(\dfrac{9x-5x}{2}\right)\\ &= 2\cos 7x\sin 2x\end{aligned}$

23. $\sin x = \dfrac{5}{13};\quad \cos x = \sqrt{1-\sin^2 x} = \sqrt{1-\left(\dfrac{5}{13}\right)^2} = \dfrac{12}{13}$

$\sin 2x = 2\sin x\cos x = 2\left(\dfrac{5}{13}\right)\left(\dfrac{12}{13}\right) = \dfrac{120}{169}$

$\cos 2x = 1 - 2\sin^2 x = 1 - 2\left(\dfrac{5}{13}\right)^2 = \dfrac{119}{169}$

$\tan 2x = \dfrac{\sin 2x}{\cos 2x} = \dfrac{120/169}{119/169} = \dfrac{120}{119}$

25. $\cos x = -\dfrac{3}{5};\quad \sin x = -\sqrt{1-\cos^2 x} = -\sqrt{1-\left(-\dfrac{3}{5}\right)^2} = -\dfrac{4}{5}$

$\sin 2x = 2\sin x\cos x = 2\left(-\dfrac{4}{5}\right)\left(-\dfrac{3}{5}\right) = \dfrac{24}{25}$

$\cos 2x = 2\cos^2 x - 1 = 2(-3/5)^2 - 1 = -7/25$

$\tan 2x = \dfrac{\sin 2x}{\cos 2x} = \dfrac{24/25}{-7/25} = -\dfrac{24}{7}$

27. $\tan x = \dfrac{3}{4};\quad \sec x = -\sqrt{1+\tan^2 x} = -\sqrt{1+\left(\dfrac{3}{4}\right)^2} = -\dfrac{5}{4};\ \cos x = -\dfrac{4}{5};\ \sin x = -\dfrac{3}{5}$

$\sin 2x = 2\sin x\cos x = 2\left(-\dfrac{3}{5}\right)\left(-\dfrac{4}{5}\right) = \dfrac{24}{25}$

$\cos 2x = 2\cos^2 x - 1 = 2\left(-\dfrac{4}{5}\right)^2 - 1 = \dfrac{7}{25}$

$\tan 2x = \dfrac{\sin 2x}{\cos 2x} = \dfrac{24/25}{7/25} = \dfrac{24}{7}$

29. $\csc x = 4;\ \sin x = \dfrac{1}{4};\ \cos x = \sqrt{1-\left(\dfrac{1}{4}\right)^2} = \dfrac{\sqrt{15}}{4}$

$\sin 2x = 2\sin x\cos x = 2\left(\dfrac{1}{4}\right)\left(\dfrac{\sqrt{15}}{4}\right) = \dfrac{\sqrt{15}}{8}$

$\cos 2x = 1 - 2\sin^2 x = 1 - 2(1/4)^2 = 7/8$

$\tan 2x = \dfrac{\sin 2x}{\cos 2x} = \dfrac{\sqrt{15}/8}{7/8} = \dfrac{\sqrt{15}}{7}$

31. $\sin\dfrac{x}{2} = \sqrt{\dfrac{1-\cos x}{2}} = \sqrt{\dfrac{1-.4}{2}} = \sqrt{.3} = \dfrac{\sqrt{30}}{10}$

$\cos\dfrac{x}{2} = \sqrt{\dfrac{1+\cos x}{2}} = \sqrt{\dfrac{1+.4}{2}} = \sqrt{.7} = \dfrac{\sqrt{70}}{10}$

$\tan\dfrac{x}{2} = \dfrac{\sin x/2}{\cos x/2} = \dfrac{\sqrt{.3}}{\sqrt{.7}} = \dfrac{\sqrt{21}}{7}$

33. First note that $\cos x = \sqrt{1-\sin^2 x} = \sqrt{1-(-3/5)^2} = 4/5$. Also
$3\pi/4 < x/2 < \pi$, hence $x/2$ lies in the second quadrant.

$\sin\dfrac{x}{2} = \sqrt{\dfrac{1-\cos x}{2}} = \sqrt{\dfrac{1-4/5}{2}} = \dfrac{1}{\sqrt{10}}$

$\cos\dfrac{x}{2} = -\sqrt{\dfrac{1+\cos x}{2}} = -\sqrt{\dfrac{1+4/5}{2}} = -\dfrac{3}{\sqrt{10}},\qquad \tan\dfrac{x}{2} = \dfrac{\sin x/2}{\cos x/2} = \dfrac{1/\sqrt{10}}{-3/\sqrt{10}} = -\dfrac{1}{3}$

35. First note that $\sec x = -\sqrt{1 + \tan^2 x} = -\sqrt{1 + (1/2)^2} = -\sqrt{5}/2$. Hence $\cos x = -2/\sqrt{5}$.

$\sin x = -\dfrac{1}{\sqrt{5}}$. Also $\dfrac{\pi}{2} < \dfrac{x}{2} < \dfrac{3\pi}{4}$ hence $\dfrac{x}{2}$ lies in the second quadrant.

$\sin \dfrac{x}{2} = \sqrt{\dfrac{1 - \cos x}{2}} = \sqrt{\dfrac{1 - \left(-2/\sqrt{5}\right)}{2}} = \sqrt{\dfrac{\sqrt{5} + 2}{2\sqrt{5}}} = \dfrac{\sqrt{50 + 20\sqrt{5}}}{10}$

$\cos \dfrac{x}{2} = -\sqrt{\dfrac{1 + \cos x}{2}} = -\sqrt{\dfrac{1 + \left(-2/\sqrt{5}\right)}{2}} = \sqrt{\dfrac{\sqrt{5} - 2}{2\sqrt{5}}} = -\dfrac{\sqrt{50 - 20\sqrt{5}}}{10}$

$\tan \dfrac{x}{2} = \dfrac{1 - \cos x}{\sin x} = \dfrac{1 - \left(-2/\sqrt{5}\right)}{-1/\sqrt{5}} = -\left(\sqrt{5} + 2\right)$

37. First note that $\cos x = \sqrt{1 - \sin^2 x} = \sqrt{1 - (.6)^2} = .8$.

$\sin 2x = 2\sin x \cos x = 2(.6)(.8) = .96$

39. $\cos 2x = 1 - 2\sin^2 x = 1 - 2(.6)^2 = .28$ **41.** $\sin \dfrac{x}{2} = \sqrt{\dfrac{1 - \cos x}{2}} = \sqrt{\dfrac{1 - .8}{2}} = \sqrt{.1} = .316$

43. $\cos 3x = \cos(2x + x)$

$\quad\quad = \cos 2x \cos x - \sin 2x \sin x$

$\quad\quad = \left(2\cos^2 x - 1\right)\cos x - \left(2\sin x \cos x\right)\sin x$

$\quad\quad = 2\cos^3 x - \cos x - 2\cos x \sin^2 x$

$\quad\quad = 2\cos^3 x - \cos x - 2\cos x\left(1 - \cos^2 x\right)$

$\quad\quad = 2\cos^3 x - \cos x - 2\cos x + 2\cos^3 x$

$\quad\quad = 4\cos^3 x - 3\cos x$

45. $\dfrac{\sin 2x}{2\sin x} = \dfrac{2\sin x \cos x}{2\sin x} = \cos x$

47. Since $2\sin x \cos x = \sin 2x$, $2\sin 2y \cos 2y = \sin 2(2y) = \sin 4y$

49. $\left(\sin x + \cos x\right)^2 - \sin 2x = \sin^2 x + 2\sin x \cos x + \cos^2 x - \sin 2x$

$\quad\quad\quad\quad\quad\quad = \sin^2 x + \cos^2 x + 2\sin x \cos x - 2\sin x \cos = 1$

51. This is an identity.

$\sin 16x = \sin 2(8x) = 2\sin 8x \cos 8x$

53. This is an identity.

$\cos^4 x - \sin^4 x = \left(\cos^2 x + \sin^2 x\right)\left(\cos^2 x - \sin^2 x\right)$

$\quad\quad\quad\quad\quad = 1\left(\cos^2 x - \sin^2 x\right) = \cos 2x$

55. This is not an identity. For example, let $x = \pi/8$.

$$\cos 4x = \cos 4\left(\frac{\pi}{8}\right) = 0 \text{ but } 2\cos 2x - 1 = 2\cos 2\left(\frac{\pi}{8}\right) - 1 = \sqrt{2} - 1$$

57. This is an identity.

$$\frac{1 + \cos 2x}{\sin 2x} = \frac{1 + 2\cos^2 x - 1}{2\sin x \cos x}$$

$$= \frac{2\cos x \cos x}{2\sin x \cos x}$$

$$= \frac{\cos x}{\sin x} = \cot x$$

59. This is an identity.

$$\sin 3x = \sin(2x + x)$$

$$= \sin 2x \cos x + \cos 2x \sin x$$

$$= 2\sin x \cos x \cos x + (1 - 2\sin^2 x)\sin x$$

$$= \sin x(2 - 2\sin^2 x + 1 - 2\sin^2 x)$$

$$= \sin x(3 - 4\sin^2 x)$$

61. This is not an identity. For example, let $x = 0$.

$$\cos 2x = \cos 2(0) = 1 \text{ but } \frac{2\tan x}{\sec^2 x} = \frac{2\tan 0}{\sec^2 0} = 0$$

63. This is an identity.

$$\csc^2\left(\frac{x}{2}\right) = \frac{1}{\sin^2\left(\frac{x}{2}\right)} = \frac{1}{\left(\pm\sqrt{\frac{1 - \cos x}{2}}\right)^2} = \frac{1}{\frac{1 - \cos x}{2}} = \frac{2}{1 - \cos x}$$

65. $\dfrac{\sin x - \sin 3x}{\cos x + \cos 3x} = \dfrac{2\cos\left(\dfrac{x + 3x}{2}\right)\sin\left(\dfrac{x - 3x}{2}\right)}{2\cos\left(\dfrac{x + 3x}{2}\right)\cos\left(\dfrac{x - 3x}{2}\right)}$

$$= \frac{\sin(-x)}{\cos(-x)} = \frac{-\sin x}{\cos x} = -\tan x$$

67. $\dfrac{\sin 4x + \sin 6x}{\cos 4x - \cos 6x} = \dfrac{2\sin\left(\dfrac{4x + 6x}{2}\right)\cos\left(\dfrac{4x - 6x}{2}\right)}{-2\sin\left(\dfrac{4x + 6x}{2}\right)\sin\left(\dfrac{4x - 6x}{2}\right)}$

$$= \frac{\cos(-x)}{-\sin(-x)} = \frac{\cos x}{\sin x} = \cot x$$

69. $\dfrac{\sin x + \sin y}{\cos x - \cos y} = \dfrac{2\sin\left(\dfrac{x + y}{2}\right)\cos\left(\dfrac{x - y}{2}\right)}{-2\sin\left(\dfrac{x + y}{2}\right)\sin\left(\dfrac{x - y}{2}\right)} = \dfrac{\cos\left(\dfrac{x - y}{2}\right)}{-\sin\left(\dfrac{x - y}{2}\right)} = -\cot\left(\dfrac{x - y}{2}\right)$

71. **a.** Use Strategy 4: Prove that $(1-\cos x)(1+\cos x)=\sin x \sin x$

$$(1-\cos x)(1+\cos x)=1-\cos^2 x$$
$$=\sin^2 x$$
$$=\sin x \sin x$$

Therefore $\dfrac{1-\cos x}{\sin x}=\dfrac{\sin x}{1+\cos x}$

b. This follows precisely as stated.

7.4 Inverse Trigonometric Functions

1. Since $\sin\dfrac{\pi}{2}=1$ and $-\dfrac{\pi}{2}\le\dfrac{\pi}{2}\le\dfrac{\pi}{2}$, $\sin^{-1}1=\dfrac{\pi}{2}$.

3. Since $\tan\left(-\dfrac{\pi}{4}\right)=-1$ and $-\dfrac{\pi}{2}<-\dfrac{\pi}{4}<\dfrac{\pi}{2}$, $\tan^{-1}(-1)=-\dfrac{\pi}{4}$.

5. Since $\cos 0=1$ and $0\le 0\le\pi$, $\cos^{-1}1=0$.

7. Since $\tan\dfrac{\pi}{6}=\dfrac{\sqrt{3}}{3}$ and $-\dfrac{\pi}{2}<\dfrac{\pi}{6}<\dfrac{\pi}{2}$, $\tan^{-1}\left(\dfrac{\sqrt{3}}{3}\right)=\dfrac{\pi}{6}$.

9. Since $\sin\left(-\dfrac{\pi}{4}\right)=-\dfrac{\sqrt{2}}{2}$ and $-\dfrac{\pi}{2}\le-\dfrac{\pi}{4}\le\dfrac{\pi}{2}$, $\sin^{-1}\left(-\dfrac{\sqrt{2}}{2}\right)=-\dfrac{\pi}{4}$.

11. Since $\tan\left(-\dfrac{\pi}{3}\right)=-\sqrt{3}$ and $-\dfrac{\pi}{2}<-\dfrac{\pi}{3}<\dfrac{\pi}{2}$, $\tan^{-1}\left(-\sqrt{3}\right)=-\dfrac{\pi}{3}$.

13. Since $\cos\left(\dfrac{2\pi}{3}\right)=-\dfrac{1}{2}$ and $0\le\dfrac{2\pi}{3}\le\pi$, $\cos^{-1}\left(-\dfrac{1}{2}\right)=\dfrac{2\pi}{3}$.

15. .3576

17. −1.2728

19. $\sin^{-1}(\sin 7)=\sin^{-1}(.65698\ldots)=.7168$

21. $\tan^{-1}(\tan(-4))=\tan^{-1}(-1.1578\ldots)=-.8584$

23. $\cos^{-1}(\cos(-8.5))=\cos^{-1}(-.602011\ldots)=2.2168$

25. Since $u=-\dfrac{\pi}{3}$, $\cos u=\cos\left(-\dfrac{\pi}{3}\right)=\dfrac{1}{2}$ and $\tan u=\tan\left(-\dfrac{\pi}{3}\right)=-\sqrt{3}$.

27. $\sin^{-1}(\cos 0)=\sin^{-1}1=\pi/2$

29. $\cos^{-1}\left(\sin\dfrac{4\pi}{3}\right)=\cos^{-1}\left(-\dfrac{\sqrt{3}}{2}\right)=\dfrac{5\pi}{6}$

31. $\sin^{-1}\left(\cos\dfrac{7\pi}{6}\right) = \sin^{-1}\left(-\dfrac{\sqrt{3}}{2}\right) = -\dfrac{\pi}{3}$ **33.** $\sin^{-1}\left(\sin\dfrac{2\pi}{3}\right) = \sin^{-1}\left(\dfrac{\sqrt{3}}{2}\right) = \dfrac{\pi}{3}$

35. $\cos^{-1}\left(\cos\left(-\dfrac{\pi}{6}\right)\right) = \cos^{-1}\left(\dfrac{\sqrt{3}}{2}\right) = \dfrac{\pi}{6}$

37. Let $u = \cos^{-1}\dfrac{3}{5}$. Then $\cos u = \dfrac{3}{5}$, $0 \le u \le \pi$. Then

$$\sin\left(\cos^{-1}\dfrac{3}{5}\right) = \sin u = \sqrt{1 - \cos^2 u} = \sqrt{1 - \left(\dfrac{3}{5}\right)^2} = \dfrac{4}{5}$$

39. Let $u = \tan^{-1}\left(-\dfrac{3}{4}\right)$. Then $\tan u = -\dfrac{3}{4}$, $-\dfrac{\pi}{2} < u < \dfrac{\pi}{2}$. Then

$$\cos\left(\tan^{-1}\left(-\dfrac{3}{4}\right)\right) = \cos u = \dfrac{1}{\sec u} = \dfrac{1}{\sqrt{1 + \tan^2 u}} = \dfrac{1}{\sqrt{1 + (-3/4)^2}} = \dfrac{1}{5/4} = \dfrac{4}{5}.$$

41. Let $u = \sin^{-1}\dfrac{5}{13}$. Then $\sin u = \dfrac{5}{13}$, $-\dfrac{\pi}{2} \le u \le \dfrac{\pi}{2}$. Then

$$\tan\left(\sin^{-1}\dfrac{5}{13}\right) = \tan u = \dfrac{\sin u}{\cos u} = \dfrac{\sin u}{\sqrt{1 - \sin^2 u}} = \dfrac{5/13}{\sqrt{1 - (5/13)^2}} = \dfrac{5/13}{12/13} = \dfrac{5}{12}.$$

43. $\sin^{-1} v = u$, where $\sin u = v$ and $-\dfrac{\pi}{2} \le u \le \dfrac{\pi}{2}$. Hence,

$$\cos\left(\sin^{-1} v\right) = \cos u = \sqrt{1 - \sin^2 u} = \sqrt{1 - v^2}\,.$$

45. $\sin^{-1} v = u$, where $\sin u = v$ and $-\dfrac{\pi}{2} \le u \le \dfrac{\pi}{2}$. Hence,

$$\tan\left(\sin^{-1} v\right) = \tan u = \dfrac{\sin u}{\cos u} = \dfrac{\sin u}{\sqrt{1 - \sin^2 u}} = \dfrac{v}{\sqrt{1 - v^2}}.$$

47.   **49.**

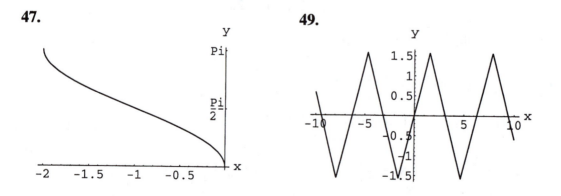

51. a. $V = V_{max} \sin 2\pi ft$

$$\sin 2\pi ft = \frac{V}{V_{max}}$$

$$2\pi ft = \sin^{-1} \frac{V}{V_{max}} + 2n\pi \quad (n \text{ any integer})$$

$$t = \frac{1}{2\pi f} \sin^{-1} \frac{V}{V_{max}} + \frac{n}{f}$$

b. The smallest positive value of t occurs when $n = 0$.

$$\frac{1}{2\pi f} \sin^{-1} \frac{V}{V_{max}} = \frac{1}{2\pi \cdot 120} \sin^{-1} \frac{8.5}{20} = 5.8219 \times 10^{-4} \text{ sec}$$

53. a. Let α = the angle formed by the road and the left edge of the light beam. Then

$$\tan \alpha = \frac{10}{x}; \ \tan(\alpha + \theta) = \frac{10 + 15}{x}.$$

Hence $\alpha = \tan^{-1} \frac{10}{x}$, $\alpha + \theta = \tan^{-1} \frac{25}{x}$, and $\theta = \tan^{-1} \frac{25}{x} - \tan^{-1} \frac{10}{x}$.

b. Graphing, we find that θ has maximum value at $x = 15.8$ ft.

55. The graph of $y = \sec(x)$ on this restricted domain is shown on the left. Since the graph passes the horizontal line test, we may conclude that the function is one-to-one and therefore has an inverse function. The graph of the inverse function is shown on the right.

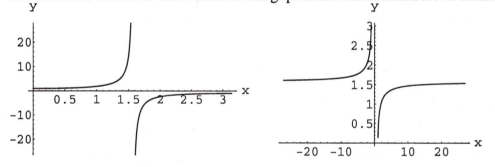

57. The graph of $y = \cot(x)$ on this restricted domain is shown on the left. Since the graph passes the horizontal line test, we may conclude that the function is one-to-one and therefore has an inverse function. The graph of the inverse function is shown on the right.

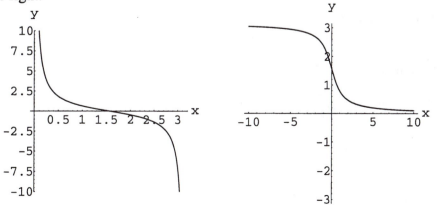

59. Let $u = \sin^{-1}(-x)$. Then

$$-x = \sin u \qquad -\pi/2 \le -u \le \pi/2$$

$$x = -\sin u$$

$$x = \sin(-u)$$

$$\sin^{-1} x = -u \qquad -\pi/2 \le -u \le \pi/2$$

$$\sin^{-1} x = -\sin^{-1}(-x)$$

$$\sin^{-1}(-x) = -\sin^{-1} x$$

61. Let $u = \cos^{-1}(-x)$. Then

$$-x = \cos u \qquad 0 \le u \le \pi, \text{ hence } -\pi \le -u \le 0, \ 0 \le \pi - u \le \pi$$

$$x = -\cos u$$

$$x = \cos(\pi - u) \quad \text{(from the stated identity)}$$

$$\pi - u = \cos^{-1} x$$

$$u = \pi - \cos^{-1} x$$

$$\cos^{-1}(-x) = \pi - \cos^{-1} x$$

63. Let $u = \cot x$. Then

$$\tan^{-1}(\pi/2 - x) = \cot x = u \qquad 0 < u < \pi, \text{ hence } -\pi/2 < \pi/2 - u < \pi/2$$

$$\pi/2 - x = \tan^{-1}(\cot x)$$

65. Let $u = \sin^{-1} x$. Then

$$\sin u = x \qquad -\pi/2 < u < \pi/2$$

$$\tan u = \frac{\sin u}{\cos u} = \frac{\sin u}{\sqrt{1 - \sin^2 u}}$$

$$= \frac{x}{\sqrt{1 - x^2}} \quad \text{(positive square root, since } u \text{ lies in quadrant I or IV)}$$

$$u = \tan^{-1}\left(\frac{x}{\sqrt{1 - x^2}}\right)$$

$$\sin^{-1} x = \tan^{-1}\left(\frac{x}{\sqrt{1 - x^2}}\right)$$

67. The statement is false. For example, let $x = 1/2$.

$$\tan^{-1}\frac{1}{2} \approx .4636 \text{ but } \frac{\sin^{-1}(1/2)}{\cos^{-1}(1/2)} = \frac{\pi/6}{\pi/3} = \frac{1}{2}$$

7.5 Trigonometric Equations

1. $\sin x = -.465$. Since $\sin^{-1}(-.465) = -.4836$, and $\pi - (-.4836) = 3.6252$, all solutions can be written as $-.4836 + 2k\pi$ and $3.6252 + 2k\pi$.

3. $\cos x = -.564$. Since $\cos^{-1}(-.564) = 2.1700$, all solutions can be written as $2.1700 + 2k\pi$ and $-2.1700 + 2k\pi$.

5. $\tan x = -.237$. Since $\tan^{-1}(-.237) = -.2327$, all solutions can be written as $-.2327 + k\pi$.

7. $\cot x = 2.3$

$\dfrac{1}{\tan x} = 2.3$

$\tan x = \dfrac{1}{2.3}$

Since $\tan^{-1}\left(\frac{1}{2.3}\right) = .4101$, all solutions can be written as $.4101 + k\pi$.

9. $\sec x = -2.65$

$\dfrac{1}{\cos x} = -2.65$

$\cos x = -\dfrac{1}{2.65}$

Since $\cos^{-1}\left(-\frac{1}{2.65}\right) = 1.9577$, all solutions can be written as $1.9577 + 2k\pi$ and $-1.9577 + 2k\pi$.

11. $\sin x = .119$. Since $\sin^{-1}(.119) = .1193$, and $\pi - (.1193) = 3.0223$, these are all solutions in $[0, 2\pi)$.

13. $\tan x = 5$. Since $\tan^{-1} 5 = 1.3734$, and $1.3734 + \pi = 4.5150$, these are all solutions in $[0, 2\pi)$.

15. $\sin x = \dfrac{\sqrt{3}}{2}$. Since $\sin^{-1}\left(\dfrac{\sqrt{3}}{2}\right) = \dfrac{\pi}{3}$, and $\pi - \dfrac{\pi}{3} = \dfrac{2\pi}{3}$, all solutions can be written as $\dfrac{\pi}{3} + 2k\pi$ and $\dfrac{2\pi}{3} + 2k\pi$.

17. $\tan x = -\sqrt{3}$. Since $\tan^{-1}\left(-\sqrt{3}\right) = -\pi/3$, all solutions can be written as $-\pi/3 + k\pi$.

19. $2\cos x = -\sqrt{3}$

$\cos x = -\sqrt{3}/2$. Since $\cos^{-1}\left(-\sqrt{3}/2\right) = 5\pi/6$, all solutions can be written as $5\pi/6 + 2k\pi$ and $-5\pi/6 + 2k\pi$.

21. $2\sin x + 1 = 0$

$\sin x = -1/2$. Since $\sin^{-1}(-1/2) = -\pi/6$, and $\pi - (-\pi/6) = 7\pi/6$, all solutions can be written as $-\dfrac{\pi}{6} + 2k\pi$ and $\dfrac{7\pi}{6} + 2k\pi$.

23. $\sin 2x = -\dfrac{\sqrt{3}}{2}$

$2x = -\dfrac{\pi}{3} + 2k\pi$ and $2x = \pi - \left(-\dfrac{\pi}{3}\right) + 2k\pi = \dfrac{4\pi}{3} + 2k\pi$

$\quad x = -\dfrac{\pi}{6} + k\pi \qquad\qquad x = \dfrac{2\pi}{3} + k\pi$

25. $2\cos\dfrac{x}{2} = \sqrt{2}$

$\quad \cos\dfrac{x}{2} = \dfrac{\sqrt{2}}{2}$

$\dfrac{x}{2} = \dfrac{\pi}{4} + 2k\pi \quad$ and $\quad \dfrac{x}{2} = -\dfrac{\pi}{4} + 2k\pi$

$\quad x = \dfrac{\pi}{2} + 4k\pi \qquad x = -\dfrac{\pi}{2} + 4k\pi$

27. $\tan 3x = -\sqrt{3}$

$\quad 3x = -\dfrac{\pi}{3} + k\pi$

$\quad x = -\dfrac{\pi}{9} + k\dfrac{\pi}{3}$

29. $5\cos 3x = -3$

$\quad \cos 3x = -\dfrac{3}{5}$

$3x = 2.2143 + 2k\pi \quad$ and $\quad \dfrac{x}{2} = -2.2143 + 2k\pi$

$\quad x = .7381 + 2k\dfrac{\pi}{3} \qquad x = -.7381 + 2k\dfrac{\pi}{3}$

31. $4\tan\dfrac{x}{2} = 8$

$\quad \tan\dfrac{x}{2} = 2$

$\quad \dfrac{x}{2} = 1.10171 + k\pi$

$\quad x = 2.2143 + 2k\pi$

Calculator in degree mode for Exercises **33-40**

33. $m = 1.1$

$\quad \sin\alpha = \dfrac{1}{1.1}$

$\quad \alpha = \sin^{-1}\dfrac{1}{1.1}$

$\quad \alpha = 65.4°$

35. $m = 2$

$\quad \sin\alpha = \dfrac{1}{2}$

$\quad \alpha = \sin^{-1}\dfrac{1}{2}$

$\quad \alpha = 30°$

37. Use $\dfrac{\sin \theta_1}{\sin \theta_2} = \dfrac{v_1}{v_2}$

with $\dfrac{v_1}{v_2} = 1.33,\ \theta_1 = 38°.$

$$\dfrac{\sin 38°}{\sin \theta_2} = 1.33$$

$$\sin \theta_2 = \dfrac{\sin 38°}{1.33}$$

$$\theta_2 = \sin^{-1}\!\left(\dfrac{\sin 38°}{1.33}\right)$$

$$\theta_2 = 27.6°$$

39. Use $\dfrac{\sin \theta_1}{\sin \theta_2} = \dfrac{v_1}{v_2}$

with $\dfrac{v_1}{v_2} = 1.66,\ \theta_1 = 24°.$

$$\dfrac{\sin 24°}{\sin \theta_2} = 1.66$$

$$\sin \theta_2 = \dfrac{\sin 24°}{1.66}$$

$$\theta_2 = \sin^{-1}\!\left(\dfrac{\sin 24°}{1.66}\right)$$

$$\theta_2 = 14.2°$$

41. $3\sin^2 x - 8\sin x - 3 = 0$
$(3\sin x + 1)(\sin x - 3) = 0$

$$\sin x = -\dfrac{1}{3} \qquad\qquad \sin x = 3$$

$\sin^{-1}\!\left(-\tfrac{1}{3}\right) = -.3398 \qquad$ impossible

The two solutions in $[0, 2\pi)$ are $-.3398 + 2\pi = 5.9433$ and $\pi - (-.3398) = 3.4814.$

43. $2\tan^2 x + 5\tan x + 3 = 0$
$(2\tan x + 3)(\tan x + 1) = 0$

$$\tan x = -\dfrac{3}{2} \qquad\qquad \tan x = -1$$

$\tan^{-1}\!\left(-\dfrac{3}{2}\right) = -.9828 \qquad \tan^{-1}(-1) = -\dfrac{\pi}{4}$

The solutions in $[0, 2\pi)$ are $-.9828 + \pi = 2.1588$ and $-.9828 + 2\pi = 5.3004,$

$-\dfrac{\pi}{4} + \pi = \dfrac{3\pi}{4},$ and $-\dfrac{\pi}{4} + 2\pi = \dfrac{7\pi}{4}.$

45. $\cot x \cos x = \cos x$
$\cot x \cos x - \cos x = 0$

$\cos x(\cot x - 1) = 0$

$\cos x = 0 \qquad \cot x = 1$

$x = \dfrac{\pi}{2}, \dfrac{3\pi}{2} \qquad x = \dfrac{\pi}{4}, \dfrac{5\pi}{4}$

47. $\cos x \csc x = 2\cos x$
$\cos x \csc x - 2\cos x = 0$

$\cos x(\csc x - 2) = 0$

$\cos x = 0 \qquad \csc x = 2$

$\qquad\qquad\qquad \sin x = 1/2$

$x = \dfrac{\pi}{2}, \dfrac{3\pi}{2} \qquad x = \dfrac{\pi}{6}, \dfrac{5\pi}{6}$

49. $4\sin x \tan x - 3\tan x + 20\sin x - 15 = 0$

$\tan x(4\sin x - 3) + 5(4\sin x - 3) = 0$

$(4\sin x - 3)(\tan x + 5) = 0$

$\sin x = 3/4 \qquad\qquad \tan x = -5$

$\sin^{-1}(3/4) = .8481 \qquad \tan^{-1}(-5) = -1.3734$

The solutions in $[0, 2\pi)$ are $.8481$, $\pi - .8481 = 2.2935$, $-1.3734 + \pi = 1.7682$,
and $-1.3734 + 2\pi = 4.9098$.

51. $\sin^2 x + 2\sin x - 2 = 0$

$\sin x = \dfrac{-2 \pm \sqrt{12}}{2}$

$\sin x = \dfrac{-2 - \sqrt{12}}{2} \qquad \sin x = \dfrac{-2 + \sqrt{12}}{2}$

impossible $\qquad\qquad \sin^{-1}\left(\dfrac{-2 + \sqrt{12}}{2}\right) = .8213$

The solutions in $[0, 2\pi)$ are $.8213$, $\pi - .8213 = 2.3203$.

53. $\tan^2 x + 1 = 3\tan x$

$\tan^2 x - 3\tan x + 1 = 0$

$\tan x = \dfrac{3 \pm \sqrt{5}}{2}$

$\tan^{-1}\left(\dfrac{3 + \sqrt{5}}{2}\right) = 1.2059 \quad \tan^{-1}\left(\dfrac{3 - \sqrt{5}}{2}\right) = .3649$

The solutions in $[0, 2\pi)$ are $1.2059, 1.2059 + \pi = 4.3475, .3649$,
and $.3649 + \pi = 3.5065$.

55. $2\tan^2 x - 1 = 3\tan x$

$2\tan^2 x - 3\tan x - 1 = 0$

$\tan x = \dfrac{3 \pm \sqrt{17}}{4}$

$\tan^{-1}\left(\dfrac{3 + \sqrt{17}}{4}\right) = 1.0591 \quad \tan^{-1}\left(\dfrac{3 - \sqrt{17}}{4}\right) = -.2737$

The solutions in $[0, 2\pi)$ are $1.0591, 1.0591 + \pi = 4.2007, -.2737 + \pi = 2.8679$,
and $-.2737 + 2\pi = 6.0095$.

57. $\sin^2 x + 3\cos^2 x = 0$ can have no solution, since the sum of two squares of non-zero
numbers is never zero, and $\sin x$ and $\cos x$ are never both zero.

59. $\sin 2x + \cos x = 0$

$2\sin x \cos x + \cos x = 0$

$\cos x(2\sin x + 1) = 0$

$\cos x = 0 \qquad \sin x = -1/2$

$x = \dfrac{\pi}{2}, \dfrac{3\pi}{2} \qquad x = \dfrac{7\pi}{6}, \dfrac{11\pi}{6}$

61. $9 - 12\sin x = 4\cos^2 x$

$9 - 12\sin x = 4(1 - \sin^2 x)$

$4\sin^2 x - 12\sin x + 5 = 0$

$(2\sin x - 5)(2\sin x - 1) = 0$

$\sin x = 5/2 \qquad \sin x = 1/2$

impossible $\qquad x = \dfrac{\pi}{6}, \dfrac{5\pi}{6}$

63. $\cos^2 x - \sin^2 x + \sin x = 0$

$1 - \sin^2 x - \sin^2 x + \sin x = 0$

$2\sin^2 x - \sin x - 1 = 0$

$(2\sin x + 1)(\sin x - 1) = 0$

$\sin x = -1/2 \qquad \sin x = 1$

$x = \dfrac{7\pi}{6}, \dfrac{11\pi}{6} \qquad x = \dfrac{\pi}{2}$

65. $\sin \dfrac{x}{2} = 1 - \cos x$

$\sin \dfrac{x}{2} = 1 - \left(1 - 2\sin^2 \dfrac{x}{2}\right)$

$2\sin^2 \dfrac{x}{2} - \sin \dfrac{x}{2} = 0$

$\sin \dfrac{x}{2}\left(2\sin \dfrac{x}{2} - 1\right) = 0$

$\sin \dfrac{x}{2} = 0 \qquad \sin \dfrac{x}{2} = \dfrac{1}{2}$

$\dfrac{x}{2} = 0 \qquad \dfrac{x}{2} = \dfrac{\pi}{6}, \dfrac{5\pi}{6}$

$x = 0 \qquad x = \dfrac{\pi}{3}, \dfrac{5\pi}{3}$

Calculator in degree mode for Exercises **67-76**

67. $\tan \theta = 7.95$. Since $\tan^{-1}(7.95) = 82.83°$, the solutions in the required interval are $82.83°$ and $82.83° + 180° = 262.83°$.

69. $\cos \theta = -.42$. Since $\cos^{-1}(-.42) = 114.83°$, the solutions in the required interval are $114.83°$ and $360° - 114.83° = 245.17°$.

71. $2\sin^2 \theta + 3\sin \theta + 1 = 0$

$(2\sin \theta + 1)(\sin \theta + 1) = 0$

$\sin \theta = -1/2 \qquad \sin \theta = -1$

$\theta = 210°, 330° \qquad \theta = 270°$

73. $\tan^2 \theta - 3 = 0$

$\tan \theta = \pm\sqrt{3}$

If $\tan \theta = \sqrt{3}$, $\theta = 60°$ or $240°$.

If $\tan \theta = -\sqrt{3}$, $\theta = 120°$ or $300°$.

75. $4\cos^2 \theta + 4\cos \theta + 1 = 0$

$(2\cos \theta + 1)^2 = 0$

$2\cos \theta + 1 = 0$

$\cos \theta = -1/2, \quad \theta = 120°, 240°$

77. Graph the function $f(x) = 4\sin 2x - 3\cos 2x - 2$ using the window $0 \le x \le 2\pi$, $-8 \le y \le 3$ and apply the zero (root) routine to obtain $x = .5275$ and $x = 1.6868$. Since the function is periodic with period π, all solutions are given by $x = .5275 + n\pi$ and $x = 1.6868 + n\pi$, where n is any integer.

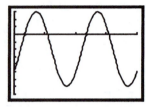

79. Graph the function $f(x) = 3\sin^3 2x - 2\cos x$ using the window $0 \le x \le 2\pi$, $-5 \le y \le 5$ and apply the zero (root) routine to obtain $x = .4959$, 1.2538, 1.5708, 1.8877, 2.6457, and 4.7124. Since the function is periodic with period 2π, all solutions are given by $x = .4959 + 2n\pi$, where n is any integer, and so on.

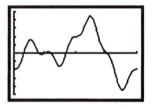

81. Graph the function $f(x) = \tan x + 5\sin x - 1$ using the window $0 \le x \le 2\pi$, $-5 \le y \le 5$, ignoring the erroneous vertical lines at the asymptotes, and apply the zero (root) routine to obtain $x = .1671$, 1.8256, 2.8867, and 4.5453. Since the function is periodic with period 2π, all solutions are given by $x = .1671 + 2n\pi$, where n is any integer, and so on.

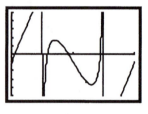

83. Graph the function $f(x) = \cos^3 x - 3\cos x + 1$ using the window $0 \le x \le 2\pi$, $-2 \le y \le 4$ and apply the zero (root) routine to obtain $x = 1.2161$ and $x = 5.0671$. Since the function is periodic with period 2π, all solutions are given by $x = 1.2161 + 2n\pi$ and $x = 5.0671 + 2n\pi$, where n is any integer.

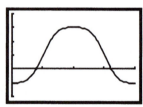

85. Graph the function $f(x) = \cos^4 x - 3\cos^3 x + \cos x - 1$ using the window $0 \le x \le 2\pi$, $-3 \le y \le 3$ and apply the zero (root) routine to obtain $x = 2.4620$ and $x = 3.8212$. Since the function is periodic with period 2π, all solutions are given by $x = 2.4620 + 2n\pi$ and $x = 3.8212 + 2n\pi$, where n is any integer.

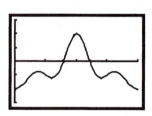

87. Graph the function $f(x) = \sin^3 x + 2\sin^2 x - 3\cos x + 2$ using the window $0 \le x \le 2\pi$, $-2 \le y \le 6$ and apply the zero (root) routine to obtain $x = .5166$ and $x = 5.6766$. Since the function is periodic with period 2π, all solutions are given by $x = .5166 + 2n\pi$ and $x = 5.6766 + 2n\pi$, where n is any integer.

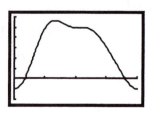

89. a. Graph $d(x) = 3\sin\left(\dfrac{2\pi}{365}(x-80)\right) + 12$

and the horizontal line $y = 11$ and use the intersection routine twice to obtain days 60 and 282 (an algebraic solution is also possible, with the same result.)

b. Graph $d(x) = 3\sin\left(\dfrac{2\pi}{365}(x-80)\right) + 12$

and use the maximum routine to obtain day 171.

91. Use $d = \dfrac{v^2}{32}\sin 2\theta$ with

$d = 2500, v = 300$.

$2500 = \dfrac{300^2}{32}\sin 2\theta$

$\sin 2\theta = \dfrac{2500 \cdot 32}{300^2}$

$\sin 2\theta = 8/9$

$2\theta = 62.74°$ or $180° - 62.74°$

$\theta = 31.37°$ or $58.63°$

93. Use $d = \dfrac{v^2}{32}\sin 2\theta$ with

$d = 288, v = 98 \times \dfrac{5280}{3600}\left(\dfrac{\text{ft}}{\text{sec}} \div \dfrac{\text{mi}}{\text{hr}}\right)$.

$288 = \dfrac{1}{32}\left(98 \times \dfrac{5280}{3600}\right)^2 \sin 2\theta$

$\sin 2\theta = 288 \cdot 32 \div \left(98 \times \dfrac{5280}{3600}\right)^2$

$\sin 2\theta = .4461$

$2\theta = 26.49°$ or $180° - 26.49°$

$\theta = 13.25°$ or $76.75°$

95. The basic equation will have no solutions if $c > 1$ or $c < -1$.

97. In dividing both sides by $\sin x$, the solutions such as 0 and π which correspond to $\sin x = 0$ are lost.

Chapter 7 Review Exercises

1. $\dfrac{\sin^2 t + \left(\tan^2 t + 2\tan t - 4\right) + \cos^2 t}{3\tan^2 t - 3\tan t} = \dfrac{\tan^2 t + 2\tan t - 4 + 1}{3\tan^2 t - 3\tan t}$

$= \dfrac{\tan^2 t + 2\tan t - 3}{3\tan^2 t - 3\tan t}$

$= \dfrac{(\tan t + 3)(\tan t - 1)}{3\tan t(\tan t - 1)}$

$= \dfrac{\tan t + 3}{3\tan t}$

$= \tfrac{1}{3} + \cot t$

3. $\dfrac{\tan^2 x - \sin^2 x}{\sec^2 x} = \left(\dfrac{\sin^2 x}{\cos^2 x} - \sin^2 x\right) \div \dfrac{1}{\cos^2 x}$

$$= \dfrac{\sin^2 x - \sin^2 x \cos^2 x}{1}$$

$$= \sin^2 x \left(1 - \cos^2 x\right)$$

$$= \sin^2 x \sin^2 x$$

$$= \sin^4 x$$

5. This is an identity.

$$\sin^4 t - \cos^4 t = \left(\sin^2 t + \cos^2 t\right)\left(\sin^2 t - \cos^2 t\right)$$

$$= 1\left(\sin^2 t - \left(1 - \sin^2 t\right)\right)$$

$$= \sin^2 t - 1 + \sin^2 t$$

$$= 2\sin^2 t - 1$$

7. This is an identity. Use Strategy 4. Prove
$\sin t \sin t = (1 - \cos t)(1 + \cos t)$ is an identity.

$$\sin t \sin t = \sin^2 t$$

$$= 1 - \cos^2 t$$

$$= (1 - \cos t)(1 + \cos t)$$

Therefore $\dfrac{\sin t}{1 - \cos t} = \dfrac{1 + \cos t}{\sin t}$

9. This is not an identity.
For example, let $t = \pi/6$.

$$\dfrac{\cos^2(\pi + \pi/6)}{\sin^2(\pi + \pi/6)} - 1 = 2,$$

but $\dfrac{1}{\sin^2(\pi/6)} = 4$

11. This is an identity.

$$(\sin x + \cos x)^2 - \sin 2x = \sin^2 x + 2\sin x \cos x + \cos^2 x - \sin 2x$$

$$= \sin^2 x + \cos^2 x + 2\sin x \cos x - 2\sin x \cos x$$

$$= 1 + 0 = 1$$

13. $\dfrac{\tan x - \sin x}{2 \tan x} = \dfrac{\sin x/\cos x - \sin x}{2(\sin x/\cos x)}$

$$= \dfrac{\sin x - \sin x \cos x}{2\sin x}$$

$$= \dfrac{\sin x(1 - \cos x)}{2\sin x}$$

$$= \dfrac{1 - \cos x}{2}$$

$$= \sin^2(x/2)$$

15. $\cos(x+y)\cos(x-y) = (\cos x\cos y - \sin x\sin y)(\cos x\cos y + \sin x\sin y)$

$$= \cos x^2\cos^2 y - \sin^2 x\sin^2 y$$

$$= \cos^2 x\left(1-\sin^2 y\right) - \left(1-\cos^2 x\right)\sin^2 y$$

$$= \cos^2 x - \cos^2 x\sin^2 y - \sin^2 y + \cos^2 x\sin^2 y$$

$$= \cos^2 x - \sin^2 y$$

17. Use Strategy 4. Prove that $(\sec x+1)(\sec x-1) = \tan x\tan x$ is an identity.

$$(\sec x+1)(\sec x-1) = \sec^2 x - 1$$

$$= \tan^2 x$$

$$= \tan x\tan x$$

Therefore $\dfrac{\sec x+1}{\tan x} = \dfrac{\tan x}{\sec x-1}$ is an identity.

19. $\dfrac{1+\tan^2 x}{\tan^2 x} = \dfrac{1}{\tan^2 x} + \dfrac{\tan^2 x}{\tan^2 x}$

$$= \cot^2 x + 1$$

$$= \csc^2 x$$

21. $\tan^2 x - \sec^2 x = \sec^2 x - 1 - \sec^2 x$

$$= -1$$

$$= \csc^2 x - 1 - \csc^2 x$$

$$= \cot^2 x - \csc^2 x$$

23. $\tan x = \dfrac{5}{12}$; $\sec x = \sqrt{1+\tan^2 x} = \sqrt{1+\left(\dfrac{5}{12}\right)^2} = \dfrac{13}{12}$; $\cos x = \dfrac{12}{13}$, $\sin x = \dfrac{5}{13}$.

$\sin 2x = 2\sin x\cos x = 2(5/13)(12/13) = 120/169$

25. First note that $\sin x$, $\cos x$, and $\sin y$ are negative and $\cos y$ is positive.

$\sec x = -\sqrt{1+\tan^2 x} = -\sqrt{1+\left(\tfrac{4}{3}\right)^2} = -\tfrac{5}{3}$ $\csc y = -\sqrt{1+\cot^2 y} = -\sqrt{1+\left(-\tfrac{5}{12}\right)^2} = -\tfrac{13}{12}$

$\cos x = \dfrac{1}{\sec x} = \dfrac{1}{-5/3} = -\dfrac{3}{5}$ $\sin y = \dfrac{1}{\csc y} = -\dfrac{12}{13}$

$\sin x = \tan x\cos x = \dfrac{4}{3}\left(-\dfrac{3}{5}\right) = -\dfrac{4}{5}$ $\cos y = \cot y\sin y = \left(-\dfrac{5}{12}\right)\left(-\dfrac{12}{13}\right) = \dfrac{5}{13}$

Then $\sin(x-y) = \sin x\cos y - \cos x\sin y$

$$= \left(-\dfrac{4}{5}\right)\left(\dfrac{5}{13}\right) - \left(-\dfrac{3}{5}\right)\left(-\dfrac{12}{13}\right)$$

$$= \dfrac{-20-36}{65}$$

$$= -\dfrac{56}{65}$$

27. First note that $\cos x = \sqrt{1 - \sin^2 x} = \sqrt{1 - \left(\dfrac{1}{4}\right)^2} = \dfrac{\sqrt{15}}{4}$

Then $\sin\left(\dfrac{\pi}{3} + x\right) = \sin\dfrac{\pi}{3}\cos x + \cos\dfrac{\pi}{3}\sin x$

$$= \left(\dfrac{\sqrt{3}}{2}\right)\left(\dfrac{\sqrt{15}}{4}\right) + \left(\dfrac{1}{2}\right)\left(\dfrac{1}{4}\right)$$

$$= \dfrac{\sqrt{45} + 1}{8} \quad \text{or} \quad \dfrac{3\sqrt{5} + 1}{8}$$

29. Yes. If $\sin x = 0$, $\sin 2x = 2\sin x \cos x = 2 \cdot 0 \cos x = 0$.

31. Using the half-angle identity,

$$\cos\dfrac{\pi}{12} = \cos\dfrac{1}{2}\left(\dfrac{\pi}{6}\right) = \sqrt{\dfrac{1 + \cos(\pi/6)}{2}} = \sqrt{\dfrac{1 + \sqrt{3}/2}{2}} = \dfrac{\sqrt{2 + \sqrt{3}}}{2}$$

Using the subtraction identity for cosine,

$$\cos\dfrac{\pi}{12} = \cos\left(\dfrac{\pi}{3} - \dfrac{\pi}{4}\right) = \cos\dfrac{\pi}{3}\cos\dfrac{\pi}{4} - \sin\dfrac{\pi}{3}\sin\dfrac{\pi}{4} = \dfrac{1}{2}\cdot\dfrac{\sqrt{2}}{2} + \dfrac{\sqrt{3}}{2}\cdot\dfrac{\sqrt{2}}{2} = \dfrac{\sqrt{2} + \sqrt{6}}{4}$$

Hence $\dfrac{\sqrt{2 + \sqrt{3}}}{2} = \dfrac{\sqrt{2} + \sqrt{6}}{4}$ and $\sqrt{2 + \sqrt{3}} = \dfrac{\sqrt{2} + \sqrt{6}}{2}$.

33. $\sin\dfrac{5\pi}{12} = \sin\left(\dfrac{\pi}{4} + \dfrac{\pi}{6}\right) = \sin\dfrac{\pi}{4}\cos\dfrac{\pi}{6} + \cos\dfrac{\pi}{4}\sin\dfrac{\pi}{6}$

$$= \dfrac{\sqrt{2}}{2}\cdot\dfrac{\sqrt{3}}{2} + \dfrac{\sqrt{2}}{2}\cdot\dfrac{1}{2}$$

$$= \dfrac{\sqrt{6} + \sqrt{2}}{4}$$

35. (a)

37. First note that $\cos x = \sqrt{1 - \sin^2 x} = \sqrt{1 - (.6)^2} = .8$.
Then $\sin 2x = \sin x \cos x = 2(.6)(.8) = .96$.

39. The slope of the line is given by $\dfrac{y_2 - y_1}{x_2 - x_1} = \dfrac{2 - 6}{-2 - 2} = 1$.

Therefore the inclination is the acute angle solution of $\tan\theta = 1$, that is, $45°$.

41. $\dfrac{\pi}{4}$ **43.** $\dfrac{\pi}{3}$ **45.** $\cos^{-1}\left(\sin\dfrac{5\pi}{3}\right) = \cos^{-1}\left(-\dfrac{\sqrt{3}}{2}\right) = \dfrac{5\pi}{6}$

47. Since $-\dfrac{\pi}{2} \le .75 \le \dfrac{\pi}{2}$, $\sin^{-1}(\sin .75) = .75$

49. $\sin^{-1}\left(\sin\dfrac{8\pi}{3}\right) = \sin^{-1}\dfrac{\sqrt{3}}{2} = \dfrac{\pi}{3}$

51.

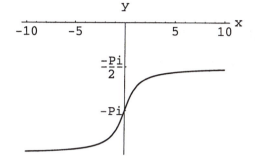

53. Let $u = \cos^{-1}(1/4)$. Then $\cos u = 1/4$, $0 \le u \le \pi$. Then

$$\sin\left(\cos^{-1}(1/4)\right) = \sin u = \sqrt{1 - \cos^2 u} = \sqrt{1 - (1/4)^2} = \sqrt{15}/4$$

55. $2\sin x = 1$

$\sin x = 1/2$

Since $\sin^{-1}\left(\dfrac{1}{2}\right) = \dfrac{\pi}{6}$, and $\pi - \dfrac{\pi}{6} = \dfrac{5\pi}{6}$, all solutions can be written as

$\dfrac{\pi}{6} + 2k\pi$ and $\dfrac{5\pi}{6} + 2k\pi$.

57. $\tan x = -1$. Since $\tan^{-1}(-1) = -\pi/4$, all solutions can be written as $-\pi/4 + k\pi$.

59. $\sin x = .7$

Since $\sin^{-1}.7 = .7754$, and $\pi - .7754 = 2.3662$, all solutions can be written as $.7754 + 2k\pi$ and $2.3662 + 2k\pi$.

61. $\tan x = 13$. Since $\tan^{-1}(13) = 1.4940$, all solutions can be written as $1.4940 + k\pi$.

63. $2\sin^2 x + 5\sin x = 3$

$2\sin^2 x + 5\sin x - 3 = 0$

$(2\sin x - 1)(\sin x + 3) = 0$

$\sin x = 1/2 \qquad\qquad \sin x = -3$

$\sin^{-1}(1/2) = \pi/6 \qquad\quad$ impossible

$x = \pi/6 + 2k\pi$ and $5\pi/6 + 2k\pi$

65. $2\sin^2 x - 3\sin x = 2$

$2\sin^2 x - 3\sin x - 2 = 0$

$(2\sin x + 1)(\sin x - 2) = 0$

$\sin x = -1/2 \qquad\qquad \sin x = 2$

$\sin^{-1}(-1/2) = -\pi/6 \qquad$ impossible

$x = -\dfrac{\pi}{6} + 2k\pi$ and $\dfrac{7\pi}{6} + 2k\pi$

67. $\sec^2 x + 3\tan^2 x = 13$

$\tan^2 x + 1 + 3\tan^2 x = 13$

$4\tan^2 x = 12$

$\tan^2 x = 3$

$\tan x = \sqrt{3}$ or $\tan x = -\sqrt{3}$

$x = \pi/3 + k\pi \qquad x = -\pi/3 + k\pi$

69. $2\sin^2 x + \sin x - 2 = 0$

$\sin x = \dfrac{-1 \pm \sqrt{17}}{4}$

$\sin x = \dfrac{-1 + \sqrt{17}}{4} \qquad\qquad \sin x = \dfrac{-1 - \sqrt{17}}{4}$

$\sin x = .7808 \qquad\qquad$ impossible

$x = \sin^{-1} .7808 = .8959 + 2k\pi$

$x = \pi - \sin^{-1} .7808 = 2.2457 + 2k\pi$

71. Graph the function $f(x) = 5\tan x - 2\sin 2x$ using the window $0 \le x \le 2\pi$, $-5 \le y \le 5$, ignoring the erroneous vertical lines at the asymptotes. Solutions are observed precisely at $x = 0, \pi, 2\pi$ and, since the function is periodic with period π, all solutions are given by $x = n\pi$, where n is any integer.

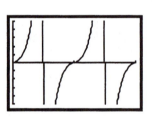

73. Graph the function $f(x) = \sin x + \sec^2 x - 3$ using the window $0 \le x \le 2\pi$, $-5 \le y \le 5$ and apply the zero (root) routine to obtain $x = .8419$, 2.2997, 4.1784, and 5.2463. Since the function is periodic with period 2π, all solutions are given by $x = .8419 + 2n\pi$, where n is any integer, and so on.

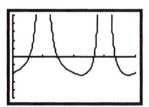

75. $\sin\theta = -.7133$. Since $\sin^{-1}(-.7133) = -45.5°$, all solutions in the required interval are given by $-45.5° + 360° = 314.5°$ and $180° - (-45.5°) = 225.5°$.

77. Use $d = \left(v^2/32\right)\sin 2\theta$ with $d = 3500$, $v = 600$.

$3500 = \left(600^2/32\right)\sin 2\theta$

$\sin 2\theta = \dfrac{3500 \cdot 32}{600^2} = \dfrac{14}{45}$

$2\theta = 18.13°$ or $180° - 18.13°$

$\theta = 9.06°$ or $80.94°$

Chapter 8
Triangle Trigonometry

8.1 Trigonometric Functions of Angles

1. $r = \sqrt{x^2 + y^2} = \sqrt{2^2 + 3^2} = \sqrt{13}$

$\sin\theta = \dfrac{y}{r} = \dfrac{3}{\sqrt{13}} \qquad \cot\theta = \dfrac{x}{y} = \dfrac{2}{3}$

$\cos\theta = \dfrac{x}{r} = \dfrac{2}{\sqrt{13}} \qquad \sec\theta = \dfrac{r}{x} = \dfrac{\sqrt{13}}{2}$

$\tan\theta = \dfrac{y}{x} = \dfrac{3}{2} \qquad \csc\theta = \dfrac{r}{y} = \dfrac{\sqrt{13}}{3}$

3. $r = \sqrt{x^2 + y^2} = \sqrt{(-5)^2 + 6^2} = \sqrt{61}$

$\sin\theta = \dfrac{y}{r} = \dfrac{6}{\sqrt{61}} \qquad \cot\theta = \dfrac{x}{y} = -\dfrac{5}{6}$

$\cos\theta = \dfrac{x}{r} = -\dfrac{5}{\sqrt{61}} \qquad \sec\theta = \dfrac{r}{x} = -\dfrac{\sqrt{61}}{5}$

$\tan\theta = \dfrac{y}{x} = -\dfrac{6}{5} \qquad \csc\theta = \dfrac{r}{y} = \dfrac{\sqrt{61}}{6}$

5. $r = \sqrt{x^2 + y^2} = \sqrt{(-3)^2 + (-\sqrt{2})^2} = \sqrt{11}$

$\sin\theta = \dfrac{y}{r} = -\dfrac{\sqrt{2}}{\sqrt{11}} \qquad \cot\theta = \dfrac{x}{y} = \dfrac{3}{\sqrt{2}}$

$\cos\theta = \dfrac{x}{r} = -\dfrac{3}{\sqrt{11}} \qquad \sec\theta = \dfrac{r}{x} = -\dfrac{\sqrt{11}}{3}$

$\tan\theta = \dfrac{y}{x} = \dfrac{\sqrt{2}}{3} \qquad \csc\theta = \dfrac{r}{y} = -\dfrac{\sqrt{11}}{\sqrt{2}}$

7. $\sin\theta = \dfrac{\text{opposite}}{\text{hypotenuse}} = \dfrac{\sqrt{2}}{\sqrt{11}}$

$\cos\theta = \dfrac{\text{adjacent}}{\text{hypotenuse}} = \dfrac{3}{\sqrt{11}}$

$\tan\theta = \dfrac{\text{opposite}}{\text{adjacent}} = \dfrac{\sqrt{2}}{3}$

9. $\sin\theta = \dfrac{\text{opposite}}{\text{hypotenuse}} = \dfrac{\sqrt{3}}{\sqrt{7}}$

$\cos\theta = \dfrac{\text{adjacent}}{\text{hypotenuse}} = \dfrac{2}{\sqrt{7}}$

$\tan\theta = \dfrac{\text{opposite}}{\text{adjacent}} = \dfrac{\sqrt{3}}{2}$

11. $\sin\theta = \dfrac{h}{m}$

$\cos\theta = \dfrac{d}{m}$

$\tan\theta = \dfrac{h}{d}$

13. $\cos A = \dfrac{c}{b}$

$\dfrac{12}{13} = \dfrac{c}{39}$

$c = 36$

15. $\tan A = \dfrac{a}{c}$

$\dfrac{5}{12} = \dfrac{15}{c}$

$c = 36$

17. $\cot A = \dfrac{c}{a}$

$6 = \dfrac{c}{1.4}$

$c = 8.4$

19. $\cos 45° = \dfrac{h}{25}$

$h = 25\cos 45°$

$h = \dfrac{25\sqrt{2}}{2}$

21. $\sin 30° = \dfrac{150}{h}$

$h = 150/\sin 30°$

$h = \dfrac{150}{1/2} = 300$

23. $\cos 30° = \dfrac{h}{100}$

$h = 100\cos 30°$

$h = 100\dfrac{\sqrt{3}}{2} = 50\sqrt{3}$

25. $a/c = \tan 60°$

$$c = \frac{a}{\tan 60°}$$

$$c = 4/\sqrt{3} = 4\sqrt{3}/3$$

27. $a/c = \tan 30°$

$$a = c \tan 30°$$

$$a = 10 \cdot \frac{1}{\sqrt{3}} = \frac{10}{\sqrt{3}} = \frac{10\sqrt{3}}{3}$$

29. $\angle A = 90° - \angle C = 90° - 50° = 40°$

$$\sin C = \frac{c}{b}, \ \sin 50° = \frac{c}{10}, \ c = 10 \sin 50° = 7.7$$

$$\cos C = \frac{a}{b}, \ \cos 50° = \frac{a}{10}, \ a = 10 \cos 50° = 6.4$$

31. $\angle C = 90° - \angle A = 90° - 14° = 76°$

$$\sin A = \frac{a}{b}, \ \sin 14° = \frac{6}{b}, \ b = \frac{6}{\sin 14°} = 24.8$$

$$\tan A = \frac{a}{c}, \ \tan 14° = \frac{6}{c}, \ c = \frac{6}{\tan 14°} = 24.1$$

33. $\angle C = 90° - \angle A = 90° - 65° = 25°$

$$\cos A = \frac{c}{b}, \ \cos 65° = \frac{5}{b}, \ b = \frac{5}{\cos 65°} = 11.8$$

$$\tan A = \frac{a}{c}, \ \tan 65° = \frac{a}{5}, \ a = 5 \tan 65° = 10.7$$

35. $\angle C = 90° - \angle A = 90° - 72° = 18°$

$$\cos A = \frac{c}{b}, \ \cos 72° = \frac{c}{3.5}, \ c = 3.5 \cos 72° = 1.1$$

$$\sin A = \frac{a}{b}, \ \sin 72° = \frac{a}{3.5}, \ a = 3.5 \sin 72° = 3.3$$

37. $\sin \theta = 3/4, \ \theta = \sin^{-1}(3/4) = 48.6°$ **39.** $\cos \theta = 2/3, \ \theta = \cos^{-1}(2/3) = 48.2°$

41. $\tan A = a/c = 4/6, \ A = \tan^{-1}(4/6) = 33.7°, \ C = 90° - A = 56.3°$

43. $\sin A = a/b = 7/10, \ A = \sin^{-1}(7/10) = 44.4°, \ C = 90° - A = 45.6°$

45. $\cos A = b/c = 12/18, \ A = \cos^{-1}(12/18) = 48.2°, \ C = 90° - A = 41.8°$

47. $\tan A = a/c = 2.5/1.4, \ A = \tan^{-1}(2.5/1.4) = 60.8°, \ C = 90° - A = 29.2°$

49. **a.** $\sin \theta = a/c = \cos \alpha$
    b. Since $\theta + \alpha + 90° = 180°$, $\theta + \alpha = 90°$.
    c. Since $\theta + \alpha = 90°$, $\alpha = 90° - \theta$ thus $\sin \theta = \cos(90° - \theta)$.

8.2 Applications of Right Triangle Trigonometry

1. Use $\tan 40° = \dfrac{\text{opposite}}{\text{adjacent}} = \dfrac{w}{120}$

$w = 120 \tan 40° = 100.7$ ft

3. Use $\cos \theta = \dfrac{\text{adjacent}}{\text{hypotenuse}} = \dfrac{9}{24}$

$\theta = \cos^{-1}(9/24) = 68.0°$

5. Use $\dfrac{c}{240} = \dfrac{\text{opposite}}{\text{adjacent}} = \tan 72.9°$, $\dfrac{d}{240} = \dfrac{\text{opposite}}{\text{adjacent}} = \tan 71.3°$

height $= c - d = 240 \tan 72.9° - 240 \tan 71.3° = 71.1$ ft

7. Let d = distance required. Then $\sin 50° = \dfrac{\text{opposite}}{\text{hypotenuse}} = \dfrac{d}{20}$.

$d = 20 \sin 50° = 15.3$ ft

9. Let θ = angle required. Then $\sin \theta = \dfrac{\text{opposite}}{\text{hypotenuse}} = \dfrac{8}{16} = \dfrac{1}{2}$, hence $\theta = 30°$.

11. Let h = height. 1 mile = 5280 feet.

$\sin 5° = \dfrac{\text{opposite}}{\text{hypotenuse}} = \dfrac{h}{5280}$

$h = 5280 \sin 5° = 460.2$ feet

13. Let x = length.

$\sin 3° = \dfrac{\text{opposite}}{\text{hypotenuse}} = \dfrac{450}{x}$

$x = 450/\sin 3° = 8598.3$ feet $(1.6$ mi.$)$

15. The angle θ between the vertical person and the mountain is $90° - 62° = 28°$.

Then $\tan \theta = \dfrac{\text{arm length}}{\text{shoulder height}}$. Shoulder height $= 5(12) = 60$ inches.

$\tan 28° = x/60$

$x = 60 \tan 28° = 31.9$ inches.

Since the person's arm length is only 27 inches, (s)he cannot touch the mountain.

17. Let x = width. Then $\tan 54° = \dfrac{\text{opposite}}{\text{adjacent}} = \dfrac{w}{20}$.

$w = 20 \tan 54° = 27.5$ ft. It is not safe for him to jump.

19. Let y = height of the streetlight. In the similar triangles shown,

$\dfrac{5.5}{4} = \dfrac{y}{10 + 4}$

$y = 14(5.5/4) = 19.25$ ft.

$\tan \theta = \dfrac{\text{opposite}}{\text{adjacent}} = \dfrac{5.5}{4}$ (using the smaller triangle)

$\theta = \tan^{-1}(5.5/4) = 54.0°$

21. Sketch a figure.

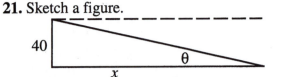

Note: θ is equal to the angle of depression, $6.5°$.
$\tan 6.5° = 40/x;\ \ x = 40/\tan 6.5° = 351.1$ meters

23. Sketch a figure.

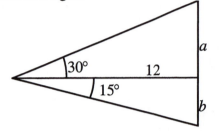

The height of the statue is $a + b$. Then
$\tan 30° = a/12 \qquad\qquad \tan 15° = b/12$

$\qquad a = 12\tan 30° \qquad\qquad b = 12\tan 15°$
$\qquad h = a + b = 12\tan 30° + 12\tan 15° = 10.1$ feet.

25. Sketch a figure.

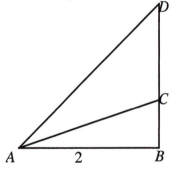

The rise of the rocket $= x = BD - BC$.
$$\tan CAB = \frac{BC}{2} \qquad\qquad \tan DAB = \frac{BD}{2}$$
$BC = 2\tan 3.5° \qquad\qquad BD = 2\tan 41°$
$x = BD - BC = 2\tan 41° - 2\tan 3.5° = 1.6$ mi.

27. Abstracting the essential information, sketch a figure.

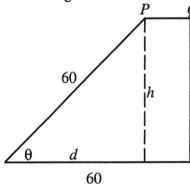

a. $\sin\theta = h/60 \quad h = 60\sin\theta = 60\sin 33° = 32.7$
    Then the height of P above the water $= 24 + h = 24 + 32.7 = 56.7$ feet.
b. $\cos\theta = d/60 \quad d = 60\cos\theta = 60\cos 33° = 50.3$
    $PQ = 60 - d = 60 - 50.3 = 9.7$ feet

29. Abstracting the essential information, sketch a figure.

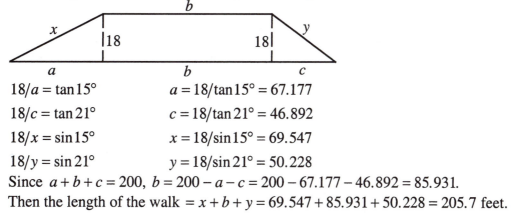

$18/a = \tan 15°$ $a = 18/\tan 15° = 67.177$
$18/c = \tan 21°$ $c = 18/\tan 21° = 46.892$
$18/x = \sin 15°$ $x = 18/\sin 15° = 69.547$
$18/y = \sin 21°$ $y = 18/\sin 21° = 50.228$
Since $a + b + c = 200$, $b = 200 - a - c = 200 - 67.177 - 46.892 = 85.931$.
Then the length of the walk $= x + b + y = 69.547 + 85.931 + 50.228 = 205.7$ feet.

31. Since the angle between the path and West is $30°$, write $d =$ distance,
$d/200 = \cos 30°$, $d = 200 \cos 30° = 173.2$ miles.

33. Sketch a figure.

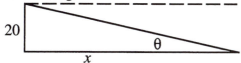

Note: θ is equal to the angle of depression, $7.4°$; x is measured in feet.
$\tan \theta = 20/x$; $x = 20/\tan \theta = 20/\tan 7.4° = 153.99$ feet

Then the rate of the car $= \dfrac{x \text{ feet}}{2 \text{ sec}} \cdot \dfrac{60 \text{ miles}}{\text{hour}} \div 88 \dfrac{\text{feet}}{\text{sec}} = \dfrac{153.99 \cdot 60}{2 \cdot 88} = 52.5$ mph.

35. a. Let $x =$ half-width of the rectangle, $y =$ height. Then Area $= A = 2xy$
Since $x/10 = \cos t$ and $y/10 = \sin t$, $x = 10 \cos t$ and $y = 10 \sin t$, hence
$A = 2(10 \sin t)(10 \cos t) = 200 \sin t \cos t$
b. Graph A using the window $0° \le t \le 90°$, $0 \le y \le 200$ and apply the maximum routine to find $t = 45°$. Then
$2x = 2 \cdot 10 \cos 45° = 14.14$ feet, $y = 10 \sin 45° = 7.07$ feet.

37. Abstracting the essential information, sketch a figure.

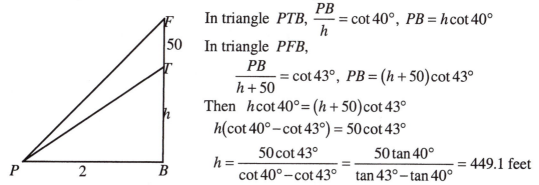

In triangle PTB, $\dfrac{PB}{h} = \cot 40°$, $PB = h \cot 40°$
In triangle PFB,
$\dfrac{PB}{h + 50} = \cot 43°$, $PB = (h + 50) \cot 43°$
Then $h \cot 40° = (h + 50) \cot 43°$
$h(\cot 40° - \cot 43°) = 50 \cot 43°$
$h = \dfrac{50 \cot 43°}{\cot 40° - \cot 43°} = \dfrac{50 \tan 40°}{\tan 43° - \tan 40°} = 449.1$ feet

8.3 The Law of Cosines

As noted in the text, all decimals are printed in rounded-off form for reading convention, but no rounding is done in the actual computation until the final answer is stored.

1. $a^2 = b^2 + c^2 - 2bc\cos A$; $\;a^2 = 10^2 + 7^2 - 2(10)(7)\cos 20°$; $\;a = 4.2$

$\cos B = \dfrac{a^2 + c^2 - b^2}{2ac}$; $\;B = 125.0°$; $\;C = 180° - A - B = 35.0°$

3. $c^2 = a^2 + b^2 - 2ab\cos C$; $\;c^2 = 6^2 + 10^2 - 2(6)(10)\cos 118°$; $\;c = 13.9$

$\cos B = \dfrac{a^2 + c^2 - b^2}{2ac}$; $\;B = 39.5°$; $\;A = 180° - B - C = 22.5°$

5. $a^2 = b^2 + c^2 - 2bc\cos A$; $\;a^2 = 12^2 + 14^2 - 2(12)(14)\cos 140°$; $\;a = 24.4$

$\cos B = \dfrac{a^2 + c^2 - b^2}{2ac}$; $\;B = 18.4°$; $\;C = 180° - A - B = 21.6°$

7. $c^2 = a^2 + b^2 - 2ab\cos C$; $\;c = 21.5$

$\cos B = \dfrac{a^2 + c^2 - b^2}{2ac}$; $\;B = 67.9°$; $\;A = 180° - B - C = 33.5°$

9. $\cos B = \dfrac{a^2 + c^2 - b^2}{2ac}$; $\;B = 21.8°$; $\;\cos C = \dfrac{a^2 + b^2 - c^2}{2ab}$; $\;C = 38.2°$

$A = 180° - B - C = 120°$

11. $\cos B = \dfrac{a^2 + c^2 - b^2}{2ac}$; $\;B = 30.8°$; $\;\cos C = \dfrac{a^2 + b^2 - c^2}{2ab}$; $\;C = 125.1°$

$A = 180° - B - C = 24.1°$

13. $\cos B = \dfrac{a^2 + c^2 - b^2}{2ac}$; $\;B = 34.5°$; $\;\cos C = \dfrac{a^2 + b^2 - c^2}{2ab}$; $\;C = 106.7°$

$A = 180° - B - C = 38.8°$

15. $\cos B = \dfrac{a^2 + c^2 - b^2}{2ac}$; $\;B = 50.5°$; $\;\cos C = \dfrac{a^2 + b^2 - c^2}{2ab}$; $\;C = 95.4°$

$A = 180° - B - C = 34.1°$

17. The sides of the triangle are given by the distance formula as:

$a = \sqrt{(5-0)^2 + (-2-0)^2} = \sqrt{29}$; $\;b = \sqrt{(1-0)^2 + (-4-0)^2} = \sqrt{17}$

$c = \sqrt{(1-5)^2 + [-4-(-2)]^2} = \sqrt{20}$

Then $\cos A = \dfrac{b^2 + c^2 - a^2}{2bc}$; $\;A = 77.5°$; $\;\cos B = \dfrac{a^2 + c^2 - b^2}{2ac}$; $\;B = 48.4°$

$C = 180° - A - B = 54.1°$

19. Regard this situation as a triangle with $a = 22 \cdot 3 = 66$ miles, $b = 31 \cdot 3 = 93$ miles and $C = 38°$. Then $c^2 = a^2 + b^2 - 2ab\cos C$, $c = 57.7$ miles.

21. The length of the shorter cable can be obtained by solving a right triangle.
$400/a = \sin 70°$

$a = 400/\sin 70° = 425.7$ feet.

Then the length c of the longer cable can be obtained by applying the Law of Cosines to the triangle with a as above, $b = 100$, and $C = 180° - 70° = 110°$.
$c^2 = a^2 + b^2 - 2ab\cos C$, $c = 469.4$ feet.

23. Apply the Law of Cosines to the triangle with $a = 60.5$, $b = 90$, and $C = 45°$.
$c^2 = a^2 + b^2 - 2ab\cos C$, $c = 63.7$ feet.

25. Apply the Alternate Law of Cosines to the triangle shown, with $a = b = 8$, $c = 10.8$.
$\cos C = \dfrac{a^2 + b^2 - c^2}{2ab}$, $C = 84.9°$

27. Regard this situation as a triangle with $a = 3$, $b = 6$, and $C = 180° - 45° = 135°$.
Then $c^2 = a^2 + b^2 - 2ab\cos C$, $c = 8.4$ km.

29. Apply the Law of Cosines to the triangle with $a = 120$, $c = 74$, and $B = 103°$.
$b^2 = a^2 + c^2 - 2ac\cos B$, $b = 154.5$ feet.

31. Sketch a figure.

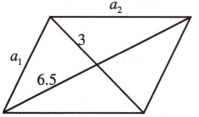

To find a_1 apply the Law of Cosines to the triangle with $b = 3, c = 6.5$, and $A_1 = 42°$. $a^2_1 = b^2 + c^2 - 2bc\cos A_1$, $a_1 = 4.7$ cm.
To find a_2 apply the Law of Cosines to the triangle with $b = 3, c = 6.5$, and $A_2 = 180° - 42° = 138°$. $a_2^2 = b^2 + c^2 - 2bc\cos A_2$, $a_2 = 9.0$ cm.

33. Apply the Law of Cosines to the triangle with $a = 350$, $b = 7 \cdot 18 = 126$ miles, and $C = 22°$. $c^2 = a^2 + b^2 - 2ab\cos C$, $c = 237.9$ miles. Now find the angle A of this triangle from the Alternate Law of Cosines. $\cos A = \dfrac{b^2 + c^2 - a^2}{2bc}$, $A = 146.6°$.
The ship should turn through $180° - A = 33.4°$.

35. Regard this situation as a triangle with A = the position of the satellite at 2:05 PM, B = the tracking station, and C = the center of the earth. Then, using standard notation,
$a = 3980$, $b = 900 + 3980 = 4880$, and angle $C = 5/360 \cdot 360° = 5°$.
$c^2 = a^2 + b^2 - 2ab\cos C$, $c = 978.7$ miles.

37. Clearly $a =$ distance from B to $C = 11.27 + 8.23 = 19.5$. Similarly
$b = 8.23 + 13 = 21.23$, $c = 11.27 + 13 = 24.27$. Apply the Alternate Law of Cosines.

$$\cos A = \frac{b^2 + c^2 - a^2}{2bc}, \ A = 50.2°, \ \cos B = \frac{a^2 + c^2 - b^2}{2ac}, \ B = 56.8°,$$
$$C = 180° - A - B = 73.0°$$

39. From the Pythagorean theorem, $AD = \sqrt{8^2 - 1^2} = \sqrt{63}$, $BE = \sqrt{7^2 - 1^2} = \sqrt{48}$.
Then $\sin ACD = AD/8$, angle $ACD = 82.8°$.
   $\sin BCE = BE/7$, angle $BCE = 81.8°$.
In triangle ABC, apply the Alternate Law of Cosines.

$$\cos ACB = \frac{8^2 + 7^2 - 9^2}{2 \cdot 8 \cdot 7}, \ \angle ACB = 73.4°$$

Then $\angle DCE = 360° - \angle ACD - \angle BCE - \angle ACB = 122°$. Arc $\overset{\frown}{DE}$ is then

$\frac{122}{360} \cdot 2\pi(1) = 2.1$ meters. The length of the rope $= AD + \overset{\frown}{DE} + BE = 17.0$ meters.

8.4 The Law of Sines

1. $b = \dfrac{a\sin B}{\sin A} = \dfrac{5\sin 22°}{\sin 48°} = 2.5$, $C = 180° - A - B = 110°$, $c = \dfrac{a\sin C}{\sin A} = \dfrac{5\sin 110°}{\sin 48°} = 6.3$

3. $B = 180° - A - C = 14°$, $b = \dfrac{a\sin B}{\sin A} = \dfrac{8\sin 14°}{\sin 116°} = 2.2$, $c = \dfrac{a\sin C}{\sin A} = \dfrac{8\sin 50°}{\sin 116°} = 6.8$

5. $A = 180° - B - C = 88°$, $a = \dfrac{b\sin A}{\sin B} = \dfrac{12\sin 88°}{\sin 44°} = 17.3$, $c = \dfrac{b\sin C}{\sin B} = \dfrac{12\sin 48°}{\sin 44°} = 12.8$

7. $C = 180° - A - B = 41.5°$, $b = \dfrac{a\sin B}{\sin A} = \dfrac{16\sin 36.2°}{\sin 102.3°} = 9.7$,

$c = \dfrac{a\sin C}{\sin A} = \dfrac{16\sin 41.5°}{\sin 102.3°} = 10.9$

9. This is the SSA case. $\sin C = \dfrac{c\sin B}{b} = \dfrac{25\sin 47°}{15} = 1.2189$

$\sin C > 1$ hence no such triangle exists.

11. This is the SSA case. $\sin A = \dfrac{a\sin B}{b} = \dfrac{12\sin 20°}{5} = .8208$

$A = 55.2°$ or $A = 180° - 55.2° = 124.8°$.
Case 1: $A = 55.2°$, $C = 104.8°$, $c = a\sin C/\sin A = 14.1$
Case 2: $A = 124.8°$, $C = 35.2°$, $c = a\sin C/\sin A = 8.4$

13. This is the SSA case. $\sin C = \dfrac{c \sin A}{a} = \dfrac{12 \sin 102°}{5} = 2.3476$

$\sin C > 1$ hence no such triangle exists.

15. This is the SSA case. $\sin B = \dfrac{b \sin C}{c} = \dfrac{11 \sin 56°}{10} = .9119$

$B = 65.8°$ or $B = 180° - 65.8° = 114.2°$.

Case 1: $B = 65.8°$, $A = 58.2°$, $a = \dfrac{b \sin A}{\sin B} = 10.3$

Case 2: $B = 114.2°$, $A = 9.8°$, $a = 2.1$

17. $b = \dfrac{a \sin B}{\sin A} = \dfrac{10.5 \sin 67°}{\sin 41°} = 14.7$, $C = 180° - A - B = 72°$,

$c = \dfrac{a \sin C}{\sin A} = \dfrac{10.5 \sin 72°}{\sin 41°} = 15.2$

19. $a^2 = b^2 + c^2 - 2bc \cos A$, $a = 9.8$, $\sin B = \dfrac{b \sin A}{a}$, $B = 23.3°$

$C = 180° - A - B = 81.7°$

21. $\cos A = \dfrac{b^2 + c^2 - a^2}{2bc}$, $A = 18.6°$, $\cos B = \dfrac{a^2 + c^2 - b^2}{2ac}$, $B = 39.6°$

$C = 180° - A - B = 121.8°$

23. $c^2 = a^2 + b^2 - 2ab \cos C$, $c = 13.9$, $\sin A = \dfrac{a \sin C}{c}$, $A = 60.1°$

$B = 180° - A - C = 72.9°$

25. This is the SSA case. $\sin C = \dfrac{c \sin B}{b} = \dfrac{12.4 \sin 62.5°}{17.2} = .6395$

$C = 39.8°$ or $C = 180° - 39.8° = 140.2°$.

Case 1: $C = 39.8°$, $A = 77.7°$, $a = b \sin A / \sin B = 18.9$

Case 2: $C = 140.2°$, impossible, because $B + C = 202.7° > 180°$

27. This is the SSA case. $\sin B = \dfrac{b \sin A}{a} = \dfrac{18.2 \sin 50.7°}{10.1} = 1.3944$

$\sin B > 1$ hence no such triangle exists.

29. $a^2 = b^2 + c^2 - 2bc \cos A$, $a = 18.5$, $\cos B = \dfrac{a^2 + c^2 - b^2}{2ac}$, $B = 36.7°$

$C = 180° - A - B = 78.3°$

31. $b = \dfrac{a \sin B}{\sin A} = \dfrac{110 \sin 35°}{\sin 19°} = 193.8$, $C = 180° - A - B = 126°$,

$c = \dfrac{a \sin C}{\sin A} = \dfrac{110 \sin 126°}{\sin 19°} = 273.3$

33. First determine angle C: $C = 180° - 57° - 42° = 81°$

Then distance $= b = \dfrac{c \sin B}{\sin C} = \dfrac{200 \sin 42°}{\sin 81°} = 135.5\,\text{meters}.$

35. Let A be the angle at the top of the tower. Then

$\sin A = \dfrac{40 \sin 57°}{54} = 2.5,\ A = 38.4°,\ B = 180° - 38.4° - 57° = 84.6°$

The required angle $\alpha = 90° - B = 5.4°.$

37. The third angle of the triangle $= 180° - 37.25° - 34.85° = 107.9°$. Clearly the vertical

side of the triangle $= 8 - 5 = 3$. Then $d = \dfrac{3 \sin 107.9°}{\sin 34.85°} = 5.0$ feet.

39. Sketch a figure.

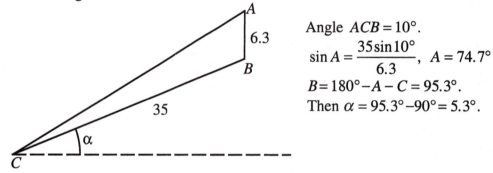

Angle $ACB = 10°.$

$\sin A = \dfrac{35 \sin 10°}{6.3},\ A = 74.7°$

$B = 180° - A - C = 95.3°.$

Then $\alpha = 95.3° - 90° = 5.3°.$

41. Call Harville vertex A, Eastview vertex B, Wellstone vertex C.

$\sin C = \dfrac{18 \sin 40°}{20},\ C = 35.3°,\ B = 180° - A - C = 104.7°.$

Required distance $= b = 20 \sin 104.7° / \sin 40° = 30.1$ km.

43. Label the text figure: A = Ray's position, B = Tom's position, C = plane's position, D = base of tower. Triangle ABD is a right triangle, hence $\cos 8.6° = 5280/c$, $c = 5340$ ft. (Or use the Pythagorean Theorem) $C = 180° - 81° - 67° = 32°$. $b = c \sin B / \sin C = 9761$ ft. The height of the plane is $\sin 81° = h/9761,\ h = 9642$ ft.

45. a. Solve triangle ABC to find $\angle BAC$, the angle at A. $180° - \angle BAC = \angle EAB$.
Solve triangle ABD to find $\angle ABD$, the angle at B. $180° - \angle ABD = \angle EBA$.
Now solve triangle EAB using the two angles and included side to find EA.
b. Follow the steps in part **a.**

$\cos BAC = \dfrac{25^2 + 75^2 - 80^2}{2(25)(75)},\ \angle BAC = 92.3°,\ \angle EAB = 87.7°.$

$\cos ABD = \dfrac{75^2 + 22^2 - 90^2}{2(75)(22)},\ \angle ABD = 127.1°,\ \angle EBA = 52.9°.$

$\angle AEB = 180° - 87.7° - 52.9° = 39.4°$

$AE = \dfrac{75 \sin 52.9°}{\sin 39.4°} = 94$ ft

47. Sketch a figure.

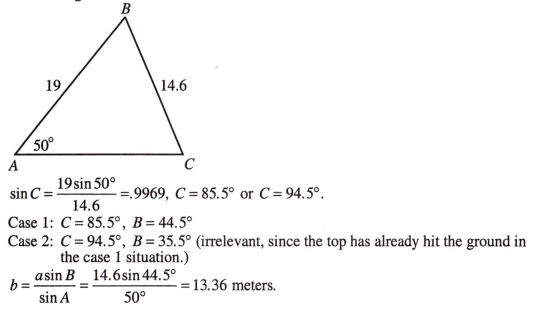

$\sin C = \dfrac{19\sin 50°}{14.6} = .9969,\ C = 85.5°\text{ or }C = 94.5°.$

Case 1: $C = 85.5°,\ B = 44.5°$

Case 2: $C = 94.5°,\ B = 35.5°$ (irrelevant, since the top has already hit the ground in the case 1 situation.)

$b = \dfrac{a\sin B}{\sin A} = \dfrac{14.6\sin 44.5°}{50°} = 13.36$ meters.

8.4A The Area of a Triangle

1. $\frac{1}{2}ab\sin C = \frac{1}{2}(4)(8)\sin 27° = 7.3$ square units

3. $\frac{1}{2}ab\sin C = \frac{1}{2}(10)(7)\sin 68° = 32.5$ square units

5. Use Heron's formula: $s = \frac{1}{2}(a+b+c) = 22$

$A = \sqrt{s(s-a)(s-b)(s-c)} = \sqrt{22(22-11)(22-15)(22-18)} = 82.3$ square units

7. Use Heron's formula: $s = \frac{1}{2}(a+b+c) = 13.5$

$A = \sqrt{s(s-a)(s-b)(s-c)} = \sqrt{13.5(13.5-7)(13.5-9)(13.5-11)} = 31.4$ square units

9. Use the distance formula to find side lengths of $\sqrt{29}, \sqrt{10}, \sqrt{61}$.

Then use Heron's formula: $s = \frac{1}{2}(a+b+c) = 8.1788$

$A = \sqrt{s(s-a)(s-b)(s-c)} = 6.5$ square units

11. As in the hint, separate the region into two triangular regions. If this is done by drawing a diagonal from the vertex with the 89° angle to the vertex with the 72° angle, then the areas of the two triangles are given by the formula Area $= \frac{1}{2}ab\sin C$. The entire area is therefore $\frac{1}{2}(55)(135)\sin 96° + \frac{1}{2}(120)(68.4)\sin 103 = 7691$ square units.

13. From the Law of Sines, the angle A between the 24-foot side and the third side is given by $\sin A = \dfrac{20\sin 44°}{24} = .5789$, $A = 35.3°$. The third angle is then given by $180° - (44° + A) = 100.6°$. The area is then $\frac{1}{2}(20)(24)\sin 100.6° = 235.9$ square feet.

15. Use Heron's formula to find the area of the deck: $s = \frac{1}{2}(65 + 72 + 88) = 112.5$.
$$A = \sqrt{s(s-a)(s-b)(s-c)} = \sqrt{112.5(112.5 - 65)(112.5 - 72)(112.5 - 88)}$$
$A = 2302.7$ square feet
Divide by 400 square feet per gallon to obtain 5.8 gallons.

17. Let A', B', C' be the points directly below A, B, C.
In right triangle $A'B'B$, $A'B = \sqrt{61}$, $\angle BA'B' = 39.806°$.
In triangle $BA'A$, $\angle BA'A = 90° - 39.806° = 50.194°$, $AB = 6.1$ from the Law of Cosines.
In right triangle $A'C'C$, $A'C = \sqrt{34}$, $\angle CA'C' = 30.964°$.
In triangle $CA'A$, $\angle CA'A = 90° - 30.964° = 59.036°$, $AC = 5.1$.
In right triangle $B'C'C$, $B'C = 5$, $\angle CB'C' = 36.870°$.
In triangle CBB', $\angle CBB' = 90° - 36.870° = 53.130°$, $BC = 4.5$.
Now use Heron's formula to find the area of triangle ABC, with sides $a = 4.5$, $b = 5.1$, $c = 6.1$, obtaining 11.18 square units.

19. Since $12 + 20 < 36$, this violates the Triangle Inequality that states that the sum of two sides of a triangle is greater than the third side. This is not a triangle and the area is undefined.

Chapter 8 Review Exercises

1. (d)   **3.** Since the hypotenuse is $\sqrt{65}$, **(e)** is true.

5. $\tan A = 12/13$
$\quad\quad A = \tan^{-1} 12/13 = 42.7°$, $C = 90° - A = 47.3°$
$\quad\quad b = \sqrt{12^2 + 13^2} = 17.7$

7. $\angle A = 90° - \angle C = 90° - 35° = 55°$
$\quad\quad \cos C = \dfrac{a}{b}$, $\cos 35° = \dfrac{12}{b}$, $b = \dfrac{12}{\cos 35°} = 14.65$

$\quad\quad \tan C = \dfrac{c}{a}$, $\tan 35° = \dfrac{c}{12}$, $c = 12\tan 35° = 8.40$

9. Let h be the height of the tower.
$\quad\quad \tan 57.3° = \dfrac{h}{145}$, $h = 145\tan 57.3° = 225.9$ feet

11. $\tan\theta = \dfrac{140}{5280}, \theta = \tan^{-1}\dfrac{140}{5280} = 1.52°$

13. $\cos A = \dfrac{b^2 + c^2 - a^2}{2bc}; A = 52.9°; \cos B = \dfrac{a^2 + c^2 - b^2}{2ac}; B = 41.6°$
$C = 180° - A - B = 85.5°$

15. $b^2 = a^2 + c^2 - 2ac\cos B; b = 21.8; \cos A = \dfrac{b^2 + c^2 - a^2}{2bc}; A = 20.6°$
$C = 180° - A - B = 29.4°$

17. Let c be the distance between the trains after 3 hours.
Train 1 has gone $45 \cdot 3 = 135$ miles $= a$.
Train 2 has gone $70 \cdot 3 = 210$ miles $= b$.
$c^2 = 135^2 + 210^2 - 2 \cdot 135 \cdot 210\cos120° = 301$ miles

19. $A = 180° - B - C = 25°, a = c\sin A / \sin C = 2.9, b = c\sin B / \sin C = 5.6$

21. This is the SSA case.
$\sin A = \dfrac{a\sin C}{c} = \dfrac{75\sin62°}{84} = .7883$
$A = 52.0°$ or $A = 180° - 52.0° = 128.0°$
Case 1: $A = 52.0°, B = 66.0°, b = c\sin B / \sin C = 86.9$
Case 2: is impossible because $A + C = 190° > 180°$.

23. This is the SSA case.
$\sin B = \dfrac{b\sin A}{a} = \dfrac{4\sin60°}{3.5} = .9897$
$B = 81.8°$ or $B = 180° - 81.8° = 98.2°$
Case 1: $B = 81.8°, C = 38.2°, c = b\sin C / \sin B = 2.5$
Case 2: $B = 98.2°, C = 21.8°, c = b\sin C / \sin B = 1.5$

25. $\frac{1}{2}bc\sin A = \frac{1}{2}(24)(15)\sin55° = 147.4$ square units.

27. Sketch a figure.

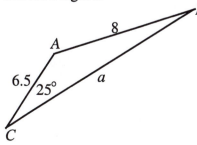

$\sin B = \dfrac{6.5\sin25°}{8}, B = 20.1°,$

$A = 134.9°, a = \dfrac{8\sin134.9°}{\sin25°} = 13.4$ km.

(The other value for B is too large for this triangle.)

29. Use the Law of Sines.

$C = 180° - A - B = 75°$, $a = b\sin A/\sin B = 41.6$, $c = b\sin C/\sin B = 54.1$

31. Use the Law of Cosines.

$b^2 = a^2 + c^2 - 2ac\cos B$; $b = 8.3$; $\cos A = \dfrac{b^2 + c^2 - a^2}{2bc}$; $A = 35.5°$

$C = 180° - A - B = 68.5°$

33. Let Alice be vertex A, Joe vertex B, and the flagpole vertex C. Thus $\angle C = 63°$.

$a = \dfrac{240\sin 54°}{\sin 63°} = 217.9$ meters is the distance from Joe to the flagpole.

$b = \dfrac{240\sin 63°}{\sin 63°} = 240$ meters is the distance from Alice to the flagpole.

35. Sketch a figure.

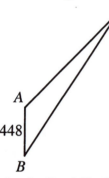

Angle $C = 65° - 62° = 3°$. Angle $B = 90° - 65° = 25°$. Angle $A = 152°$.

a. $b = \dfrac{448\sin 25°}{\sin 3°} = 3618$ feet.   **b.** $a = \dfrac{448\sin 152°}{\sin 3°} = 4019$ feet.

c. $\sin 65° = h/4019$, $h = 3642$ feet.

37. Use the Pythagorean Theorem to obtain the sides of triangle ABC.

$a = \sqrt{18^2 + 10^2} = \sqrt{424}$

$b = \sqrt{18^2 + 12^2} = \sqrt{468}$

$c = \sqrt{10^2 + 12^2} = \sqrt{244}$

Then angle $ABC = \cos^{-1}\left(\dfrac{a^2 + c^2 - b^2}{2ac}\right) = \cos^{-1}\left(\dfrac{424 + 244 - 468}{2\sqrt{424}\sqrt{244}}\right) = 71.89°$.

Therefore, $\dfrac{a+b}{c} = \dfrac{\sin A + \sin B}{\sin C}$ and $\dfrac{a-b}{c} = \dfrac{\sin A - \sin B}{\sin C}$.

39. $A = \tfrac{1}{2}(5)(8)\sin 30° = 10$ square units

41. Use Heron's formula: $s = \tfrac{1}{2}(7 + 11 + 14) = 16$

$A = \sqrt{16(16 - 7)(16 - 11)(16 - 14)} = 37.95$ square units

Chapter 9
Applications of Trigonometry

9.1 The Complex Plane and Polar Form for Complex Numbers

1-7. (graphs)

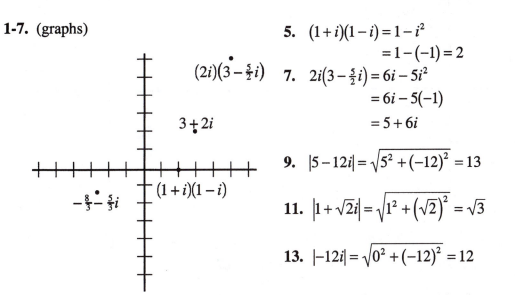

5. $(1+i)(1-i) = 1 - i^2$
$$= 1 - (-1) = 2$$

7. $2i\left(3 - \frac{5}{2}i\right) = 6i - 5i^2$
$$= 6i - 5(-1)$$
$$= 5 + 6i$$

9. $|5 - 12i| = \sqrt{5^2 + (-12)^2} = 13$

11. $\left|1 + \sqrt{2}i\right| = \sqrt{1^2 + \left(\sqrt{2}\right)^2} = \sqrt{3}$

13. $|-12i| = \sqrt{0^2 + (-12)^2} = 12$

15. Let $z = 1$, $w = i$. Then $|z + w| = |1 + i| = \sqrt{2}$ but $|z| + |w| = |1| + |i| = 1 + 1 = 2$.

17.

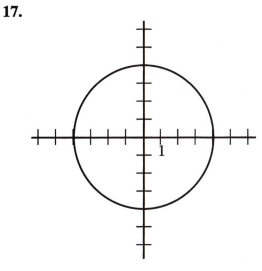

19. The graph consists of all points that are 10 units away from $(1, 0)$.

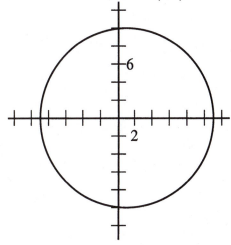

21. The graph consists of all points that are 4 units away from $2i$ or $(0,2)$.

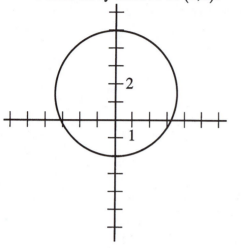

23. The graph consists of all points with x-coordinate equal to 2.

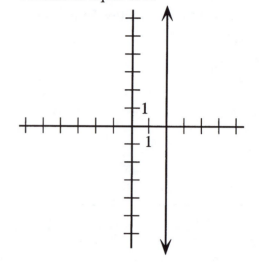

25. $r = \sqrt{3^2 + 4^2} = 5$, $\tan\theta = 4/3$, z lies in the first quadrant, hence $\theta = \tan^{-1}4/3 = .9273$.

$3 + 4i = 5(\cos\theta + i\sin\theta) = 5(\cos(.9273) + i\sin(.9273))$

27. $r = \sqrt{5^2 + (-12)^2} = 13$, $\tan\theta = -12/5$, z lies in the fourth quadrant, hence $\theta = 2\pi + \tan^{-1}(-12/5) = 5.1072$.

$5 - 12i = 13(\cos\theta + i\sin\theta) = 13(\cos(5.1072) + i\sin(5.1072))$

29. $r = \sqrt{1^2 + 2^2} = \sqrt{5}$, $\tan\theta = 2/1 = 2$, z lies in the first quadrant, hence $\theta = \tan^{-1}2 = 1.1071$.

$1 + 2i = \sqrt{5}(\cos\theta + i\sin\theta) = \sqrt{5}(\cos(1.1071) + i\sin(1.1071))$

31. $r = \sqrt{\left(-\dfrac{5}{2}\right)^2 + \left(\dfrac{7}{2}\right)^2} = \sqrt{\dfrac{74}{4}}$, $\tan\theta = \dfrac{7}{2} \div \left(-\dfrac{5}{2}\right) = -\dfrac{7}{5}$, z lies in the second quadrant,

hence $\theta = \pi + \tan^{-1}(-7/5) = 2.191$.

$-\dfrac{5}{2} + \dfrac{7}{2}i = \dfrac{\sqrt{74}}{2}(\cos\theta + i\sin\theta) = \dfrac{\sqrt{74}}{2}(\cos(2.191) + i\sin(2.191))$

33. $\left(\cos\dfrac{\pi}{12} + i\sin\dfrac{\pi}{12}\right) \cdot 2\left(\cos\dfrac{7\pi}{12} + i\sin\dfrac{7\pi}{12}\right) = 2\left(\cos\dfrac{8\pi}{12} + i\sin\dfrac{8\pi}{12}\right)$

$= 2\left(\cos\dfrac{2\pi}{3} + i\sin\dfrac{2\pi}{3}\right)$

$= 2\left(-\dfrac{1}{2} + i\dfrac{\sqrt{3}}{2}\right)$

$= -1 + i\sqrt{3}$

35. $12\left(\cos\dfrac{11\pi}{12}+i\sin\dfrac{11\pi}{12}\right)\cdot\dfrac{7}{2}\left(\cos\dfrac{\pi}{4}+i\sin\dfrac{\pi}{4}\right)=42\left(\cos\dfrac{14\pi}{12}+i\sin\dfrac{14\pi}{12}\right)$

$$=42\left(\cos\dfrac{7\pi}{6}+i\sin\dfrac{7\pi}{6}\right)$$

$$=42\left(-\dfrac{\sqrt{3}}{2}-\dfrac{1}{2}i\right)$$

$$=-21\sqrt{3}-21i$$

37. $\dfrac{6\left(\cos\dfrac{7\pi}{20}+i\sin\dfrac{7\pi}{20}\right)}{4\left(\cos\dfrac{\pi}{10}+i\sin\dfrac{\pi}{10}\right)}=\dfrac{6}{4}\left(\cos\dfrac{5\pi}{20}+i\sin\dfrac{5\pi}{20}\right)$

$$=\dfrac{3}{2}\left(\cos\dfrac{\pi}{4}+i\sin\dfrac{\pi}{4}\right)$$

$$=\dfrac{3}{2}\left(\dfrac{\sqrt{2}}{2}+i\dfrac{\sqrt{2}}{2}\right)$$

$$=\dfrac{3\sqrt{2}}{4}+\dfrac{3\sqrt{2}}{4}i$$

39. $(1+i)\left(1+\sqrt{3}i\right)=\sqrt{2}\left(\cos\dfrac{\pi}{4}+i\sin\dfrac{\pi}{4}\right)2\left(\cos\dfrac{\pi}{3}+i\sin\dfrac{\pi}{3}\right)=2\sqrt{2}\left(\cos\dfrac{7\pi}{12}+i\sin\dfrac{7\pi}{12}\right)$

41. $\dfrac{1+i}{1-i}=\dfrac{\sqrt{2}\left(\cos\dfrac{\pi}{4}+i\sin\dfrac{\pi}{4}\right)}{\sqrt{2}\left(\cos\left(-\dfrac{\pi}{4}\right)+i\sin\left(-\dfrac{\pi}{4}\right)\right)}=\cos\dfrac{\pi}{2}+i\sin\dfrac{\pi}{2}$

43. $3i\left(2\sqrt{3}+2i\right)=3\left(\cos\dfrac{\pi}{2}+i\sin\dfrac{\pi}{2}\right)4\left(\cos\dfrac{\pi}{6}+i\sin\dfrac{\pi}{6}\right)=12\left(\cos\dfrac{2\pi}{3}+i\sin\dfrac{2\pi}{3}\right)$

45. $i(i+1)\left(-\sqrt{3}+i\right)=\left(\cos\dfrac{\pi}{2}+i\sin\dfrac{\pi}{2}\right)\sqrt{2}\left(\cos\dfrac{\pi}{4}+i\sin\dfrac{\pi}{4}\right)2\left(\cos\dfrac{5\pi}{6}+i\sin\dfrac{5\pi}{6}\right)$

$$=2\sqrt{2}\left(\cos\left(\dfrac{\pi}{2}+\dfrac{\pi}{4}+\dfrac{5\pi}{6}\right)+i\sin\left(\dfrac{\pi}{2}+\dfrac{\pi}{4}+\dfrac{5\pi}{6}\right)\right)$$

$$=2\sqrt{2}\left(\cos\dfrac{19\pi}{12}+i\sin\dfrac{19\pi}{12}\right)$$

47. Since $z=r(\cos\theta+i\sin\theta)$, $i=\cos\dfrac{\pi}{2}+i\sin\dfrac{\pi}{2}$, $iz=r\left(\cos\left(\theta+\dfrac{\pi}{2}\right)+i\sin\left(\theta+\dfrac{\pi}{2}\right)\right)$

the argument of iz is 90° more than that of z, hence iz is equivalent to rotating z through 90° in the complex plane.

49. a. $\text{slope} = \dfrac{b-0}{a-0} = \dfrac{b}{a}$

b. $\text{slope} = \dfrac{d-0}{c-0} = \dfrac{d}{c}$

c. Line L has slope $= \dfrac{d}{c}$; from the point-slope form, the equation is $y - b = \dfrac{d}{c}(x-a)$.

d. Line M has slope $= \dfrac{b}{a}$; from the point-slope form, the equation is $y - d = \dfrac{b}{a}(x-c)$.

e.

f. Substituting $(a+c, b+d)$ in the equation of line L yields:

$$y - b = \dfrac{d}{c}(x-a)$$

$$b + d - b = \dfrac{d}{c}(a+c-a)$$

$$d = d$$

Substituting $(a+c, b+d)$ in the equation of line M yields:

$$y - d = \dfrac{b}{a}(x-c)$$

$$b + d - d = \dfrac{b}{a}(a+c-c)$$

$$b = b$$

Thus $(a+c, b+d)$ lies on both lines L and M. The conclusion follows as stated in the text.

51. a. $r_2\left(\cos\theta_2 + i\sin\theta_2\right)\left(\cos\theta_2 - i\sin\theta_2\right) = r_2\left(\cos^2\theta_2 + \sin^2\theta_2\right) = r_2 \cdot 1 = r_2$

b. $r_1\left(\cos\theta_1 + i\sin\theta_1\right)\left(\cos\theta_2 - i\sin\theta_2\right)$

$$= r_1\left(\cos\theta_1\cos\theta_2 - i\cos\theta_1\sin\theta_2 + i\sin\theta_1\cos\theta_2 - i^2\sin\theta_1\sin\theta_2\right)$$

$$= r_1\left(\cos\theta_1\cos\theta_2 + \sin\theta_1\sin\theta_2 + i\left(\sin\theta_1\cos\theta_2 - \cos\theta_1\sin\theta_2\right)\right)$$

$$= r_1\left(\cos\left(\theta_1 - \theta_2\right) + i\sin\left(\theta_1 - \theta_2\right)\right)$$

9.2 DeMoivre's Theorem and *n*th Roots of Complex Numbers

1. $\left(\cos\dfrac{\pi}{12}+i\sin\dfrac{\pi}{12}\right)^{6}=\cos\dfrac{6\pi}{12}+i\sin\dfrac{6\pi}{12}=\cos\dfrac{\pi}{2}+i\sin\dfrac{\pi}{2}=0+1i$ or i

3. $\left[3\left(\cos\dfrac{7\pi}{30}+i\sin\dfrac{7\pi}{30}\right)\right]^{5}=3^{5}\left(\cos\dfrac{35\pi}{30}+i\sin\dfrac{35\pi}{30}\right)=243\left(\cos\dfrac{7\pi}{6}+i\sin\dfrac{7\pi}{6}\right)$

$$=243\left(-\dfrac{\sqrt{3}}{2}-i\dfrac{1}{2}\right)=-\dfrac{243\sqrt{3}}{2}-\dfrac{243}{2}i$$

5. $(1-i)^{12}=\left[\sqrt{2}\left(\cos\dfrac{7\pi}{4}+i\sin\dfrac{7\pi}{4}\right)\right]^{12}=\left(\sqrt{2}\right)^{12}\left(\cos\dfrac{84\pi}{4}+i\sin\dfrac{84\pi}{4}\right)$

$$=64(\cos 21\pi+i\sin 21\pi)=64(-1+0i)=-64+0i \text{ or } -64$$

7. $\left(\dfrac{\sqrt{3}}{2}+\dfrac{1}{2}i\right)^{10}=\left(\cos\dfrac{\pi}{6}+i\sin\dfrac{\pi}{6}\right)^{10}=\cos\dfrac{10\pi}{6}+i\sin\dfrac{10\pi}{6}=\cos\dfrac{5\pi}{3}+i\sin\dfrac{5\pi}{3}=\dfrac{1}{2}-i\dfrac{\sqrt{3}}{2}$

9. $\left(-\dfrac{1}{\sqrt{2}}+\dfrac{i}{\sqrt{2}}\right)^{14}=\left(\cos\dfrac{3\pi}{4}+i\sin\dfrac{3\pi}{4}\right)^{14}=\cos\dfrac{42\pi}{4}+i\sin\dfrac{42\pi}{4}$

$$=\cos\dfrac{21\pi}{2}+i\sin\dfrac{21\pi}{2}=0+i1 \text{ or } i$$

11. Apply the roots-of-unity formula with $n=4,\ k=0,1,2,3$.

$k=0:\ \cos 0+i\sin 0=1$

$k=1:\ \cos\dfrac{2\pi}{4}+i\sin\dfrac{2\pi}{4}=i$

$k=2:\ \cos\dfrac{4\pi}{4}+i\sin\dfrac{4\pi}{4}=-1$

$k=3:\ \cos\dfrac{6\pi}{4}+i\sin\dfrac{6\pi}{4}=-i$

13. Apply the root formula with $n=3,\ k=0,1,2$.

$k=0:\ \sqrt[3]{64}\left[\cos\left(\dfrac{\pi/5}{3}\right)+i\sin\left(\dfrac{\pi/5}{3}\right)\right]=4\left(\cos\dfrac{\pi}{15}+i\sin\dfrac{\pi}{15}\right)$

$k=1:\ \sqrt[3]{64}\left[\cos\left(\dfrac{\pi/5+2\pi}{3}\right)+i\sin\left(\dfrac{\pi/5+2\pi}{3}\right)\right]=4\left(\cos\dfrac{11\pi}{15}+i\sin\dfrac{11\pi}{15}\right)$

$k=2:\ \sqrt[3]{64}\left[\cos\left(\dfrac{\pi/5+4\pi}{3}\right)+i\sin\left(\dfrac{\pi/5+4\pi}{3}\right)\right]=4\left(\cos\dfrac{7\pi}{5}+i\sin\dfrac{7\pi}{5}\right)$

15. Apply the root formula with $n = 4,\ k = 0,1,2,3$.

$$k = 0:\ \sqrt[4]{81}\left[\cos\left(\frac{\pi/12}{4}\right) + i\sin\left(\frac{\pi/12}{4}\right)\right] = 3\left(\cos\frac{\pi}{48} + i\sin\frac{\pi}{48}\right)$$

$$k = 1:\ \sqrt[4]{81}\left[\cos\left(\frac{\pi/12 + 2\pi}{4}\right) + i\sin\left(\frac{\pi/12 + 2\pi}{4}\right)\right] = 3\left(\cos\frac{25\pi}{48} + i\sin\frac{25\pi}{48}\right)$$

$$k = 2:\ \sqrt[4]{81}\left[\cos\left(\frac{\pi/12 + 4\pi}{4}\right) + i\sin\left(\frac{\pi/12 + 4\pi}{4}\right)\right] = 3\left(\cos\frac{49\pi}{48} + i\sin\frac{49\pi}{48}\right)$$

$$k = 3:\ \sqrt[4]{81}\left[\cos\left(\frac{\pi/12 + 6\pi}{4}\right) + i\sin\left(\frac{\pi/12 + 6\pi}{4}\right)\right] = 3\left(\cos\frac{73\pi}{48} + i\sin\frac{73\pi}{48}\right)$$

17. $-1 = 1(\cos\pi + i\sin\pi)$. Apply the root formula with $n = 5,\ k = 0,1,2,3,4$.

$$k = 0:\ \sqrt[5]{1}\left[\cos\frac{\pi}{5} + i\sin\frac{\pi}{5}\right] = \cos\frac{\pi}{5} + i\sin\frac{\pi}{5}$$

$$k = 1:\ \sqrt[5]{1}\left[\cos\left(\frac{\pi + 2\pi}{5}\right) + i\sin\left(\frac{\pi + 2\pi}{5}\right)\right] = \cos\frac{3\pi}{5} + i\sin\frac{3\pi}{5}$$

$$k = 2:\ \sqrt[5]{1}\left[\cos\left(\frac{\pi + 4\pi}{5}\right) + i\sin\left(\frac{\pi + 4\pi}{5}\right)\right] = \cos\pi + i\sin\pi$$

$$k = 3:\ \sqrt[5]{1}\left[\cos\left(\frac{\pi + 6\pi}{5}\right) + i\sin\left(\frac{\pi + 6\pi}{5}\right)\right] = \cos\frac{7\pi}{5} + i\sin\frac{7\pi}{5}$$

$$k = 4:\ \sqrt[5]{1}\left[\cos\left(\frac{\pi + 8\pi}{5}\right) + i\sin\left(\frac{\pi + 8\pi}{5}\right)\right] = \cos\frac{9\pi}{5} + i\sin\frac{9\pi}{5}$$

19. $i = 1\left(\cos\frac{\pi}{2} + i\sin\frac{\pi}{2}\right)$. Apply the root formula with $n = 5,\ k = 0,1,2,3,4$.

$$k = 0:\ \sqrt[5]{1}\left[\cos\left(\frac{\pi/2}{5}\right) + i\sin\left(\frac{\pi/2}{5}\right)\right] = \cos\frac{\pi}{10} + i\sin\frac{\pi}{10}$$

$$k = 1:\ \sqrt[5]{1}\left[\cos\left(\frac{\pi/2 + 2\pi}{5}\right) + i\sin\left(\frac{\pi/2 + 2\pi}{5}\right)\right] = \cos\frac{\pi}{2} + i\sin\frac{\pi}{2}$$

$$k = 2:\ \sqrt[5]{1}\left[\cos\left(\frac{\pi/2 + 4\pi}{5}\right) + i\sin\left(\frac{\pi/2 + 4\pi}{5}\right)\right] = \cos\frac{9\pi}{10} + i\sin\frac{9\pi}{10}$$

$$k = 3:\ \sqrt[5]{1}\left[\cos\left(\frac{\pi/2 + 6\pi}{5}\right) + i\sin\left(\frac{\pi/2 + 6\pi}{5}\right)\right] = \cos\frac{13\pi}{10} + i\sin\frac{13\pi}{10}$$

$$k = 4:\ \sqrt[5]{1}\left[\cos\left(\frac{\pi/2 + 8\pi}{5}\right) + i\sin\left(\frac{\pi/2 + 8\pi}{5}\right)\right] = \cos\frac{17\pi}{10} + i\sin\frac{17\pi}{10}$$

21. $1 + i = \sqrt{2}(\cos \pi/4 + i \sin \pi/4)$. Apply the root formula with $n = 2$, $k = 0,1$.

$k = 0$: $\sqrt{\sqrt{2}}\left[\cos\left(\dfrac{\pi/4}{2}\right) + i\sin\left(\dfrac{\pi/4}{2}\right)\right] = \sqrt[4]{2}\left(\cos\dfrac{\pi}{8} + i\sin\dfrac{\pi}{8}\right)$

$k = 1$: $\sqrt{\sqrt{2}}\left[\cos\left(\dfrac{\pi/4 + 2\pi}{2}\right) + i\sin\left(\dfrac{\pi/4 + 2\pi}{2}\right)\right] = \sqrt[4]{2}\left(\cos\dfrac{9\pi}{8} + i\sin\dfrac{9\pi}{8}\right)$

23. The solutions are the six sixth roots of -1. Write $-1 = \cos\pi + i\sin\pi$ and apply the root formula to obtain

$\cos\dfrac{\pi}{6} + i\sin\dfrac{\pi}{6} = \dfrac{\sqrt{3}}{2} + i\dfrac{1}{2}$ $\qquad$ $\cos\dfrac{7\pi}{6} + i\sin\dfrac{7\pi}{6} = -\dfrac{\sqrt{3}}{2} - i\dfrac{1}{2}$

$\cos\dfrac{\pi}{2} + i\sin\dfrac{\pi}{2} = i$ $\qquad$ $\cos\dfrac{3\pi}{2} + i\sin\dfrac{3\pi}{2} = -i$

$\cos\dfrac{5\pi}{6} + i\sin\dfrac{5\pi}{6} = -\dfrac{\sqrt{3}}{2} + i\dfrac{1}{2}$ $\qquad$ $\cos\dfrac{11\pi}{6} + i\sin\dfrac{11\pi}{6} = \dfrac{\sqrt{3}}{2} - i\dfrac{1}{2}$

25. The solutions are the three cube roots of i. Write $i = \cos\dfrac{\pi}{2} + i\sin\dfrac{\pi}{2}$ and apply the root formula to obtain

$\cos\dfrac{\pi}{6} + i\sin\dfrac{\pi}{6} = \dfrac{\sqrt{3}}{2} + i\dfrac{1}{2}$

$\cos\dfrac{5\pi}{6} + i\sin\dfrac{5\pi}{6} = -\dfrac{\sqrt{3}}{2} + i\dfrac{1}{2}$

$\cos\dfrac{3\pi}{2} + i\sin\dfrac{3\pi}{2} = -i$

27. The solutions are the three cube roots of $-27i$. Write

$-27i = 27\left(\cos\dfrac{3\pi}{2} + i\sin\dfrac{3\pi}{2}\right)$

and apply the root formula to obtain

$3\left(\cos\dfrac{\pi}{2} + i\sin\dfrac{\pi}{2}\right) = 3i$

$3\left(\cos\dfrac{7\pi}{6} + i\sin\dfrac{7\pi}{6}\right) = -\dfrac{3\sqrt{3}}{2} - \dfrac{3i}{2}$

$3\left(\cos\dfrac{11\pi}{6} + i\sin\dfrac{11\pi}{6}\right) = \dfrac{3\sqrt{3}}{2} - \dfrac{3i}{2}$

29. The solutions are the four fourth roots of $-1 + \sqrt{3}i$. Write

$-1 + \sqrt{3}i = 2\left(\cos\dfrac{2\pi}{3} + i\sin\dfrac{2\pi}{3}\right)$ and apply the root formula to obtain

$\sqrt[4]{2}\left(\cos\dfrac{\pi}{6} + i\sin\dfrac{\pi}{6}\right) = \sqrt[4]{2}\left(\dfrac{\sqrt{3}}{2} + i\dfrac{1}{2}\right)$

$\sqrt[4]{2}\left(\cos\dfrac{2\pi}{3} + i\sin\dfrac{2\pi}{3}\right) = \sqrt[4]{2}\left(-\dfrac{1}{2} + i\dfrac{\sqrt{3}}{2}\right)$

$\sqrt[4]{2}\left(\cos\dfrac{7\pi}{6} + i\sin\dfrac{7\pi}{6}\right) = \sqrt[4]{2}\left(-\dfrac{\sqrt{3}}{2} - i\dfrac{1}{2}\right)$

$\sqrt[4]{2}\left(\cos\dfrac{5\pi}{3} + i\sin\dfrac{5\pi}{3}\right) = \sqrt[4]{2}\left(\dfrac{1}{2} - i\dfrac{\sqrt{3}}{2}\right)$

31. With calculator in parametric mode, set the values $0 \leq t \leq 2\pi$, t-step $\approx .067$, $-1.5 \leq x \leq 1.5$, $-1 \leq y \leq 1$, and graph the unit circle $x = \cos t$, $y = \sin t$. Then reset the t-step to $2\pi / 7$ and graph again.

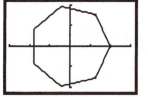

Using the trace feature, find the seven vertices to be 1, $.6235 \pm .7818i$, $-.2225 \pm .9749i$, $-.9010 \pm .4339i$.

33. With calculator in parametric mode, set the values $0 \leq t \leq 2\pi$, t-step $\approx .067$, $-1.5 \leq x \leq 1.5$, $-1 \leq y \leq 1$, and graph the unit circle $x = \cos t$, $y = \sin t$. Then reset the t-step to $2\pi / 8 = \pi / 4$ and graph again.

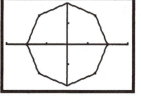

Using the trace feature, find the eight vertices to be ± 1, $\pm i$ $.7071 \pm .7071i$, $-.7071 \pm .7071i$.

35. With calculator in parametric mode, set the values $0 \leq t \leq 2\pi$, t-step $\approx .067$, $-1.5 \leq x \leq 1.5$, $-1 \leq y \leq 1$, and graph the unit circle $x = \cos t$, $y = \sin t$. Then reset the t-step to $2\pi / 9$ and graph again.

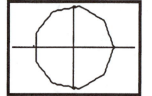

Using the trace feature, find the nine vertices to be 1, $.7660 \pm .6428i$, $.1736 \pm .9848i$, $-.5 \pm .8660i$, $-.9397 \pm .3420i$.

37. As suggested, the solutions are the solutions of $x^6 - 1 = 0$ with the exception of 1 itself, that is : $\dfrac{-1 \pm i\sqrt{3}}{2}$, $\dfrac{1 \pm i\sqrt{3}}{2}$, and -1.

39. Since the argument of v is 12 times the argument of w, when w has increased by 2π, v has increased by 24π, making 12 trips around the circle.

41. Since the u_i are distinct, clearly, the vu_i are distinct. Since v is a solution of $z^n = r(\cos\theta + i\sin\theta)$, we can write $v^n = r(\cos\theta + i\sin\theta)$ and similarly $u_i{}^n = 1$. Then $(vu_i)^n = v^n u_i{}^n = r(\cos\theta + i\sin\theta) \cdot 1 = r(\cos\theta + i\sin\theta)$.

Thus the vu_i are the solutions of the equation.

9.3 Vectors in the Plane

1. $\vec{PQ} = \langle 5-2, 9-3 \rangle = \langle 3,6 \rangle$; $\|\langle 3,6 \rangle\| = \sqrt{3^2 + 6^2} = 3\sqrt{5}$

3. $\vec{PQ} = \langle -4-(-7), -5-0 \rangle = \langle 3,-5 \rangle$; $\|\langle 3,-5 \rangle\| = \sqrt{3^2 + (-5)^2} = \sqrt{34}$

5. $\vec{PQ} = \langle 7-1, 11-5 \rangle = \langle 6,6 \rangle$ **7.** $\vec{PQ} = \langle -10-(-4), 2-(-8) \rangle = \langle -6,10 \rangle$

9. $\vec{PQ} = \langle \frac{17}{5} - \frac{4}{5}, -\frac{12}{5} - (-2) \rangle = \langle \frac{13}{5}, -\frac{2}{5} \rangle$

11. $\mathbf{u} + \mathbf{v} = \langle -2,4 \rangle + \langle 6,1 \rangle = \langle 4,5 \rangle$
$\mathbf{u} - \mathbf{v} = \langle -2,4 \rangle - \langle 6,1 \rangle = \langle -2-6, 4-1 \rangle = \langle -8,3 \rangle$
$3\mathbf{u} - 2\mathbf{v} = 3\langle -2,4 \rangle - 2\langle 6,1 \rangle = \langle -6,12 \rangle - \langle 12,2 \rangle = \langle -18,10 \rangle$

13. $\mathbf{u} + \mathbf{v} = \langle 3, 3\sqrt{2} \rangle + \langle 4\sqrt{2}, 1 \rangle = \langle 3+4\sqrt{2}, 3\sqrt{2}+1 \rangle$
$\mathbf{u} - \mathbf{v} = \langle 3, 3\sqrt{2} \rangle - \langle 4\sqrt{2}, 1 \rangle = \langle 3-4\sqrt{2}, 3\sqrt{2}-1 \rangle$
$3\mathbf{u} - 2\mathbf{v} = 3\langle 3, 3\sqrt{2} \rangle - 2\langle 4\sqrt{2}, 1 \rangle = \langle 9, 9\sqrt{2} \rangle - \langle 8\sqrt{2}, 2 \rangle = \langle 9-8\sqrt{2}, 9\sqrt{2}-2 \rangle$

15. $\mathbf{u} + \mathbf{v} = 2\langle -2,5 \rangle + \frac{1}{4}\langle -7,12 \rangle = \langle -4,10 \rangle + \langle -\frac{7}{4}, 3 \rangle = \langle -\frac{23}{4}, 13 \rangle$
$\mathbf{u} - \mathbf{v} = 2\langle -2,5 \rangle - \frac{1}{4}\langle -7,12 \rangle = \langle -4,10 \rangle - \langle -\frac{7}{4}, 3 \rangle = \langle -\frac{9}{4}, 7 \rangle$
$3\mathbf{u} - 2\mathbf{v} = 3\langle -4,10 \rangle - 2\langle -\frac{7}{4}, 3 \rangle = \langle -12,30 \rangle - \langle -\frac{7}{2}, 6 \rangle = \langle -\frac{17}{2}, 24 \rangle$

17. $\mathbf{u} + \mathbf{v} = 8\mathbf{i} + 2(3\mathbf{i} - 2\mathbf{j}) = 8\mathbf{i} + 6\mathbf{i} - 4\mathbf{j} = 14\mathbf{i} - 4\mathbf{j}$
$\mathbf{u} - \mathbf{v} = 8\mathbf{i} - 2(3\mathbf{i} - 2\mathbf{j}) = 8\mathbf{i} - 6\mathbf{i} + 4\mathbf{j} = 2\mathbf{i} + 4\mathbf{j}$
$3\mathbf{u} - 2\mathbf{v} = 3(8\mathbf{i}) - 2 \cdot 2(3\mathbf{i} - 2\mathbf{j}) = 24\mathbf{i} - 12\mathbf{i} + 8\mathbf{j} = 12\mathbf{i} + 8\mathbf{j}$

19. $\mathbf{u} + \mathbf{v} = -(2\mathbf{i} + \frac{3}{2}\mathbf{j}) + \frac{3}{4}\mathbf{i} = -2\mathbf{i} - \frac{3}{2}\mathbf{j} + \frac{3}{4}\mathbf{i} = -\frac{5}{4}\mathbf{i} - \frac{3}{2}\mathbf{j}$
$\mathbf{u} - \mathbf{v} = -(2\mathbf{i} + \frac{3}{2}\mathbf{j}) - \frac{3}{4}\mathbf{i} = -2\mathbf{i} - \frac{3}{2}\mathbf{j} - \frac{3}{4}\mathbf{i} = -\frac{11}{4}\mathbf{i} - \frac{3}{2}\mathbf{j}$
$3\mathbf{u} - 2\mathbf{v} = -3(2\mathbf{i} + \frac{3}{2}\mathbf{j}) - 2(\frac{3}{4}\mathbf{i}) = -6\mathbf{i} - \frac{9}{2}\mathbf{j} - \frac{3}{2}\mathbf{i} = -\frac{15}{2}\mathbf{i} - \frac{9}{2}\mathbf{j}$

21. $\mathbf{u} + 2\mathbf{w} = \mathbf{i} - 2\mathbf{j} + 2(-4\mathbf{i} + \mathbf{j}) = \mathbf{i} - 2\mathbf{j} - 8\mathbf{i} + 2\mathbf{j} = -7\mathbf{i}$ $\langle -7,0 \rangle$

23. $\frac{1}{2}\mathbf{w} = \frac{1}{2}(-4\mathbf{i} + \mathbf{j}) = -2\mathbf{i} + \frac{1}{2}\mathbf{j}$ $\langle -2, \frac{1}{2} \rangle$

25. $\frac{1}{4}(8\mathbf{u} + 4\mathbf{v} - \mathbf{w}) = 2\mathbf{u} + \mathbf{v} - \frac{1}{4}\mathbf{w} = 2(\mathbf{i} - 2\mathbf{j}) + (3\mathbf{i} + \mathbf{j}) - \frac{1}{4}(-4\mathbf{i} + \mathbf{j})$
$= 2\mathbf{i} - 4\mathbf{j} + 3\mathbf{i} + \mathbf{j} + \mathbf{i} - \frac{1}{4}\mathbf{j} = 6\mathbf{i} - \frac{13}{4}\mathbf{j}$ $\langle 6, -\frac{13}{4} \rangle$

27. $\mathbf{v} = \langle \|\mathbf{v}\|\cos\theta, \|\mathbf{v}\|\sin\theta \rangle = \langle 4\cos 0°, 4\sin 0° \rangle = \langle 4, 0 \rangle$

29. $\mathbf{v} = \langle \|\mathbf{v}\|\cos\theta, \|\mathbf{v}\|\sin\theta \rangle = \langle 10\cos 225°, 10\sin 225° \rangle = \langle -5\sqrt{2}, -5\sqrt{2} \rangle$

31. $\mathbf{v} = \langle \|\mathbf{v}\|\cos\theta, \|\mathbf{v}\|\sin\theta \rangle = \langle 6\cos 40°, 6\sin 40° \rangle = \langle 4.5963, 3.8567 \rangle$

33. $\mathbf{v} = \langle \|\mathbf{v}\|\cos\theta, \|\mathbf{v}\|\sin\theta \rangle = \langle \tfrac{1}{2}\cos 250°, \tfrac{1}{2}\sin 250° \rangle = \langle -.1710, -.4698 \rangle$

35. $\|\mathbf{v}\| = \sqrt{4^2 + 4^2} = 4\sqrt{2}$ $\tan\theta = 4/4 = 1$, $\theta = 45°$ (quadrant I)

37. $\|\mathbf{v}\| = \sqrt{(-8)^2 + 0^2} = 8$ $\tan\theta = 0/8 = 1$, $\theta = 180°$ (negative x-axis)

39. $\|6\mathbf{j}\| = \sqrt{0^2 + 6^2} = 6$ $\tan\theta = 6/0$, undefined, $\theta = 90°$ (positive y-axis)

41. $\|-2\mathbf{i} + 8\mathbf{j}\| = \sqrt{(-2)^2 + 8^2} = 2\sqrt{17}$ $\tan\theta = 8/-2 = -4$, $\theta = 104.04°$ (quadrant II)

43. $\|\mathbf{v}\| = \sqrt{4^2 + (-5)^2} = \sqrt{41}$,   unit vector $= \dfrac{1}{\sqrt{41}}\langle 4, -5 \rangle = \left\langle \dfrac{4}{\sqrt{41}}, -\dfrac{5}{\sqrt{41}} \right\rangle$

45. $\|\mathbf{v}\| = \sqrt{5^2 + 10^2} = 5\sqrt{5}$,   unit vector $= \dfrac{1}{5\sqrt{5}}(5\mathbf{i} + 10\mathbf{j}) = \dfrac{1}{\sqrt{5}}\mathbf{i} + \dfrac{2}{\sqrt{5}}\mathbf{j}$

47. $\mathbf{u} = \langle 30\cos 0°, 30\sin 0° \rangle = \langle 30, 0 \rangle$, $\mathbf{v} = \langle 90\cos 60°, 90\sin 60° \rangle = \langle 45, 45\sqrt{3} \rangle$

Resultant $= \mathbf{u} + \mathbf{v} = \langle 75, 45\sqrt{3} \rangle$, $\|\mathbf{u} + \mathbf{v}\| = \sqrt{75^2 + \left(45\sqrt{3}\right)^2} = 30\sqrt{13} = 108.2$ pounds

$\tan\theta = \dfrac{45\sqrt{3}}{75}$, $\theta = 46.1°$ (quadrant I)

49. $\mathbf{u} = \langle 12\cos 130°, 12\sin 130° \rangle = \langle -7.7135, 9.1925 \rangle$

$\mathbf{v} = \langle 20\cos 250°, 20\sin 250° \rangle = \langle -6.8404, -18.7939 \rangle$

Resultant $= \mathbf{u} + \mathbf{v} = \langle -14.5539, -9.6013 \rangle$

$\|\mathbf{u} + \mathbf{v}\| = \sqrt{(-14.5539)^2 + (-9.6013)^2} = 17.4356$ kg

$\tan\theta = \dfrac{-9.6013}{-14.5539} = .6597$, $\theta = 213.4132°$ (quadrant III)

51. $\mathbf{u}_1 + \mathbf{u}_2 + \mathbf{u}_3 + \mathbf{v} = 0$

$\mathbf{v} = -\mathbf{u}_1 - \mathbf{u}_2 - \mathbf{u}_3$

$= -\langle 2, 5 \rangle - \langle -6, 1 \rangle - \langle -4, -8 \rangle$

$= \langle 8, 2 \rangle$

53. $\mathbf{v} + \mathbf{0} = \langle c, d \rangle + \langle 0, 0 \rangle = \langle c, d \rangle = \mathbf{v}$

55. $r(\mathbf{u}+\mathbf{v}) = r(\langle a,b\rangle + \langle c,d\rangle)$

$\qquad = r\langle a+c, b+d\rangle$

$\qquad = \langle ra+rc, rb+rd\rangle$

$\qquad = \langle ra, rb\rangle + \langle rc+rd\rangle$

$\qquad = r\langle a,b\rangle + r\langle c,d\rangle$

$\qquad = r\mathbf{u} + r\mathbf{v}$

57. $(rs)\mathbf{v} = rs\langle c,d\rangle$

$\qquad = \langle rsc, rsd\rangle$

$\qquad = r\langle sc, sd\rangle = r(s\mathbf{v})$

Also

$\quad \langle rsc, rsd\rangle = \langle src, srd\rangle$

$\qquad\qquad = s\langle rc, rd\rangle = s(r\mathbf{v})$

59. Write one force as $\langle 30\cos 0°, 30\sin 0°\rangle = \langle 30,0\rangle$.

Write second force as $\langle 20\cos 28°, 20\sin 28°\rangle = \langle 17.6590, 9.3894\rangle$.

Then the resultant $= \langle 30,0\rangle + \langle 17.6590, 9.3894\rangle = \langle 47.6590, 9.3894\rangle$.

The magnitude of the resultant $= \sqrt{47.6590^2 + 9.3894^2} = 48.575$ pounds.

61. Following the example, write

$\dfrac{\left\|\overrightarrow{TP}\right\|}{50} = \sin 40°, \quad \left\|\overrightarrow{TP}\right\| = 50\sin 40° = 32.1$ pounds parallel to plane.

$\dfrac{\left\|\overrightarrow{TQ}\right\|}{50} = \cos 40°, \quad \left\|\overrightarrow{TQ}\right\| = 50\cos 40° = 38.3$ pounds perpendicular to plane.

63. Here, $\dfrac{\left\|\overrightarrow{TP}\right\|}{150} = \dfrac{60}{150} = \cos\theta, \quad \theta = \cos^{-1}\left(\dfrac{60}{150}\right) = 66.4°$.

In Exercises **65-70**, navigation angles are measured in degrees clockwise from north. First they are converted to standard position and measured clockwise. Use **p** for the plane's speed and direction; **w** for the wind's speed and direction.

65. The plane's direction is $60°$, convert to $30°$ in standard position.

$\mathbf{p} = \langle 250\cos 30°, 250\sin 30°\rangle$

The wind is blowing from $330°$ to $150°$; convert to $-60°$ in standard position.

$\mathbf{w} = \langle 40\cos(-60°), 40\sin(-60°)\rangle$

$\mathbf{p}+\mathbf{w} = \langle 250\cos 30° + 40\cos(-60°), 250\sin 30° + 40\sin(-60°)\rangle = \langle 236.5, 90.4\rangle$

This is a vector with magnitude $\sqrt{236.5^2 + 90.4^2} = 253.2$ mph and direction $\tan^{-1}(90.4/236.5) = 20.9°$. This represents a navigation angle of $90° - 20.9° = 69.1°$.

67. The plane's direction is $300°$, convert to $150°$ in standard position.
$$\mathbf{p} = \langle 300\cos 150°, 300\sin 150° \rangle$$
The wind's direction is $30°$, convert to $60°$ in standard position.
$$\mathbf{w} = \langle 50\cos 60°, 50\sin 60° \rangle$$
$$\mathbf{p} + \mathbf{w} = \langle 300\cos 150° + 50\cos 60°, 300\sin 150° + 50\sin 60° \rangle = \langle -234.8, 193.3 \rangle$$
This is a vector with magnitude $\sqrt{(-234.8)^2 + 193.3^2} = 304.1$ mph and direction
(quadrant II) $180° + \tan^{-1}(193.3/-234.8) = 140.5°$.
This represents a navigation angle of $309.5°$.

69. The plane's course is $70°$, convert to $20°$ in standard position. Hence
$$\mathbf{p} + \mathbf{w} = \langle 400\cos 20°, 400\sin 20° \rangle$$
The wind blowing south is represented by $\mathbf{w} = \langle 0, -60 \rangle$.
$$\mathbf{p} + \mathbf{w} - \mathbf{w} = \langle 400\cos 20°, 400\sin 20° + 60 \rangle = \langle 375.9, 196.8 \rangle.$$ This is a vector with
magnitude $\sqrt{375.9^2 + 196.8^2} = 424.3$ mph and direction $\tan^{-1}(196.8/375.9) = 27.6°$.
This represents a navigation angle of $62.4°$.

71. Represent the current by $\mathbf{c} = \langle -1, 0 \rangle$ and the desired direction by $\mathbf{r} = \langle 0, r \rangle$. Then she
should swim in the direction $\mathbf{s}$ given by $\mathbf{s} + \mathbf{c} = \mathbf{r}$, $\mathbf{s} = \mathbf{r} - \mathbf{c} = \langle 0, r \rangle - \langle -1, 0 \rangle = \langle 1, r \rangle$.
Since $\|\mathbf{s}\| = 2.8 = \|\langle 1, r \rangle\| = \sqrt{1 + r^2}$, $r = 2.62$. Then the direction of $\mathbf{s}$ is given by
$\tan^{-1}(r/1) = \tan^{-1}(2.62) = 69.08°$.

73. Following the hint,
$$\mathbf{v} = c\cos 65°\,\mathbf{i} + c\sin 65°\,\mathbf{j}, \quad \mathbf{u} = d\cos 148°\,\mathbf{i} + d\sin 148°\,\mathbf{j}$$
$$\mathbf{u} + \mathbf{v} = 0\mathbf{i} + 400\mathbf{j} \ \text{(to counteract gravity)}$$
Hence: $c\cos 65° + d\cos 148° = 0$
$\qquad\qquad c\sin 65° + d\sin 148° = 400$
Solve to obtain $c = 341.77$ lbs on $\mathbf{v}$, $d = 170.32$ lbs on $\mathbf{u}$.

75. Represent the force in the $28°$ rope by $\mathbf{u} = a\cos 152°\,\mathbf{i} + a\sin 152°\,\mathbf{j}$.
Represent the force in the $38°$ rope by $\mathbf{v} = b\cos 38°\,\mathbf{i} + b\sin 38°\,\mathbf{j}$.
$$\mathbf{u} + \mathbf{v} = 0\mathbf{i} + 600\mathbf{j}$$
Hence: $a\cos 152° + b\cos 38° = 0$, $a\sin 152° + b\sin 38° = 600$
Solve to obtain $a = 517.55$ lb on the $28°$ rope, $b = 579.90$ lbs on the $38°$ rope.

77. a. $\mathbf{v} = \langle x_2 - x_1, y_2 - y_1 \rangle$; $k\mathbf{v} = \langle kx_2 - kx_1, ky_2 - ky_1 \rangle$
   b. $\|\mathbf{v}\| = \sqrt{(x_2 - x_1)^2 + (y_2 - y_1)^2}$; $\|k\mathbf{v}\| = \sqrt{(kx_2 - kx_1)^2 + (ky_2 - ky_1)^2}$
   c. $\|k\mathbf{v}\| = \sqrt{k^2(x_2 - x_1)^2 + k^2(y_2 - y_1)^2} = \sqrt{k^2}\sqrt{(x_2 - x_1)^2 + (y_2 - y_1)^2}$
   $$= |k|\sqrt{(x_2 - x_1)^2 + (y_2 - y_1)^2} = |k|\|\mathbf{v}\|$$

d. $\tan \theta = \dfrac{y_2 - y_1}{x_2 - x_1} = \dfrac{k}{k} \dfrac{y_2 - y_1}{x_2 - x_1} = \dfrac{ky_2 - ky_1}{kx_2 - kx_1} = \tan \beta$

Since the angles have the same tangent, they can only differ by a multiple of π, and so are parallel. They have either the same or the opposite direction.

e. If $k>0$, the signs of the components of $k\mathbf{v}$ are the same as those of $\mathbf{v}$, so the two vectors lie in the same quadrant. Therefore, they must have the same direction. If $k<0$, then the components of $k\mathbf{v}$ and the components of $\mathbf{v}$ must have opposite signs and the two vectors do not lie in the same quadrant. Therefore, they do not have the same direction and must have opposite directions.

79. a. Since $\mathbf{u} - \mathbf{v} = \langle a-c, b-d \rangle$, $\|\mathbf{u} - \mathbf{v}\| = \sqrt{(a-c)^2 + (b-d)^2}$. The magnitude of $\mathbf{w}$ is given by the distance betwwen the points (a,b) and (c,d) . Hence, $\|\mathbf{u} - \mathbf{v}\| = \|\mathbf{w}\|$.

b. $\mathbf{u} - \mathbf{v}$ lies on the straight line through $(0,0)$ and $(a-c, b-d)$ which has slope

$\dfrac{(b-d)-0}{(a-c)-0} = \dfrac{b-d}{a-c}$. $\mathbf{w}$ lies on the line joining (a,b) and (c,d). This also has slope

$\dfrac{b-d}{a-c}$. Since the slopes are the same, the vectors, $\mathbf{u} - \mathbf{v}$ and $\mathbf{w}$ are parallel.

Therefore, they either point in the same direction or in opposite directions. We can see that they have the same direction by considering the signs of the components of $\mathbf{u} - \mathbf{v}$. If $a-c>0$ and $b-d>0$, then $a>c$, $b>d$ and $\mathbf{u} - \mathbf{v}$ and $\mathbf{w}$ both point up and right. If $a-c>0$ and $b-d<0$, both vectors will point left and up. If $a-c<0$ and $b-d<0$, both vectors will point left and down. In any case they point in the same direction.

9.4 The Dot Product

1. $\mathbf{u} \bullet \mathbf{v} = \langle 3,4 \rangle \bullet \langle -5,2 \rangle = 3(-5) + 4 \cdot 2 = -7$; $\mathbf{u} \bullet \mathbf{u} = \langle 3,4 \rangle \bullet \langle 3,4 \rangle = 3 \cdot 3 + 4 \cdot 4 = 25$
$\mathbf{v} \bullet \mathbf{v} = \langle -5,2 \rangle \bullet \langle -5,2 \rangle = (-5)(-5) + 2 \cdot 2 = 29$

3. $\mathbf{u} \bullet \mathbf{v} = (2\mathbf{i} + \mathbf{j}) \bullet 3\mathbf{i} = 2 \cdot 3 + 1 \cdot 0 = 6$; $\mathbf{u} \bullet \mathbf{u} = (2\mathbf{i} + \mathbf{j}) \bullet (2\mathbf{i} + \mathbf{j}) = 2 \cdot 2 + 1 \cdot 1 = 5$
$\mathbf{v} \bullet \mathbf{v} = 3\mathbf{i} \bullet 3\mathbf{i} = 3 \cdot 3 = 9$

5. $\mathbf{u} \bullet \mathbf{v} = (3\mathbf{i} + 2\mathbf{j}) \bullet (2\mathbf{i} + 3\mathbf{j}) = 3 \cdot 2 + 2 \cdot 3 = 12$
$\mathbf{u} \bullet \mathbf{u} = (3\mathbf{i} + 2\mathbf{j}) \bullet (3\mathbf{i} + 2\mathbf{j}) = 3 \cdot 3 + 2 \cdot 2 = 13$
$\mathbf{v} \bullet \mathbf{v} = (2\mathbf{i} + 3\mathbf{j}) \bullet (2\mathbf{i} + 3\mathbf{j}) = 2 \cdot 2 + 3 \cdot 3 = 13$

7. $\mathbf{u} \bullet (\mathbf{v} + \mathbf{w}) = \langle 2,5 \rangle \bullet (\langle -4,3 \rangle + \langle 2,-1 \rangle) = \langle 2,5 \rangle \bullet \langle -2,2 \rangle = -4 + 10 = 6$

9. $(\mathbf{u} + \mathbf{v}) \bullet (\mathbf{v} + \mathbf{w}) = (\langle 2,5 \rangle + \langle -4,3 \rangle) \bullet (\langle -4,3 \rangle + \langle 2,-1 \rangle) = \langle -2,8 \rangle \bullet \langle -2,2 \rangle = 4 + 16 = 20$

11. $(3\mathbf{u} + \mathbf{v}) \bullet (2\mathbf{w}) = (3\langle 2,5 \rangle + \langle -4,3 \rangle) \bullet (2\langle 2,-1 \rangle) = (\langle 6,15 \rangle + \langle -4,3 \rangle) \bullet \langle 4,-2 \rangle$
$= \langle 2,18 \rangle \bullet \langle 4,-2 \rangle = 8 - 36 = -28$

13. $\cos\theta = \dfrac{\langle 4,-3\rangle \bullet \langle 1,2\rangle}{\|\langle 4,-3\rangle\|\|\langle 1,2\rangle\|} = \dfrac{-2}{5\sqrt{5}};\quad \theta = \cos^{-1}\dfrac{-2}{5\sqrt{5}} = 1.75065$ rad

15. $\cos\theta = \dfrac{(2\mathbf{i}-3\mathbf{j})\bullet(-\mathbf{i})}{\|2\mathbf{i}-3\mathbf{j}\|\|-\mathbf{i}\|} = \dfrac{-2}{\sqrt{13}(1)} = \dfrac{-2}{\sqrt{13}};\quad \theta = \cos^{-1}\dfrac{-2}{\sqrt{13}} = 2.1588$ rad

17. $\cos\theta = \dfrac{\left(\sqrt{2}\mathbf{i}+\sqrt{2}\mathbf{j}\right)\bullet(\mathbf{i}-\mathbf{j})}{\left\|\sqrt{2}\mathbf{i}+\sqrt{2}\mathbf{j}\right\|\|\mathbf{i}-\mathbf{j}\|} = \dfrac{0}{2\sqrt{2}} = 0;\quad \theta = \dfrac{\pi}{2}$ rad

19. $\langle 2,6\rangle \bullet \langle 3,-1\rangle = 6-6 = 0$. Orthogonal

21. $\langle 9,-6\rangle \bullet \langle -6,4\rangle = -54-24 = -78$. Not orthogonal
However, $\langle 9,-6\rangle = -\frac{3}{2}\langle -6,4\rangle$, hence the vectors are parallel.

23. $(2\mathbf{i}-2\mathbf{j})\bullet(5\mathbf{i}+8\mathbf{j}) = 10-16 = -6$. Not orthogonal
Since $2\mathbf{i}-2\mathbf{j} \neq k(5\mathbf{i}+8\mathbf{j})$ the vectors are also not parallel.

25. The vectors will be orthogonal if their dot product is zero.
$(2\mathbf{i}+3\mathbf{j})\bullet(3\mathbf{i}-k\mathbf{j}) = 6-3k = 0$ if $k=2$

27. The vectors will be orthogonal if their dot product is zero.
$(\mathbf{i}-\mathbf{j})\bullet\left(k\mathbf{i}+\sqrt{2}\mathbf{j}\right) = k-\sqrt{2} = 0$ if $k=\sqrt{2}$.

29. $proj_{\mathbf{u}}\mathbf{v} = \left(\dfrac{\mathbf{u}\bullet\mathbf{v}}{\|\mathbf{u}\|^2}\right)\mathbf{u}$ $\qquad\qquad proj_{\mathbf{v}}\mathbf{u} = \left(\dfrac{\mathbf{u}\bullet\mathbf{v}}{\|\mathbf{v}\|^2}\right)\mathbf{v}$

$= \dfrac{(3\mathbf{i}-5\mathbf{j})\bullet(6\mathbf{i}+2\mathbf{j})}{\|3\mathbf{i}-5\mathbf{j}\|^2}(3\mathbf{i}-5\mathbf{j})$ $\qquad = \dfrac{(3\mathbf{i}-5\mathbf{j})\bullet(6\mathbf{i}+2\mathbf{j})}{\|6\mathbf{i}+2\mathbf{j}\|^2}(6\mathbf{i}+2\mathbf{j})$

$= \dfrac{8}{34}(3\mathbf{i}-5\mathbf{j})$ $\qquad\qquad\qquad = \dfrac{8}{40}(6\mathbf{i}+2\mathbf{j})$

$= \dfrac{12}{17}\mathbf{i}-\dfrac{20}{17}\mathbf{j}$ $\qquad\qquad\qquad = \dfrac{6}{5}\mathbf{i}+\dfrac{2}{5}\mathbf{j}$

31. $proj_{\mathbf{u}}\mathbf{v} = \left(\dfrac{\mathbf{u}\bullet\mathbf{v}}{\|\mathbf{u}\|^2}\right)\mathbf{u} = \dfrac{(\mathbf{i}+\mathbf{j})\bullet(\mathbf{i}-\mathbf{j})}{\|\mathbf{i}+\mathbf{j}\|^2}(\mathbf{i}+\mathbf{j}) = 0(\mathbf{i}+\mathbf{j}) = \mathbf{0}$

$proj_{\mathbf{v}}\mathbf{u} = \left(\dfrac{\mathbf{u}\bullet\mathbf{v}}{\|\mathbf{v}\|^2}\right)\mathbf{v} = \mathbf{0}$ (similarly)

33. $comp_{\mathbf{v}}\mathbf{u} = \dfrac{\mathbf{u}\bullet\mathbf{v}}{\|\mathbf{v}\|} = \dfrac{(10\mathbf{i}+4\mathbf{j})\bullet(3\mathbf{i}-2\mathbf{j})}{\|3\mathbf{i}-2\mathbf{j}\|} = \dfrac{22}{\sqrt{13}}$

35. $comp_{\mathbf{v}}\mathbf{u} = \dfrac{\mathbf{u} \bullet \mathbf{v}}{\|\mathbf{v}\|} = \dfrac{(3\mathbf{i} + 2\mathbf{j}) \bullet (-\mathbf{i} + 3\mathbf{j})}{\|-\mathbf{i} + 3\mathbf{j}\|} = \dfrac{3}{\sqrt{10}}$

37. $\begin{aligned}
\mathbf{u} \bullet (\mathbf{v} + \mathbf{w}) &= \langle a, b \rangle \bullet \left(\langle c, d \rangle + \langle r, s \rangle \right) \\
&= \langle a, b \rangle \bullet \langle c + r, d + s \rangle \\
&= a(c + r) + b(d + s) \\
&= (ac + bd) + (ar + bs) \\
&= \langle a, b \rangle \bullet \langle c, d \rangle + \langle a, b \rangle \bullet \langle r, s \rangle \\
&= \mathbf{u} \bullet \mathbf{v} + \mathbf{u} \bullet \mathbf{w}
\end{aligned}$

39. $\mathbf{0} \bullet \mathbf{u} = \langle 0, 0 \rangle \bullet \langle a, b \rangle = 0a + 0b = 0$

41. If $\theta = 0$, $\mathbf{u} = k\mathbf{v}$ for some positive k. $\quad\quad$ If $\theta = \pi$, $\mathbf{u} = k\mathbf{v}$ for some negative k.

$\cos\theta = 1$, then $\begin{aligned}[t]
\mathbf{u} \bullet \mathbf{v} &= k\mathbf{v} \bullet \mathbf{v} \\
&= k\|\mathbf{v}\|^2 \\
&= k\|\mathbf{v}\|\|\mathbf{v}\| \bullet 1 \\
&= \|\mathbf{u}\|\|\mathbf{v}\|\cos\theta
\end{aligned}$
$\quad\quad$
$\cos\theta = -1$, then $\begin{aligned}[t]
\mathbf{u} \bullet \mathbf{v} &= k\mathbf{v} \bullet \mathbf{v} \\
&= k\|\mathbf{v}\|^2 = k\|\mathbf{v}\|\|\mathbf{v}\| \\
&= -k\|\mathbf{u}\|\|\mathbf{v}\| \\
&= \|\mathbf{u}\|\|\mathbf{v}\|\cos\theta
\end{aligned}$

43. The side containing $(1,2)$ and $(3,4)$ has vector description $\langle 3 - 1, 4 - 2 \rangle = \langle 2, 2 \rangle$.
The side containing $(3,4)$ and $(5,2)$ has vector description $\langle 5 - 3, 2 - 4 \rangle = \langle 2, -2 \rangle$.
Since these two vectors are orthogonal, the line segments form a right angle.

45. Let $\mathbf{u} = \mathbf{i}$, $\mathbf{v} = \mathbf{i} + 2\mathbf{j}$, $\mathbf{w} = \mathbf{i} + 3\mathbf{j}$. Then $\mathbf{u} \bullet \mathbf{v} = 1 = \mathbf{u} \bullet \mathbf{w}$. Many examples are possible, one need only choose $\mathbf{u}$ orthogonal to $\mathbf{v} - \mathbf{w}$.

47. As in Example 7, the force has magnitude
$\left| comp_{\mathbf{v}}\mathbf{F} \right| = \|\mathbf{F}\|\cos(90° - 30°) = |600\cos 60°| = 300$ pounds.

49. $\overrightarrow{PQ} = 4\mathbf{i} + \mathbf{j}$, Work $= \mathbf{F} \bullet \overrightarrow{PQ} = (2\mathbf{i} + 5\mathbf{j}) \bullet (4\mathbf{i} + \mathbf{j}) = 13$

51. $\overrightarrow{PQ} = 3\mathbf{i} + 6\mathbf{j}$, Work $= \mathbf{F} \bullet \overrightarrow{PQ} = (2\mathbf{i} + 3\mathbf{j}) \bullet (3\mathbf{i} + 6\mathbf{j}) = 24$

53. The force is a vector $30\cos(-60°)\mathbf{i} + 30\sin(-60°)\mathbf{j} = 15\mathbf{i} - 15\sqrt{3}\mathbf{j} = \mathbf{F}$.
The displacement is a vector $75\mathbf{i} = \mathbf{D}$.
The work $= \mathbf{F} \bullet \mathbf{D} = \left(15\mathbf{i} - 15\sqrt{3}\mathbf{j}\right) \bullet (75\mathbf{i}) = 1125$ ft-lb.

55. Following the hint, the work is
$-\mathbf{F} \bullet \mathbf{d} = -(0\mathbf{i} - 40\mathbf{j}) \bullet (100\cos 20°\,\mathbf{i} + 100\sin 20°\,\mathbf{j}) = 4000\sin 20° = 1368$ ft-lb.

Chapter 9 Review Exercises

1. $|i(4+2i)|+|3-i| = |4i-2|+|3-i| = \sqrt{4^2+(-2)^2}+\sqrt{3^2+(-1)^2} = \sqrt{20}+\sqrt{10}$

3.

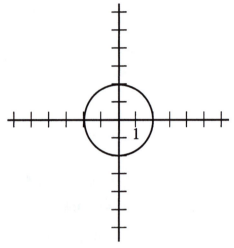

5. $r = \sqrt{1^2+\sqrt{3}^2} = 2.$ $\tan\theta = \sqrt{3}/1;\ \theta = \pi/3$ (quadrant I)

$1+\sqrt{3}i = 2(\cos\pi/3 + i\sin\pi/3)$

7. $2\left(\cos\dfrac{\pi}{12}+i\sin\dfrac{\pi}{12}\right)\cdot 4\left(\cos\dfrac{\pi}{6}+i\sin\dfrac{\pi}{6}\right) = 8\left(\cos\dfrac{\pi}{4}+i\sin\dfrac{\pi}{4}\right)$

$$= 8\left(\dfrac{\sqrt{2}}{2}+i\dfrac{\sqrt{2}}{2}\right) = 4\sqrt{2}+i4\sqrt{2}$$

9. $\dfrac{12\left(\cos\dfrac{7\pi}{12}+i\sin\dfrac{7\pi}{12}\right)}{3\left(\cos\dfrac{5\pi}{12}+i\sin\dfrac{5\pi}{12}\right)} = 4\left(\cos\dfrac{\pi}{6}+i\sin\dfrac{\pi}{6}\right) = 4\left(\dfrac{\sqrt{3}}{2}+i\dfrac{1}{2}\right) = 2\sqrt{3}+2i$

11. $\left[\sqrt[3]{3}\left(\cos\dfrac{5\pi}{36}+i\sin\dfrac{5\pi}{36}\right)\right]^{12} = \left(3^{\frac{1}{3}}\right)^{12}\left(\cos\dfrac{5\pi}{3}+i\sin\dfrac{5\pi}{3}\right)$

$$= 81\left(\dfrac{1}{2}+i\left(-\dfrac{\sqrt{3}}{2}\right)\right)$$

$$= \dfrac{81}{2}-\dfrac{81\sqrt{3}}{2}i$$

13. Apply the roots-of-unity formula with $n = 6$, $k = 0, 1, 2, 3, 4, 5$.

$k = 0$: $\cos 0 + i\sin 0$

$k = 1$: $\cos\dfrac{2\pi}{6} + i\sin\dfrac{2\pi}{6} = \cos\dfrac{\pi}{3} + i\sin\dfrac{\pi}{3}$

$k = 2$: $\cos\dfrac{4\pi}{6} + i\sin\dfrac{4\pi}{6} = \cos\dfrac{2\pi}{3} + i\sin\dfrac{2\pi}{3}$

$k = 3$: $\cos\dfrac{6\pi}{6} + i\sin\dfrac{6\pi}{6} = \cos\pi + i\sin\pi$

$k = 4$: $\cos\dfrac{8\pi}{6} + i\sin\dfrac{8\pi}{6} = \cos\dfrac{4\pi}{3} + i\sin\dfrac{4\pi}{3}$

$k = 5$: $\cos\dfrac{10\pi}{6} + i\sin\dfrac{10\pi}{6} = \cos\dfrac{5\pi}{3} + i\sin\dfrac{5\pi}{3}$

15. The solutions are the four fourth roots of $i = \cos\dfrac{\pi}{2} + i\sin\dfrac{\pi}{2}$, with $n = 4$, $k = 0, 1, 2, 3$.

$k = 0$: $\cos\left(\dfrac{\pi/2}{4}\right) + i\sin\left(\dfrac{\pi/2}{4}\right) = \cos\dfrac{\pi}{8} + i\sin\dfrac{\pi}{8}$

$k = 1$: $\cos\left(\dfrac{\pi/2 + 2\pi}{4}\right) + i\sin\left(\dfrac{\pi/2 + 2\pi}{4}\right) = \cos\dfrac{5\pi}{8} + i\sin\dfrac{5\pi}{8}$

$k = 2$: $\cos\left(\dfrac{\pi/2 + 4\pi}{4}\right) + i\sin\left(\dfrac{\pi/2 + 4\pi}{4}\right) = \cos\dfrac{9\pi}{8} + i\sin\dfrac{9\pi}{8}$

$k = 3$: $\cos\left(\dfrac{\pi/2 + 6\pi}{4}\right) + i\sin\left(\dfrac{\pi/2 + 6\pi}{4}\right) = \cos\dfrac{13\pi}{8} + i\sin\dfrac{13\pi}{8}$

17. $\mathbf{u} + \mathbf{v} = \langle 3, -2 \rangle + \langle 8, 1 \rangle = \langle 11, -1 \rangle$

19. $\|2\mathbf{v} - 4\mathbf{u}\| = \|2\langle 8, 1 \rangle - 4\langle 3, -2 \rangle\| = \|\langle 16, 2 \rangle - \langle 12, -8 \rangle\| = \|\langle 4, 10 \rangle\| = \sqrt{4^2 + 10^2} = 2\sqrt{29}$

21. $4\mathbf{u} - \mathbf{v} = 4(-2\mathbf{i} + \mathbf{j}) - (3\mathbf{i} - 4\mathbf{j}) = -8\mathbf{i} + 4\mathbf{j} - 3\mathbf{i} + 4\mathbf{j} = -11\mathbf{i} + 8\mathbf{j}$

23. $\|\mathbf{u} + \mathbf{v}\| = \|-2\mathbf{i} + \mathbf{j} + 3\mathbf{i} - 4\mathbf{j}\| = \|\mathbf{i} - 3\mathbf{j}\| = \sqrt{1^2 + (-3)^2} = \sqrt{10}$

25. $\mathbf{v} = \langle \|\mathbf{v}\|\cos\theta, \|\mathbf{v}\|\sin\theta \rangle = \langle 5\cos 45°, 5\sin 45° \rangle = \langle 5\sqrt{2}/2, 5\sqrt{2}/2 \rangle$

27. A vector in the opposite direction is given by $-3\mathbf{i} + 6\mathbf{j}$. Since the magnitude of this vector is given by $\sqrt{(-3)^2 + 6^2} = 3\sqrt{5}$, the required unit vector is $\dfrac{1}{3\sqrt{5}}(-3\mathbf{i} + 6\mathbf{j}) = -\dfrac{1}{\sqrt{5}}\mathbf{i} + \dfrac{2}{\sqrt{5}}\mathbf{j}$.

29. Let the velocity of the plane be $\mathbf{v} = \langle 300\cos(-30°), 300\sin(-30°) \rangle = \langle 150\sqrt{3}, -150 \rangle$.

Let the velocity of the wind be $\mathbf{w} = \langle 0, -40 \rangle$. Hence the course of the plane can be represented by $\mathbf{v} + \mathbf{w} = \langle 150\sqrt{3}, -150 \rangle + \langle 0, -40 \rangle = \langle 150\sqrt{3}, -190 \rangle$. Then

$$\|\mathbf{v} + \mathbf{w}\| = \sqrt{\left(150\sqrt{3}\right)^2 + (-190)^2} = \sqrt{103,600} = 321.87 \text{ mph.}$$

$\tan\theta = \dfrac{-190}{150\sqrt{3}}$, $\theta = -36.18°$ which represents a course of $126.18°$.

31. $\mathbf{u} \bullet \mathbf{v} = \langle 3, -4 \rangle \bullet \langle -2, 5 \rangle = 3(-2) + (-4)5 = -26$

33. $(\mathbf{u} + \mathbf{v}) \bullet \mathbf{w} = \left(\langle 3, -4 \rangle + \langle -2, 5 \rangle\right) \bullet \langle 0, 3 \rangle = \langle 1, 1 \rangle \bullet \langle 0, 3 \rangle = 1 \cdot 0 + 1 \cdot 3 = 3$

35. $\cos\theta = \dfrac{(5\mathbf{i} - 2\mathbf{j}) \bullet (3\mathbf{i} + \mathbf{j})}{\|5\mathbf{i} - 2\mathbf{j}\|\|3\mathbf{i} + \mathbf{j}\|} = \dfrac{13}{\sqrt{29}\sqrt{10}}$; $\theta = \cos^{-1}\dfrac{13}{\sqrt{29}\sqrt{10}} = .70$ rad

37. $proj_\mathbf{v}\mathbf{u} = \left(\dfrac{\mathbf{u} \bullet \mathbf{v}}{\|\mathbf{v}\|^2}\right)\mathbf{v} = \dfrac{(4\mathbf{i} - 3\mathbf{j}) \bullet (2\mathbf{i} + \mathbf{j})}{\|2\mathbf{i} + \mathbf{j}\|^2}(2\mathbf{i} + \mathbf{j}) = \dfrac{5}{\left(\sqrt{5}\right)^2}(2\mathbf{i} + \mathbf{j}) = 2\mathbf{i} + \mathbf{j}$

39. $(\mathbf{u} + \mathbf{v}) \bullet (\mathbf{u} - \mathbf{v}) = \mathbf{u} \bullet \mathbf{u} - \mathbf{u} \bullet \mathbf{v} + \mathbf{v} \bullet \mathbf{u} - \mathbf{v} \bullet \mathbf{v} = \|\mathbf{u}\|^2 - \mathbf{u} \bullet \mathbf{v} + \mathbf{v} \bullet \mathbf{u} - \|\mathbf{v}\|^2 = \|\mathbf{u}\|^2 - \|\mathbf{v}\|^2$

If $\mathbf{u}$ and $\mathbf{v}$ have the same magnitude, this expression equals zero, hence $\mathbf{u} + \mathbf{v}$ and $\mathbf{u} - \mathbf{v}$ are orthogonal.

41. Represent the force of gravity by $\mathbf{F} = \langle 0, -3500 \rangle$ and let $\mathbf{v}$ be a unit vector pointing down the ramp.

The force needed to hold the car against the force of gravity is $- proj_\mathbf{v}\mathbf{F}$.

The magnitude of this force is $\|\mathbf{F}\|\cos\theta = 3500\cos 60° = 3500\left(\tfrac{1}{2}\right) = 1750$ pounds.

Chapter 10
Analytic Geometry

10.1 Circles and Ellipses

Note: Throughout this chapter, windows have been chosen on the TI-83, wherever possible, to show vertices as often as possible without distorting the shapes of the graphs unreasonably.

1. Circle, center on y-axis: $x^2 + (y-3)^2 = 4$.

3. Ellipse, center at origin, major axis horizontal: $x^2 + 6y^2 = 18$

5. Circle, center in first quadrant: $(x-4)^2 + (y-3)^2 = 4$

7. Ellipse, major axis horizontal, $a = 5$, $b = 2$.
$$-9.4 \le x \le 9.4, -9.4 \le y \le 9.4$$

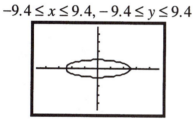

9. $4x^2 + 3y^2 = 12$
$$\frac{x^2}{3} + \frac{y^2}{4} = 1 \text{ Ellipse,}$$
major axis vertical, $a = 2, b = \sqrt{3}$.
$$-3 \le x \le 3, -2 \le y \le 2$$

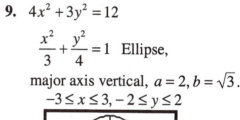

11. $\dfrac{x^2}{10} - 1 = -\dfrac{y^2}{36}$
$$\frac{x^2}{10} + \frac{y^2}{36} = 1$$
Ellipse, major axis vertical, $a = 6, b = \sqrt{10}$.
$$-2\sqrt{10} \le x \le 2\sqrt{10}, -7 \le y \le 7$$

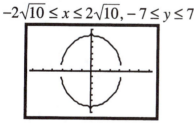

13. $4x^2 + 4y^2 = 1$
$$\frac{x^2}{1/4} + \frac{y^2}{1/4} = 1$$
Circle, center at origin, radius $\frac{1}{2}$.
$$-1.5 \le x \le 1.5, -1 \le y \le 1$$

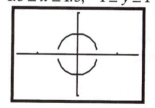

15. $a = 7, b = 2$, major axis horizontal
$$\frac{x^2}{7^2} + \frac{y^2}{2^2} = 1$$
$$\frac{x^2}{49} + \frac{y^2}{4} = 1$$

17. $2a = 12, 2b = 8$, so $a = 6, b = 4$,
major axis horizontal: $\dfrac{x^2}{6^2} + \dfrac{y^2}{4^2} = 1$
$$\frac{x^2}{36} + \frac{y^2}{16} = 1$$

19. $a = 7$, $b = 3$, major axis vertical

$$\frac{x^2}{3^2} + \frac{y^2}{7^2} = 1 \text{ or } \frac{x^2}{9} + \frac{y^2}{49} = 1$$

21. $a = 4$, $b = 2$

$$\pi ab = \pi \cdot 4 \cdot 2 = 8\pi$$

23. Rewrite: $\dfrac{x^2}{4} + \dfrac{y^2}{3} = 1$

$a = 2$, $b = \sqrt{3}$

$\pi ab = \pi \cdot 2 \cdot \sqrt{3} = 2\pi\sqrt{3}$

25. Rewrite: $\dfrac{6x^2}{14} + \dfrac{2y^2}{14} = 1$

$$\frac{x^2}{7/3} + \frac{y^2}{7} = 1$$

$a = \sqrt{\dfrac{7}{3}}$, $b = \sqrt{7}$, $\pi ab = \pi\sqrt{\dfrac{7}{3}}\sqrt{7} = \dfrac{7\pi}{\sqrt{3}}$

27. Ellipse, center $(1,5)$, major axis vertical

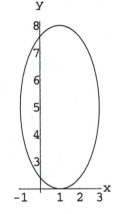

29. Ellipse, center $(-1,4)$, major axis horizontal

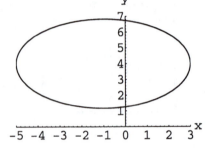

31. Complete the square:

$$9x^2 + 4y^2 + 54x - 8y + 49 = 0$$

$$9(x+3)^2 + 4(y-1)^2 = 36$$

$$\frac{(x+3)^2}{4} + \frac{(y-1)^2}{9} = 1$$

Ellipse, center $(-3,1)$, major axis vertical

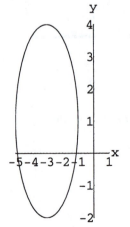

33. Complete the square:

$$x^2 + y^2 + 6x - 8y + 5 = 0$$

$$(x+3)^2 + (y-4)^2 = 20$$

Circle, center $(-3,4)$, radius $\sqrt{20}$

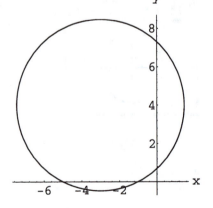

35. Complete the square:
$$4x^2 + y^2 + 24x - 4y + 36 = 0$$
$$4(x+3)^2 + (y-2)^2 = 4$$
$$\frac{(x+3)^2}{1} + \frac{(y-2)^2}{4} = 1$$
Ellipse, center $(-3,2)$, major axis vertical

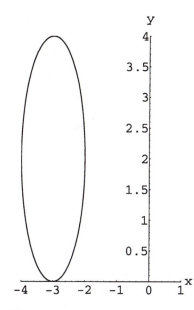

37. Rewrite as
$$\frac{x^2}{10} + \frac{y^2}{36} = 1$$
Use $x = \sqrt{10}\cos t$, $y = 6\sin t$.
$-10 \le x \le 10$, $-7 \le y \le 7$

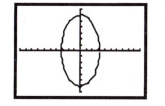

39. Rewrite as
$$\frac{x^2}{1/4} + \frac{y^2}{1/4} = 1$$
Use $x = \tfrac{1}{2}\cos t$, $y = \tfrac{1}{2}\sin t$.
$-3 \le x \le 3$, $-2 \le y \le 2$

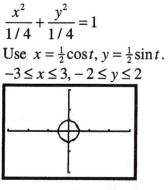

41. Use
$$x - 1 = 2\cos t, \; y - 5 = 3\sin t$$
$$x = 2\cos t + 1, \; y = 3\sin t + 5$$
$-5 \le x \le 10$, $-1 \le y \le 9$

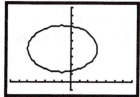

43. Use $x + 1 = 4\cos t$, $y - 4 = \sqrt{8}\sin t$
$$x = 4\cos t - 1, \; y = \sqrt{8}\sin t + 4$$
$-7 \le x \le 7$, $-1 \le y \le 9$

45. $2a = 7 - (-1)$, $a = 4$
$2b = 4 - 0$, $b = 2$
major axis vertical
$$\frac{(x-2)^2}{4} + \frac{(y-3)^2}{16} = 1$$

47. $2a = 12$, $a = 6$, $2b = 5$, $b = 5/2$
major axis vertical
$$\frac{(x-7)^2}{25/4} + \frac{(y+4)^2}{36} = 1$$

49. $a = 9 - 3 = 6$, $b = (-2) - (-6) = 4$
major axis horizontal
$$\frac{(x-3)^2}{36} + \frac{(y+2)^2}{16} = 1$$

51. $2a = 14$, $a = 7$, $2b = 8$, $b = 4$
$$\frac{(x+5)^2}{49} + \frac{(y-3)^2}{16} = 1 \quad \text{or} \quad \frac{(x+5)^2}{16} + \frac{(y-3)^2}{49} = 1$$

53. Circle, center at origin: $2x^2 + 2y^2 - 8 = 0$

55. Ellipse, major axis vertical, center in first quadrant: $2x^2 + y^2 - 8x - 6y + 9 = 0$

57. Ellipse, major axis vertical, center in third quadrant: $\dfrac{(x+3)^2}{4} + \dfrac{(y+3)^2}{8} = 1$

59. If $a = b$, the equation becomes $\dfrac{x^2}{a^2} + \dfrac{y^2}{a^2} = 1$, $x^2 + y^2 = a^2$.
This is the equation of a circle with center at the origin, radius a.

61. $2a = 477{,}736 \quad a = 238{,}868 \quad 2b = 477{,}078 \quad b = 238{,}539$
$$c = \sqrt{a^2 - b^2} = \sqrt{(238{,}868)^2 - (238{,}539)^2} = 12{,}533$$
The minimum distance is $a - c = 238{,}868 - 12{,}533 = 226{,}335$ miles.
The maximum distance is $a + c = 238{,}868 + 12{,}533 = 251{,}401$ miles.

63. $a^2 = 100 \quad b^2 = 99 \quad c = \sqrt{a^2 - b^2} = \sqrt{100 - 99} = 1$, eccentricity $= \dfrac{c}{a} = \dfrac{1}{10} = .1$

65. $a^2 = 40 \quad b^2 = 10 \quad c = \sqrt{a^2 - b^2} = \sqrt{40 - 10} = \sqrt{30}$, eccentricity $= \dfrac{c}{a} = \dfrac{\sqrt{30}}{\sqrt{40}} = \dfrac{\sqrt{3}}{2}$

67. The closer the eccentricity is to zero, the more the ellipse resembles a circle. The closer the eccentricity is to 1, the more elongated the ellipse becomes.

69. The minimum distance from the center is $a - c = 6540 + 6400 = 12{,}940$.
The maximum distance from the center is $a + c = 22{,}380 + 6400 = 28{,}780$.
Solving these two equations yields $a = 20{,}860$, $c = 7920$.
Hence, eccentricity $= \dfrac{c}{a} = \dfrac{7920}{20{,}860} = .38$

71. The guest can be anywhere inside an ellipse with foci 50 feet apart: $c = 25$, and sum of distances to foci: $150 - 50 = 100$: $a = 50$. The fence is an ellipse with major axis 100 ft, minor axis $2\left(\sqrt{50^2 - 25^2}\right) = 50\sqrt{3}$ ft.
The area is $ab = \pi \cdot 50 \cdot 50\sqrt{3} = 2500\pi\sqrt{3}$ sq ft.

10.2 Hyperbolas

1. Ellipse, center at origin, major axis horizontal: $x^2 + 4y^2 = 1$.

3. Hyperbola, focal axis on x-axis: $2x^2 - y^2 = 8$.

5. Ellipse: center at origin, major axis vertical: $6x^2 + 2y^2 = 18$.

7. Write: $\dfrac{x^2}{6} - \dfrac{y^2}{16} = 1$

$\dfrac{x^2}{\left(\sqrt{6}\right)^2} - \dfrac{y^2}{4^2} = 1$

$a = \sqrt{6}, b = 4$. Hyperbola, vertices $\left(\pm\sqrt{6}, 0\right)$, asymptotes $y = \pm\dfrac{4}{\sqrt{6}}x$.

Use window $-6 \le x \le 6, -6 \le y \le 6$.

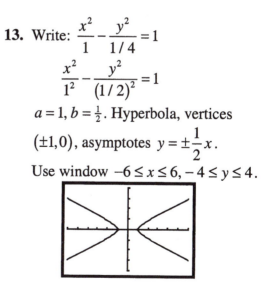

9. Write: $\dfrac{x^2}{4} - \dfrac{y^2}{16} = 1$

$\dfrac{x^2}{2^2} - \dfrac{y^2}{4^2} = 1$

$a = 2, b = 4$. Hyperbola, vertices $(\pm 2, 0)$, asymptotes $y = \pm 2x$.
Use the standard window.

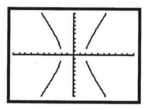

11. Write: $\dfrac{x^2}{10} - \dfrac{y^2}{36} = 1$

$\dfrac{x^2}{\left(\sqrt{10}\right)^2} - \dfrac{y^2}{6^2} = 1$

$a = \sqrt{10}, b = 6$. Hyperbola, vertices $\left(\pm\sqrt{10}, 0\right)$, asymptotes $y = \pm\dfrac{6}{\sqrt{10}}x$.
Use the standard window.

13. Write: $\dfrac{x^2}{1} - \dfrac{y^2}{1/4} = 1$

$\dfrac{x^2}{1^2} - \dfrac{y^2}{(1/2)^2} = 1$

$a = 1, b = \frac{1}{2}$. Hyperbola, vertices $(\pm 1, 0)$, asymptotes $y = \pm\dfrac{1}{2}x$.
Use window $-6 \le x \le 6, -4 \le y \le 4$.

15. Since the x-intercepts are ± 3, $a = 3$. Since an asymptote is $y = 2x$,

$\dfrac{b}{a} = 2, b = 2a, b = 6$. Equation: $\dfrac{x^2}{3^2} - \dfrac{y^2}{6^2} = 1$.

17. Since a vertex is $(2,0)$, $a = 2$. The equation must be of form $\dfrac{x^2}{2^2} - \dfrac{y^2}{b^2} = 1$. Since $\left(4, \sqrt{3}\right)$ is on the graph, its coordinates must satisfy the equation. Hence

$$\frac{4^2}{2^2} - \frac{\left(\sqrt{3}\right)^2}{b^2} = 1$$

$$b^2 = 1$$

The equation is: $\dfrac{x^2}{4} - \dfrac{y^2}{1} = 1$.

19. Hyperbola, center $(-1,-3)$, focal axis on vertical line $x = -1$, $a = 5$, $b = 4$, asymptotes $y + 3 = \pm\dfrac{5}{4}(x + 1)$.

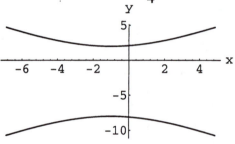

21. Hyperbola, center $(-3,2)$, focal axis on horizontal line $y = 2$, $a = 1$, $b = 2$, asymptotes $y - 2 = \pm 2(x + 3)$.

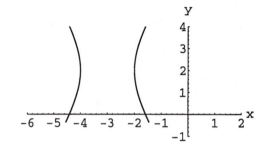

23. $(y + 4)^2 - 8(x - 1)^2 = 8$

$$\frac{(y + 4)^2}{8} - \frac{(x - 1)^2}{1} = 1$$

Hyperbola, center $(1, -4)$, focal axis on vertical line $x = 1$, $a = \sqrt{8}$, $b = 1$, asymptotes $y + 4 = \pm\sqrt{8}(x - 1)$.

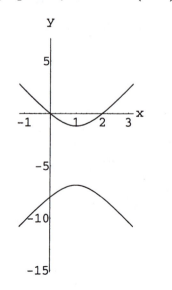

25. $4y^2 - x^2 + 6x - 24y + 11 = 0$

$$4\left(y^2 - 6y + 9\right) - \left(x^2 - 6x + 9\right)$$

$$= -11 + 36 - 9$$

$$4(y - 3)^2 - (x - 3)^2 = 16$$

$$\frac{(y - 3)^2}{4} - \frac{(x - 3)^2}{16} = 1$$

Hyperbola, center $(3,3)$, focal axis on vertical line $x = 3$, $a = 2$, $b = 4$, asymptotes $y - 4 = \pm\dfrac{1}{2}(x - 3)$.

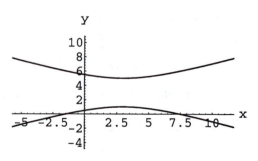

27. $2x^2 + 2y^2 - 12x - 16y + 26 = 0$

$2(x^2 - 6x + 9) + 2(y^2 - 8y + 16)$

$\qquad = -26 + 18 + 32$

$(x-3)^2 + (y-4)^2 = 12$

Circle, center $(3,4)$, radius $\sqrt{12}$.

29. $2x^2 + 3y^2 - 12x - 24y + 54 = 0$

$2(x^2 - 6x + 9) + 3(y^2 - 8y + 16)$

$\qquad = -54 + 18 + 48$

$\dfrac{(x-3)^2}{6} + \dfrac{(y-4)^2}{4} = 1$

Ellipse, center $(3,4)$,
focal axis horizontal.

31. $x^2 - 3y^2 + 4x + 12y = 20$

$(x^2 + 4x + 4) - 3(y^2 - 4y + 4) = 20 + 4 - 12$

$\dfrac{(x+2)^2}{12} - \dfrac{(y-2)^2}{4} = 1$

Hyperbola, center $(-2,2)$, focal axis on horizontal line $y = 2$, $a = \sqrt{12}$,

$b = 2$, asymptotes $y - 2 = \pm \dfrac{1}{\sqrt{3}}(x+2)$.

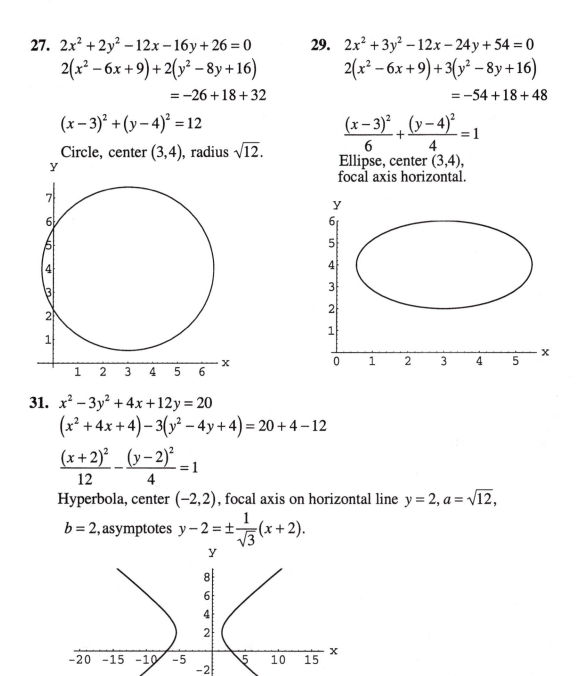

33. Since the center is $(-2,3)$ and the focal axis is vertical, the equation must be of form:

$\dfrac{(y-3)^2}{a^2} - \dfrac{(x+2)^2}{b^2} = 1$.

Since the distance from center to vertex is $3 - 1 = 2$, $a = 2$; $\dfrac{(y-3)^2}{4} - \dfrac{(x+2)^2}{b^2} = 1$

Since $\left(-2 + 3\sqrt{10}, 11\right)$ is on the graph, its coordinates satisfy the equation.

$\dfrac{(11-3)^2}{4} - \dfrac{\left(-2 + 3\sqrt{10} + 2\right)^2}{b^2} = 1$, $b^2 = 6$. The equation is $\dfrac{(y-3)^2}{4} - \dfrac{(x+2)^2}{6} = 1$.

35. Since the center is (4,2) and the focal axis is horizontal, the equation must be of

form: $\dfrac{(x-4)^2}{a^2} - \dfrac{(y-2)^2}{b^2} = 1.$

Since the distance from center to vertex is $7 - 4 = 3$, $a = 3$; $\dfrac{(x-4)^2}{9} - \dfrac{(y-2)^2}{b^2} = 1$

Since an asymptote has slope $\dfrac{4}{3}$, $\dfrac{b}{a} = \dfrac{4}{3}$, $b = 4$.

The equation is $\dfrac{(x-4)^2}{9} - \dfrac{(y-2)^2}{16} = 1.$

37. Center at origin, focal axis is vertical: $y^2 - 2x^2 = 6$

39. Center in first quadrant: $\dfrac{(y-2)^2}{4} - \dfrac{(x-3)^2}{9} = 1$

41. Center in third quadrant: $\dfrac{(x+3)^2}{3} - \dfrac{(y+3)^2}{4} = 1$

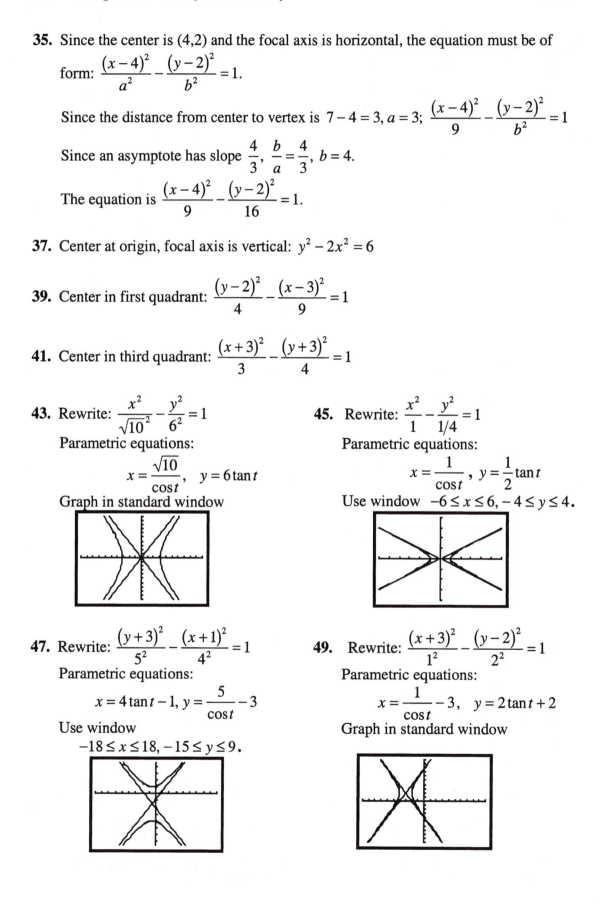

43. Rewrite: $\dfrac{x^2}{\sqrt{10}^2} - \dfrac{y^2}{6^2} = 1$
Parametric equations:
$$x = \dfrac{\sqrt{10}}{\cos t}, \quad y = 6\tan t$$
Graph in standard window

45. Rewrite: $\dfrac{x^2}{1} - \dfrac{y^2}{1/4} = 1$
Parametric equations:
$$x = \dfrac{1}{\cos t}, \quad y = \dfrac{1}{2}\tan t$$
Use window $-6 \le x \le 6, -4 \le y \le 4.$

47. Rewrite: $\dfrac{(y+3)^2}{5^2} - \dfrac{(x+1)^2}{4^2} = 1$
Parametric equations:
$$x = 4\tan t - 1, \, y = \dfrac{5}{\cos t} - 3$$
Use window
$-18 \le x \le 18, -15 \le y \le 9.$

49. Rewrite: $\dfrac{(x+3)^2}{1^2} - \dfrac{(y-2)^2}{2^2} = 1$
Parametric equations:
$$x = \dfrac{1}{\cos t} - 3, \quad y = 2\tan t + 2$$
Graph in standard window

51. The graphs are shown below on one set of coordinate axes.

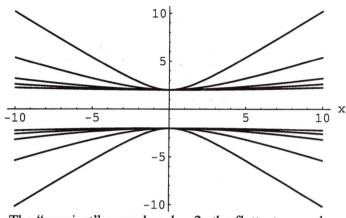

The "curviest" curve has $b = 2$; the flattest curve has $b = 20$. Clearly, as b increases, the graphs get flatter. However, although the hyperbola may look like two horizontal lines for very large b, it is still a hyperbola, with asymptotes $y = \pm 2x / b$ that have non-zero slopes.

53. The asymptotes of $\dfrac{x^2}{a^2} - \dfrac{y^2}{a^2} = 1$ are $y = \pm \dfrac{a}{a} x$ or $y = \pm x$.

Since they have slopes 1 and –1 and $1(-1) = -1$, they are perpendicular.

55. The data require the position (x,y) of the explosion to satisfy: The distances of (x,y) from two points 5280 feet apart differ by 2200 feet. This defines a hyperbola with $2c = 5280$, $2a = 2200$, thus $a = 1100$, $c = 2640$, $b^2 = 2640^2 - 1100^2 = 5{,}759{,}600$.

If the points are $(\pm 1100, 0)$, then the equation must be : $\dfrac{x^2}{1{,}210{,}000} - \dfrac{y^2}{5{,}759{,}600} = 1.$

The exact location, however, cannot be determined from only the given information.

57. $a^2 = 10$ $b^2 = 40$ $c^2 = a^2 + b^2 = 50$ $e = \dfrac{c}{a} = \dfrac{\sqrt{50}}{\sqrt{10}} = \sqrt{5}$

59. Rewrite: $\dfrac{(y-2)^2}{3} - \dfrac{(x+2)^2}{6} = 1$

$a^2 = 3$ $b^2 = 6$ $c^2 = a^2 + b^2 = 9$ $e = \dfrac{c}{a} = \dfrac{3}{\sqrt{3}} = \sqrt{3}$

61. Rewrite: $4(x^2 - 4x + 4) - 5(y^2 + 10y + 25) = -71 + 16 - 125$

$$4(x-2)^2 - 5(y+5)^2 = -180$$

$$\frac{(y+5)^2}{36} - \frac{(x-2)^2}{45} = 1$$

$a^2 = 36$ $b^2 = 45$ $c^2 = a^2 + b^2 = 81$ $e = \dfrac{c}{a} = \dfrac{9}{6} = \dfrac{3}{2}$

10.3 Parabolas

1. Parabola, opens to the right: $6x = y^2$

3. Hyperbola, focal axis horizontal: $2x^2 - y^2 = 8$

5. Ellipse, major axis horizontal: $x^2 + 6y^2 = 18$

7. Parabola, opens upward.
   Rewrite: $x^2 = \frac{1}{3}y$
   $\qquad 4p = \frac{1}{3}$
   $\qquad\quad p = \frac{1}{12}$
   Focus: $\left(0, \frac{1}{12}\right)$, Directrix $y = -\frac{1}{12}$

9. Parabola, opens upward.
   Rewrite: $x^2 = 4y$
   $\qquad 4p = 4$
   $\qquad\quad p = 1$
   Focus: $(0,1)$, Directrix $y = -1$

11. Rewrite: $x - 2 = y^2$. $4p = 1$; $p = \frac{1}{4}$. Opens to the right.
    Vertex: $(2,0)$ Focus: $\left(2\frac{1}{4}, 0\right)$ Directrix: $x = 2 - \frac{1}{4} = \frac{7}{4}$

13. Rewrite: $-1(x - 2) = (y + 1)^2$. $4p = -1$; $p = -\frac{1}{4}$. Opens to the left.
    Vertex: $(2, -1)$ Focus: $\left(\frac{7}{4}, -1\right)$ Directrix: $x = 2 + \frac{1}{4} = \frac{9}{4}$

15. Rewrite: $3\left(x - \frac{2}{3}\right) = (y + 3)^2$. $4p = 3$; $p = \frac{3}{4}$. Opens to the right.
    Vertex: $\left(\frac{2}{3}, -3\right)$ Focus: $\left(\frac{2}{3} + \frac{3}{4}, -3\right) = \left(\frac{17}{12}, -3\right)$ Directrix: $x = \frac{2}{3} - \frac{3}{4} = -\frac{1}{12}$

17. Complete the square: $x = y^2 - 9y$
    $$x + \frac{81}{4} = y^2 - 9y + \frac{81}{4}$$
    $$x + \frac{81}{4} = \left(y - \frac{9}{2}\right)^2$$
    $4p = 1$; $p = \frac{1}{4}$. Opens to the right.
    Vertex: $\left(-\frac{81}{4}, \frac{9}{2}\right)$
    Focus: $\left(-\frac{81}{4} + \frac{1}{4}, \frac{9}{2}\right) = \left(-20, \frac{9}{2}\right)$
    Directrix: $x = -\frac{81}{4} - \frac{1}{4} = -\frac{41}{2}$

19. Complete the square: $y = 3x^2 + x - 4$
    $$y + 4 + \frac{3}{36} = 3\left(x^2 + \frac{1}{3}x + \frac{1}{36}\right)$$
    $$\frac{1}{3}\left(y + \frac{49}{12}\right) = \left(x + \frac{1}{6}\right)^2$$
    $4p = \frac{1}{3}$; $p = \frac{1}{12}$. Opens upward.
    Vertex: $\left(-\frac{1}{6}, -\frac{49}{12}\right)$
    Focus: $\left(-\frac{1}{6}, -\frac{49}{12} + \frac{1}{12}\right) = \left(-\frac{1}{6}, -4\right)$
    Directrix: $x = -\frac{49}{12} - \frac{1}{12} = -\frac{25}{6}$

21. Complete the square: $y = -3x^2 + 4x + 5$
    $$y - 5 - 4/3 = -3\left(x^2 - 4/3x + 4/9\right)$$
    $$-1/3(y - 19/3) = (x - 2/3)^2$$
    $4p = -\frac{1}{3}$; $p = -\frac{1}{12}$. Opens downward.
    Vertex: $\left(\frac{2}{3}, \frac{19}{3}\right)$, Focus: $\left(\frac{2}{3}, \frac{19}{3} - \frac{1}{12}\right) = \left(\frac{2}{3}, \frac{25}{4}\right)$, Directrix: $x = \frac{19}{3} + \frac{1}{12} = \frac{77}{12}$

23. Vertex: (1, 2)

25. Vertex: (0, 2)

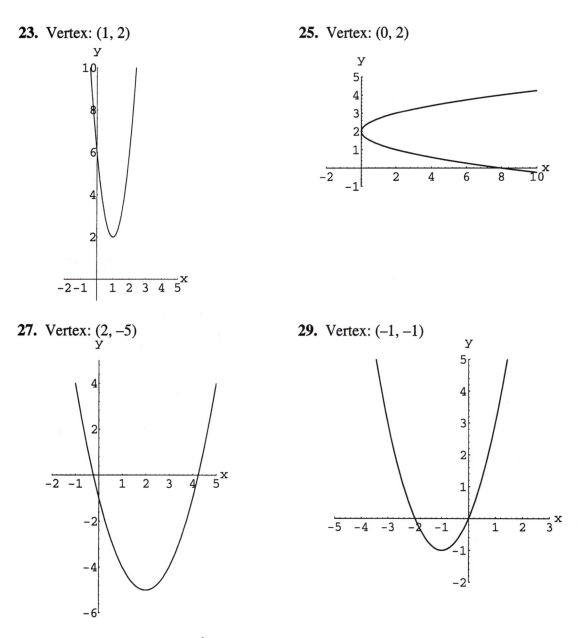

27. Vertex: (2, –5)

29. Vertex: (–1, –1)

31. The equation has form $x^2 = 4py$. Since (2, 12) is on the graph, its coordinates satisfy the equation, hence $2^2 = 4p \cdot 12$, $4p = \frac{1}{3}$.
$x^2 = \frac{1}{3}y$ or $y = 3x^2$

33. The equation has form $(x-1)^2 = 4py$. Since (2, 13) is on the graph, its coordinates satisfy the equation, hence $(2-1)^2 = 4p \cdot 13$, $4p = \frac{1}{13}$.
$(x-1)^2 = \frac{1}{13}y$ or $y = 13(x-1)^2$

35. The equation has form $(y-1)^2 = 4p(x-2)$. Since $(5,0)$ is on the graph, its coordinates satisfy the equation, hence $(0-1)^2 = 4p(5-2)$, $4p = \frac{1}{3}$.
$(y-1)^2 = \frac{1}{3}(x-2)$ or $(x-2) = 3(y-1)^2$

37. Since the focus is to the right of the vertex a distance $\frac{1}{16}$, $p = \frac{1}{16}$, and the parabola opens to the right. The equation is $4\left(\frac{1}{16}\right)(x+3) = (y+2)^2$ or $(x+3) = 4(y+2)^2$.

39. Since the focus is above the vertex a distance $\frac{1}{8}$, $p = \frac{1}{8}$, and the parabola opens upward. The equation is $4\left(\frac{1}{8}\right)(y-1) = (x-1)^2$ or $(y-1) = 2(x-1)^2$.

41. A rough sketch (not shown) indicates that the parabola must open to the right. The equation has form $(y-3)^2 = 4p(x+1)$. Since $(8,0)$ is on the graph, its coordinates satisfy the equation, hence

$$(0-3)^2 = 4p(8+1)$$
$$4p = 1$$

The equation is $(y-3)^2 = (x+1)$. Confirm that $(0,4)$ is on the graph, since its coordinates also satisfy the equation.

43. Rewrite: $y^2 = 4x$

$$x = \frac{t^2}{4} \quad y = t$$

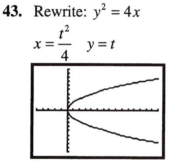

Use window: $-10 \le t \le 10$,
$-5 \le x \le 15, -10 \le y \le 10$

45. $x = t \quad y = 4(t-1)^2 + 2$

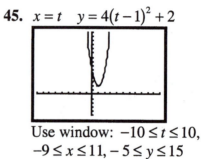

Use window: $-10 \le t \le 10$,
$-9 \le x \le 11, -5 \le y \le 15$

47. $x = 2(t-2)^2 \quad y = t$

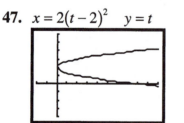

Use window: $-10 \le t \le 10$,
$-2 \le x \le 10, -4 \le y \le 6$

49. Complete the square:

$$x^2 = 6x - y - 5$$
$$x^2 - 6x + 9 = -y - 5 + 9$$
$$(x-3)^2 = -(y-4)$$

Parabola, vertex $(3,4)$

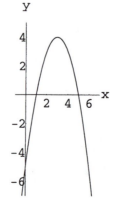

51. Complete the square:

$$3y^2 - 2y = x - 1$$

$$3\left(y^2 - \tfrac{2}{3}y + \tfrac{1}{9}\right) = x - 1 + \tfrac{1}{3}$$

$$\left(y - \tfrac{1}{3}\right)^2 = \tfrac{1}{3}\left(x - \tfrac{2}{3}\right)$$

Parabola, vertex $\left(\tfrac{2}{3}, \tfrac{1}{3}\right)$

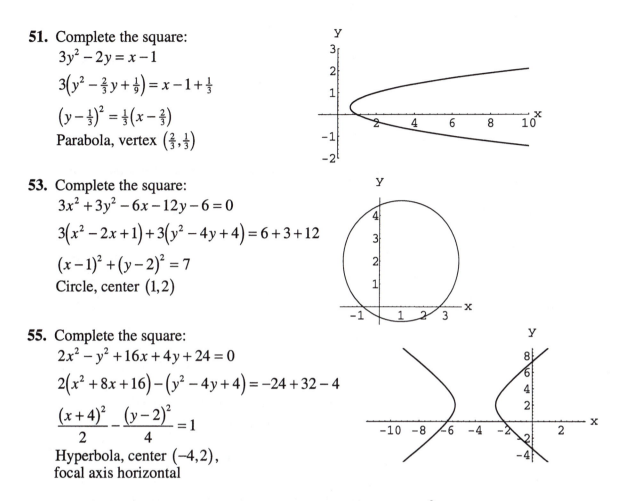

53. Complete the square:

$$3x^2 + 3y^2 - 6x - 12y - 6 = 0$$

$$3\left(x^2 - 2x + 1\right) + 3\left(y^2 - 4y + 4\right) = 6 + 3 + 12$$

$$\left(x - 1\right)^2 + \left(y - 2\right)^2 = 7$$

Circle, center $(1, 2)$

55. Complete the square:

$$2x^2 - y^2 + 16x + 4y + 24 = 0$$

$$2\left(x^2 + 8x + 16\right) - \left(y^2 - 4y + 4\right) = -24 + 32 - 4$$

$$\frac{\left(x + 4\right)^2}{2} - \frac{\left(y - 2\right)^2}{4} = 1$$

Hyperbola, center $(-4, 2)$,
focal axis horizontal

57. Parabola, vertex in first quadrant, opens upward: $y = \left(x - 4\right)^2 + 2$

59. Parabola, vertex in third quadrant, opens downward: $y = -x^2 - 8x - 18$

61. Parabola, vertex in third quadrant, opens upward: $y = \left(x + 5\right)^2 - 3$

63. Clearly, 0. The vertex of $y = x^2 + c$ is at $(0, c)$.

65. Replace x with 9 and solve $4y^2 + 4y = 5 \cdot 9 - 12$ to obtain

$$y = \frac{-1 \pm \sqrt{34}}{2} : \left(9, \frac{-1 \pm \sqrt{34}}{2}\right)$$ are the points.

67. By appropriate choice of axes the parabola can be viewed as opening upward with vertex at the origin. The equation is then $x^2 = 4py$. When $x = \tfrac{1}{2}(4) = 2$, $y = 1.5$, hence $2^2 = 4p(1.5)$

$$p = \tfrac{2}{3}$$

The receiver should be placed at the focus, $\tfrac{2}{3}$ feet or 8 inches from the vertex.

69. If the parabola is viewed as having equation $y^2 = 4px$, when $y = \frac{1}{2}(3) = 1.5$, $x = 1.5$
hence $(1.5)^2 = 4p(1.5)$

$$p = \frac{1.5}{4} = .375$$

The bulb should be placed at the focus, .375 inches from the vertex.

71. Choose the origin at the center of the bridge. The equation of the parabolas is then
$x^2 = 4py$. When $x = \frac{1}{2}(420) = 210$, $y = 100$, hence

$$210^2 = 4p \cdot 100$$

$$4p = 441$$

The equation is then $x^2 = 441y$. One hundred feet from one of the towers, take

$x = 210 - 100 = 110$. Then $y = \dfrac{110^2}{441} = 27.44$ feet.

10.4 Rotations and Second-Degree Equations

1. $A = 1$, $B = -2$, $C = 3$. $B^2 - 4AC = (-2)^2 - 4(1)(3) = -8 < 0$. Ellipse

3. $A = 1$, $B = 2$, $C = 1$. $B^2 - 4AC = 2^2 - 4(1)(1) = 0$. Parabola

5. $A = 17$, $B = -48$, $C = 31$. $B^2 - 4AC = (-48)^2 - 4(17)(31) = 196 > 0$. Hyperbola

7. $A = 9$, $B = 0$, $C = 4$. $B^2 - 4AC = 0^2 - 4(9)(4) = -144 < 0$. Ellipse

Solve for y to obtain $y = \frac{1}{2}\left(2 \pm 3\sqrt{-5 - 6x - x^2}\right)$.

Use window $-6 \le x \le 0$, $-3 \le y \le 4$

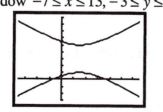

9. $A = -1$, $B = 0$, $C = 4$. $B^2 - 4AC = 0^2 - 4(-1)(4) = 16 > 0$. Hyperbola

Solve for y to obtain $y = \frac{1}{2}\left(6 \pm \sqrt{25 - 6x + x^2}\right)$.

Use window $-7 \le x \le 13$, $-3 \le y \le 9$

11. $A = 0$, $B = 0$, $C = 3$.

$B^2 - 4AC = 0^2 - 4(0)(3) = 0$. Parabola

Solve for y to obtain

$y = \frac{1}{3}\left(1 \pm \sqrt{3x - 2}\right)$.

Use window $-1 \le x \le 8$, $-3 \le y \le 3$

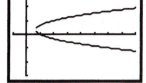

13. $A = 41$, $B = -24$, $C = 34$.

$B^2 - 4AC = (-24)^2 - 4(41)(34)$

$= -5000 < 0$

Ellipse. Solve for y to obtain

$y = \frac{1}{34}\left(12x \pm 5\sqrt{34 - 50x^2}\right)$.

Use window $-1.5 \le x \le 1.5$, $-1 \le y \le 1$

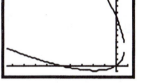

15. $A = 17$, $B = -48$, $C = 31$.

$B^2 - 4AC = (-48)^2 - 4(17)(31)$

$= 196 > 0$

Hyperbola Solve for y to obtain

$y = \frac{1}{31}\left(24x \pm 7\sqrt{x^2 - 31}\right)$. Use window

$-15 \le x \le 15$, $-10 \le y \le 10$

17. $A = 9$, $B = 24$, $C = 16$.

$B^2 - 4AC = 24^2 - 4(9)(16) = 0$.

Parabola. Solve for y to obtain

$y = \frac{1}{16}\left(65 - 12x \pm 5\sqrt{169 - 120x}\right)$.

Use window $-19 \le x \le 2$, $-1 \le y \le 13$

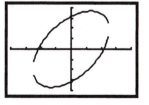

19. $A = 23$, $B = 26\sqrt{3}$, $C = -3$.

$B^2 - 4AC = \left(26\sqrt{3}\right)^2 - 4(23)(-3)$

$= 2304 > 0$

Hyperbola. Solve for y to obtain

$y = \frac{1}{3}\left(8\sqrt{3} + 13\sqrt{3}x \pm 24\sqrt{1 + x + x^2}\right)$.

Use window $15 \le x \le 15$, $-15 \le y \le 15$

21. $A = 17$, $B = -12$, $C = 8$.

$B^2 - 4AC = (-12)^2 - 4(17)(8)$

$= -400 < 0$

Ellipse. Solve for y to obtain

$y = \frac{1}{4}\left(3x \pm \sqrt{160 - 25x^2}\right)$.

Use window $-4 \le x \le 4$, $-4 \le y \le 4$

23. $A = 3$, $B = 2\sqrt{3}$, $C = 1$.

$B^2 - 4AC = \left(2\sqrt{3}\right)^2 - 4(3)(1) = 0$. Parabola.

Solve for y to obtain $y = (2 - x)\sqrt{3} \pm 2\sqrt{7 - 4x}$.

Use window $-9 \le x \le 4$, $-2 \le y \le 10$

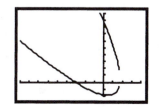

10.4A Rotation of Axes

1. $x = 3, y = 2$
Using the rotation equations, write:
$$3 = u\cos 45° - v\sin 45° = (u - v)/\sqrt{2}$$
$$2 = u\sin 45° + v\cos 45° = (u + v)/\sqrt{2}$$
Solving, we obtain $u = \dfrac{5\sqrt{2}}{2}, v = -\dfrac{\sqrt{2}}{2}$.
$$\left(\frac{5\sqrt{2}}{2}, -\frac{\sqrt{2}}{2}\right)$$

3. $x = 1, y = 0$
Using the rotation equations, write:
$$1 = u\cos 30° - v\sin 30° = (u\sqrt{3} - v)/2$$
$$0 = u\sin 30° + v\cos 30° = (u + v\sqrt{3})/2$$
Solving, we obtain $u = \dfrac{\sqrt{3}}{2}, v = -\dfrac{1}{2}$.
$$\left(\frac{\sqrt{3}}{2}, -\frac{1}{2}\right)$$

5. Using the rotation equations, write: $x = u\cos 45° - v\sin 45° = (u - v)/\sqrt{2}$
$$y = u\sin 45° + v\cos 45° = (u + v)/\sqrt{2}$$
Substituting, we obtain $\dfrac{(u - v)}{\sqrt{2}}\dfrac{(u + v)}{\sqrt{2}} = 1$.
$$u^2 - v^2 = 2$$

7. Using the rotation equations, write:
$$x = u\cos 30° - v\sin 30° = (u\sqrt{3} - v)/2$$
$$y = u\sin 30° + v\cos 30° = (u + v\sqrt{3})/2$$
Substituting, we obtain
$$7\left(\frac{u\sqrt{3} - v}{2}\right)^2 - 6\sqrt{3}\left(\frac{u\sqrt{3} - v}{2}\right)\left(\frac{u + v\sqrt{3}}{2}\right) + 13\left(\frac{u + v\sqrt{3}}{2}\right)^2 - 16 = 0.$$
$$7\left(3u^2 - 2\sqrt{3}uv + v^2\right) - 6\sqrt{3}\left(\sqrt{3}u^2 + 3uv - uv - \sqrt{3}v^2\right) + 13\left(u^2 + 2\sqrt{3}uv + 3v^2\right) = 4 \cdot 16$$
$$16u^2 + 64v^2 = 4 \cdot 16$$
$$u^2 + 4v^2 = 4$$

9. $A = 41 \quad B = -24 \quad C = 34$
$$\cot 2\theta = \frac{A - C}{B} = \frac{41 - 34}{-24} = -\frac{7}{24}, \quad \cos 2\theta = -\frac{7}{25}$$
$$\sin \theta = \sqrt{\frac{1 - \cos 2\theta}{2}} = \sqrt{\frac{1 - (-7/25)}{2}} = \sqrt{\frac{1 + 7/25}{2}} = \frac{4}{5}, \quad \cos \theta = \frac{3}{5}, \quad \theta = \sin^{-1}\frac{3}{5}$$
The rotation equations are: $x = \dfrac{3u - 4v}{5}, \quad y = \dfrac{4u + 3v}{5}$

11. $A = 17$, $B = -48$, $C = 31$, so $\cot 2\theta = \dfrac{A-C}{B} = \dfrac{17-31}{-48} = \dfrac{7}{24}$ and $\cos 2\theta = \dfrac{7}{25}$.

$\sin = \sqrt{\dfrac{1-\cos 2\theta}{2}} = \sqrt{\dfrac{1-7/25}{2}} = \dfrac{3}{5}$, $\cos \theta = \dfrac{4}{5}$, $\theta = 36.87°$

The rotation equations are: $x = \dfrac{4u-3v}{5}$, $y = \dfrac{3u+4v}{5}$

13. a. Substitute the rotation equations: $x = u\cos\theta - v\sin\theta$, $y = u\sin\theta + v\cos\theta$

to obtain: $A(u\cos\theta - v\sin\theta)^2 + B(u\cos\theta - v\sin\theta)(u\sin\theta + v\cos\theta) +$

$\qquad C(u\sin\theta + v\cos\theta)^2 + D(u\cos\theta - v\sin\theta) + E(u\sin\theta + v\cos\theta) + F = 0$

Multiply out and collect terms to obtain: $\left(A\cos^2\theta + B\sin\theta\cos\theta + C\sin^2\theta\right)u^2 +$

$\qquad\qquad \left(-2A\sin\theta\cos\theta + B\left(\cos^2\theta - \sin^2\theta\right) + 2C\sin\theta\cos\theta\right)uv +$

$\qquad\qquad \left(A\sin^2\theta - B\sin\theta\cos\theta + C\cos^2\theta\right)v^2 + (D\cos\theta + E\sin\theta)u +$

$\qquad\qquad (-D\sin\theta + E\cos\theta)v + F = 0$

b. Indeed. $B' = 2(C-A)\sin\theta\cos\theta + B\left(\cos^2\theta - \sin^2\theta\right)$

c. Immediately upon applying the double-angle identities,

$\quad B' = (C-A)\sin 2\theta + B\cos 2\theta$.

d. Then if $\cot 2\theta = (A-C)/B$

$\qquad \dfrac{\cos 2\theta}{\sin 2\theta} = \dfrac{A-C}{B}$

$\qquad (C-A)\sin 2\theta + B\cos 2\theta = 0$.   Hence $B' = 0$.

15. a. Using the results of Exercise **13(a)**, write

$\quad B'^2 - 4A'C' = \left[2(C-A)\sin\theta\cos\theta + B\left(\cos^2\theta - \sin^2\theta\right)\right]^2 -$

$\qquad 4\left(A\cos^2\theta + B\sin\theta\cos\theta + C\sin^2\theta\right)\left(A\sin^2\theta - B\sin\theta\cos\theta + C\cos^2\theta\right)$

Expanding and collecting terms yields

$\quad B'^2 - 4A'C' = A^2\left(4\cos^2\theta\sin^2\theta - 4\cos^2\theta\sin^2\theta\right) +$

$\qquad B^2\left(\cos^4\theta - 2\cos^2\theta\sin^2\theta + \sin^4\theta + 4\cos^2\theta\sin^2\theta\right) +$

$\qquad C^2\left(4\cos^2\theta\sin^2\theta - 4\cos^2\theta\sin^2\theta\right) +$

$\qquad AB\left(-4\cos^3\theta\sin\theta + 4\cos\theta\sin^3\theta + 4\cos^3\theta\sin\theta - 4\cos\theta\sin^3\theta\right) +$

$\qquad AC\left(-8\sin^2\theta\cos^2\theta - 4\cos^4\theta - 4\sin^4\theta\right) +$

$\qquad BC\left(4\cos^3\theta\sin\theta - 4\cos\theta\sin^3\theta - 4\cos^3\theta\sin\theta + 4\cos\theta\sin^3\theta\right)$

The coefficients of A^2, C^2, AB, and BC are seen to reduce to zero, and the right side reduces to: $B'^2 - 4A'C' = B^2\left(\cos^4\theta + 2\cos^2\theta\sin^2\theta + \sin^4\theta\right) -$

$\qquad\qquad 4AC\left(\cos^4\theta + 2\cos^2\theta\sin^2\theta + \sin^4\theta\right)$

$\qquad = \left(B^2 - 4AC\right)\left(\cos^2\theta + \sin^2\theta\right) = B^2 - 4AC$

b. $B^2 - 4AC = B'^2 - 4A'C' = -4A'C'$ if $B' = 0$.
Therefore if $B^2 - 4AC < 0$, $-4A'C' < 0$, $A'C' > 0$, and the graph is an ellipse.
Similarly if $B^2 - 4AC = 0$, $-4A'C' = 0$, $A'C' = 0$, and the graph is a parabola.
And if $B^2 - 4AC > 0$, $-4A'C' > 0$, $A'C' < 0$, and the graph is a hyperbola.

10.5 Plane Curves and Parametric Equations

1. $-5 \le x \le 6, -2 \le y \le 2$

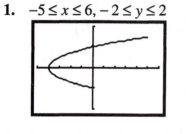

3. $-3 \le x \le 4, -2 \le y \le 3$

5. $0 \le x \le 14, -15 \le y \le 0$

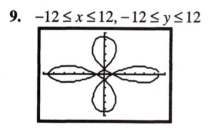

7. $-2 \le x \le 20, -11 \le y \le 11$

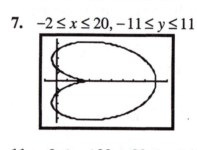

9. $-12 \le x \le 12, -12 \le y \le 12$

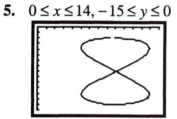

11. $-2 \le x \le 20, -20 \le y \le 4$

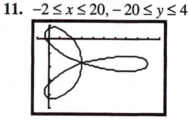

13. $-25 \le x \le 22, -25 \le y \le 26$

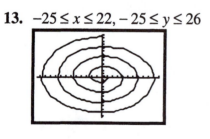

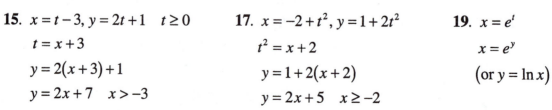

15. $x = t - 3, y = 2t + 1 \quad t \ge 0$

$t = x + 3$

$y = 2(x + 3) + 1$

$y = 2x + 7 \quad x > -3$

17. $x = -2 + t^2, y = 1 + 2t^2$

$t^2 = x + 2$

$y = 1 + 2(x + 2)$

$y = 2x + 5 \quad x \ge -2$

19. $x = e^t$

$x = e^y$

(or $y = \ln x$)

21. $x = 3\cos t,\ y = 3\sin t$

$x^2 + y^2 = (3\cos t)^2 + (3\sin t)^2$

$x^2 + y^2 = 9$

23. The graphs look identical on the page (and hence are not shown). Both give a straight line segment between $P = (-4, 7)$ and $Q = (2, -5)$. However, the graph in (a) moves from P to Q, while the graph in (b) moves from Q to P.

25. $(x-7)^2 + (y+4)^2 = 6^2$ is the equation. Parametrize as:

$x - 7 = 6\cos t \qquad\qquad y + 4 = 6\sin t$

$\quad x = 7 + 6\cos t \qquad\qquad y = -4 + 6\sin t$

Use window $0 \le t \le 2\pi,\ -3 \le x \le 17,\ -11 \le y \le 3$

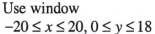

27. Complete the square to obtain $(x-7)^2 + (y+4)^2 = 36$. The parametrization of Exercise **25** applies (and the graph is identical).

29. a. $\dfrac{d-b}{c-a}$ **b.** $y - b = \dfrac{d-b}{c-a}(x-a)$ or $\dfrac{y-b}{d-b} = \dfrac{x-a}{c-a}$

c. Solving both equations for t, we obtain $\dfrac{x-a}{c-a} = t,\ \dfrac{y-b}{d-b} = t$.

Thus $\dfrac{y-b}{d-b} = \dfrac{x-a}{c-a}$ results.

31. In Exercise **30**, the parametrization

$x = a + (c-a)t,\ y = b + (d-b)t,\ 0 \le t \le 1$

was developed. Here:

$x = -6 + (12 - (-6))t \quad y = 12 + (-10 - 12)t$

$x = 18t - 6 \qquad\qquad y = 12 - 22t,\ 0 \le t \le 1$

Use window

$-8 \le x \le 14,\ -12 \le y \le 14$

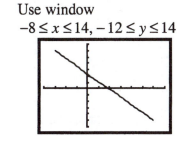

33. In Exercise **30**, the parametrization

$x = a + (c-a)t,\ y = b + (d-b)t,\ 0 \le t \le 1$

was developed. Here:

$x = 18 + (-16 - 18)t \quad y = 4 + (14 - 4)t$

$x = 18 - 34t \qquad\qquad y = 10t + 4,\ 0 \le t \le 1$

Use window

$-20 \le x \le 20,\ 0 \le y \le 18$

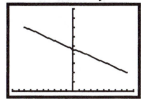

35. a. $k = 1$ $k = 2$

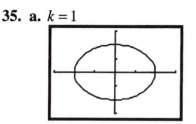   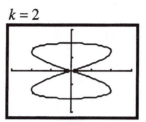

$k = 3$ $k = 4$

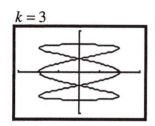   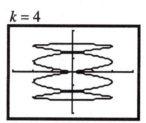

b. $k = 5$ $k = 6$ (The prediction was left to the student.)

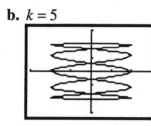

37. Graph using the window
$-15 \le x \le 5, 0 \le y \le 20$.

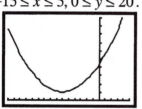

There appears to be a local minimum at $(-6, 2)$.
Since $y = 2$ must be the minimum value for y,
and this corresponds to $t = 0$, therefore $x = -6$,
this is confirmed.

39. Graph using the window
$0 \le x \le 8, 0 \le y \le 6$.

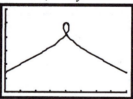

There appears to be a local maximum at $(4, 5)$.
Since $y = 5$ must be the maximum value for y,
and this corresponds to $t = 0$, therefore $x = 4$,
this is confirmed.

41. a. Apply the text formula for projectile
motion to obtain parametric equations
$x = (110 \cos 28°)t$, $y = (110 \sin 28°)t - 16t^2$.

Graph using the window
$0 \le t \le 3.5, 0 \le x \le 350, 0 \le y \le 50$
(calculator in degree mode).

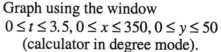

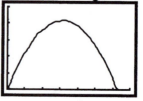

b. Find that $y = 0$ when $t = 3.23$. This is the time that the skeet is in the air.

c. The highest point occurs midway in the flight when $t = 1.61$ seconds; this yields on substitution $y = 41.67$ feet.

43. a. Apply the text formula for projectile motion to obtain parametric equations
$$x = (88\cos 48°)t, \ y = (88\sin 48°)t - 16t^2 + 4$$
Graph using the window $0 \le t \le 4.5, 0 \le x \le 250, 0 \le y \le 100$.

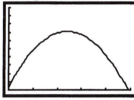

b. Trace to find that when $x = 200$, y is greater than 40, so the arrow will go over the wall.

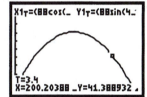

45. Apply the text formula for projectile motion to obtain parametric equations
$$x = (v\cos 30°)t \quad y = (v\sin 30°)t - 16t^2$$
$$x = v\frac{\sqrt{3}}{2}t \qquad y = \frac{v}{2}t - 16t^2$$
When $x = 300, y = 0$, hence
$$300 = v\frac{\sqrt{3}}{2}t \quad 0 = \frac{v}{2}t - 16t^2$$
Thus $t = \dfrac{600}{v\sqrt{3}}$. Substituting into the second equation gives $0 = \dfrac{v}{2}\left(\dfrac{600}{v\sqrt{3}}\right) - 16\left(\dfrac{600}{v\sqrt{3}}\right)^2$.

The positive solution of this equation is $v = 105.29$ ft/sec.

47. a. The four paths are shown in the window $0 \le t \le 8, 0 \le x \le 350, 0 \le y \le 200$.

b. The ball went farthest for the 40° angle.

c. The ball would go farthest of all for a 45° angle. The 40° and 45° angle paths are shown below.

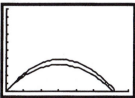

49. a. The diagram shows that $\dfrac{\pi}{2} + \theta = t,\ \theta = t - \dfrac{\pi}{2}$.

In right triangle PQC,

$\cos\theta = CQ/3$ $\qquad\qquad\qquad$ $\sin\theta = PQ/3$

$CQ = 3\cos\theta$ $\qquad\qquad\qquad$ $PQ = 3\sin\theta$

$x = OT - CQ = 3t - 3\cos(t - \pi/2)$ $\qquad$ $y = CT + PQ = 3 + 3\sin(t - \pi/2)$.

b. $\cos(t - \pi/2) = \cos t \cos\pi/2 + \sin t \sin\pi/2 = \sin t$

$\sin(t - \pi/2) = \sin t \cos\pi/2 - \cos t \sin\pi/2 = -\cos t$

Hence $x = 3t - 3\sin t = 3(t - \sin t)$

$\qquad y = 3 - 3\cos t = 3(1 - \cos t)$.

51. a. The diagram shows that $\dfrac{3\pi}{2} + \theta = t,\ \theta = t - \dfrac{3\pi}{2}$.

In right triangle PQC,

$\cos\theta = CQ/3$ $\qquad\qquad\qquad$ $\sin\theta = PQ/3$

$CQ = 3\cos\theta$ $\qquad\qquad\qquad$ $PQ = 3\sin\theta$

$x = OT + CQ = 3t + 3\cos(t - 3\pi/2)$ $\qquad$ $y = CT - PQ = 3 - 3\sin(t - 3\pi/2)$.

b. $\cos(t - 3\pi/2) = \cos t \cos 3\pi/2 + \sin t \sin 3\pi/2 = -\sin t$

$\sin(t - 3\pi/2) = \sin t \cos 3\pi/2 - \cos t \sin 3\pi/2 = \cos t$

Hence $x = 3t - 3\sin t = 3(t - \sin t),\ y = 3 - 3\cos t = 3(1 - \cos t)$.

53. a. Graphs A and B do not collide. A passes through the point of intersection when $t = 1.2$ and B passes through the point when $t = 0.95$.

b. A and B are closest together near $t = 1.10$.

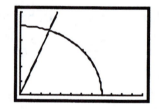

c. A and C very nearly collide near $t = 1.13$.

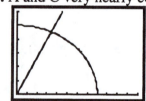

d. The given function is obtained by applying the distance formula to the points $(8\cos t, 5\sin t)$ from A and $(3t, 4t)$ from C. In function mode, graph

$$y = \sqrt{(8\cos x - 3x)^2 + (5\sin x - 4x)^2}$$ and use trace or the minimum routine to show that there is a minimum at $x = 1.13$, however, the minimum is not zero.

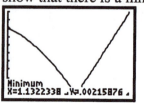

55. a. Examine the figures below:

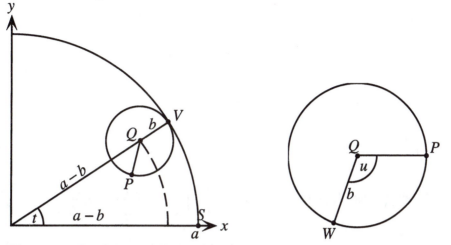

The center Q of the small circle is always at distance $a - b$ from the origin. Let t be the angle that the line from the origin to Q makes with the x-axis. Then the coordinates of Q are $x = (a - b)\cos t$, $y = (a - b)\sin t$. The change in x-coordinate on the small circle from Q to P is $b\cos u$, where u is the angle that PQ makes with the positive x-axis. Similarly, the change in y-coordinate from Q to P is $-b\sin u$. Therefore, the coordinates of P are

$$x = (a - b)\cos t + b\cos u$$

$$y = (a - b)\sin t - b\sin u$$

Since the inner circle rolls without slipping, the arc length that P moves around the inner circle from P to W must equal the arc length that the inner circle has moved along the circumference of the larger circle from S to V. Since the length of a circular arc is the radius times the angle, this means $bu = (a - b)t$, or $u = (a - b)t / b$. Therefore

$$x = (a - b)\cos t + b\cos\left(\frac{a - b}{b}t\right)$$

$$y = (a - b)\sin t - b\sin\left(\frac{a - b}{b}t\right)$$

b. In this case

$x = 4\cos t + \cos 4t$

$y = 4\sin t - \sin 4t$

$0 \le t \le 2\pi, -6 \le x \le 7, -5 \le y \le 5$

c. In this case

$x = 3\cos t + 2\cos\frac{3}{2}t$

$y = 3\sin t - 2\sin\frac{3}{2}t$

$0 \le t \le 4\pi, -6 \le x \le 7, -5 \le y \le 5$

10.6 Polar Coordinates

1. $P - \left(2, \dfrac{\pi}{4}\right), R - (5,\pi), T - \left(4, \dfrac{3\pi}{2}\right), V - (7,0), Q - \left(3, \dfrac{2\pi}{3}\right),$

$S - \left(7, \dfrac{7\pi}{6}\right), U - \left(6, -\dfrac{\pi}{3}\right)$

3. $(-5,3\pi), (-5,-\pi), (5,2\pi), (5,-2\pi)$ **5.** $\left(-1, \dfrac{11\pi}{6}\right), \left(-1, -\dfrac{13\pi}{6}\right), \left(1, \dfrac{5\pi}{6}\right), \left(1, -\dfrac{7\pi}{6}\right)$

7. $r = 3, \theta = \dfrac{\pi}{3}, x = 3\cos\dfrac{\pi}{3} = \dfrac{3}{2}, y = 3\sin\dfrac{\pi}{3} = \dfrac{3\sqrt{3}}{2}, \left(\dfrac{3}{2}, \dfrac{3\sqrt{3}}{2}\right)$

9. $r = -1, \theta = \dfrac{5\pi}{6}, x = -1\cos\dfrac{5\pi}{6} = \dfrac{\sqrt{3}}{2}, y = -1\sin\dfrac{5\pi}{6} = -\dfrac{1}{2}, \left(\dfrac{\sqrt{3}}{2}, -\dfrac{1}{2}\right)$

11. $r = \sqrt{x^2 + y^2} = \sqrt{\left(3\sqrt{3}\right)^2 + (-3)^2} = 6, \tan\theta = \dfrac{y}{x} = \dfrac{-3}{3\sqrt{3}} = -\dfrac{1}{\sqrt{3}}$

$\theta = -\dfrac{\pi}{6}$ (Quadrant IV), $\left(6, -\dfrac{\pi}{6}\right)$

13. $r = \sqrt{2^2 + 4^2} = 2\sqrt{5}, \tan\theta = \dfrac{4}{2} = 2, \theta = \tan^{-1}2$ (Quadrant I),

$\left(2\sqrt{5}, \tan^{-1}2\right) = \left(2\sqrt{5}, 1.107\right)$

15. $r = \sqrt{(-5)^2 + (2.5)^2} = \sqrt{31.25} = \dfrac{5}{2}\sqrt{5}, \tan\theta = \dfrac{2.5}{-5} = -\dfrac{1}{2}$

$\theta = \pi + \tan^{-1}\left(-\dfrac{1}{2}\right),$ $\left(\dfrac{5}{2}\sqrt{5}, \pi + \tan^{-1}\left(-\dfrac{1}{2}\right)\right) = (5.59, 2.6679)$

17.

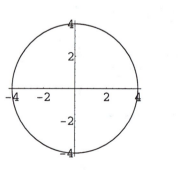

19.

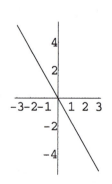

21.

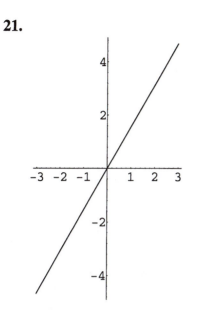

23.

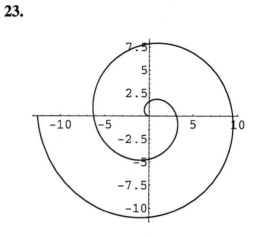

25.

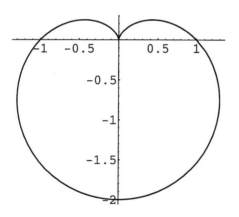

27.

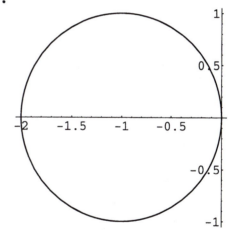

29.

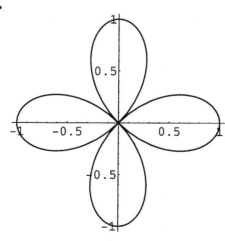

31.

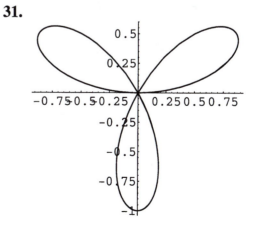

33.

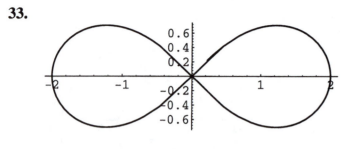

35.

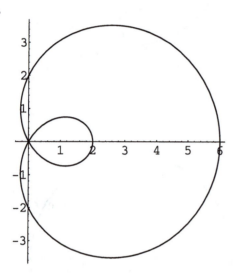

37.

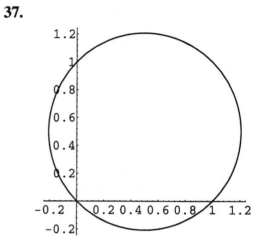

39.

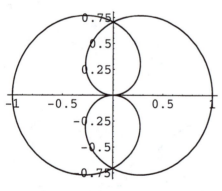

41.

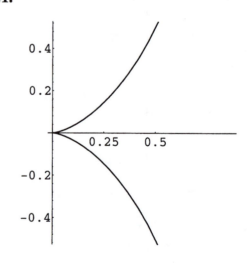

43.

45.

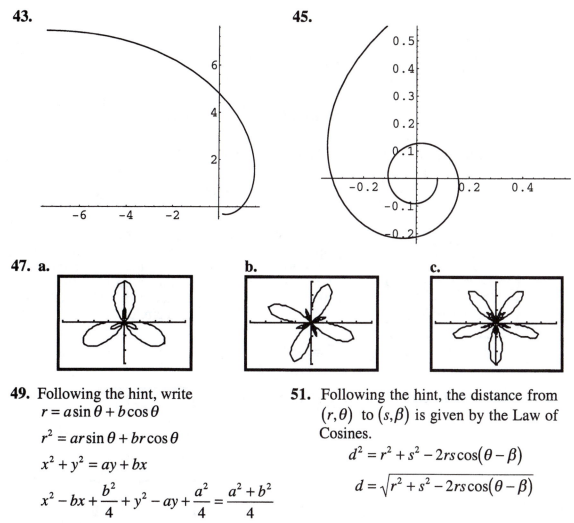

47. a.　　　　　　　　　　**b.**　　　　　　　　　　**c.**

49. Following the hint, write
$r = a\sin\theta + b\cos\theta$

$r^2 = ar\sin\theta + br\cos\theta$

$x^2 + y^2 = ay + bx$

$x^2 - bx + \dfrac{b^2}{4} + y^2 - ay + \dfrac{a^2}{4} = \dfrac{a^2 + b^2}{4}$

$\left(x - \dfrac{b}{2}\right)^2 + \left(y - \dfrac{a}{2}\right)^2 = \dfrac{a^2 + b^2}{4}$

This is the equation of a circle.

51. Following the hint, the distance from
(r, θ) to (s, β) is given by the Law of
Cosines.

$$d^2 = r^2 + s^2 - 2rs\cos(\theta - \beta)$$

$$d = \sqrt{r^2 + s^2 - 2rs\cos(\theta - \beta)}$$

10.7 Polar Equations of Conics

1. $e = 1$, parabola, vertex at $\theta = \pi$. Graph (**d**)

3. $r = \dfrac{3}{1 - 2\sin\theta}$: $e = 2$, hyperbola, vertices at $\theta = \dfrac{3\pi}{2}, \dfrac{\pi}{2}$. Graph (**c**)

5. $r = \dfrac{2}{1 - \frac{2}{3}\sin\theta}$: $e = \dfrac{2}{3}$, ellipse, vertices at $\theta = \dfrac{\pi}{2}, \dfrac{3\pi}{2}$. Graph (**a**)

7. $r = \dfrac{4}{1 + \frac{4}{3}\sin\theta}$: $e = \dfrac{4}{3}$, hyperbola.　　**9.** $r = \dfrac{8/3}{1 + \sin\theta}$: $e = 1$, parabola.

11. $r = \dfrac{1/3}{1 - \frac{2}{3}\cos\theta}$: $e = \dfrac{2}{3}$, ellipse.

13. $a^2 = 100, b^2 = 99, c^2 = a^2 - b^2 = 1, e = c/a = 1/10 = .1$

15. $a^2 = 10, b^2 = 40, c^2 = a^2 + b^2 = 50, e = c/a = \sqrt{50}/\sqrt{10} = \sqrt{5}$

17. Complete the square: $16(x^2 - 2x + 1) - 9(y^2 - 4y + 4) = -124 + 16 - 36$

$$\dfrac{(y-2)^2}{16} - \dfrac{(x-1)^2}{9} = 1; \quad a^2 = 16, b^2 = 9, c^2 = a^2 + b^2 = 25; \ e = \dfrac{c}{a} = \dfrac{5}{4}$$

19. a.

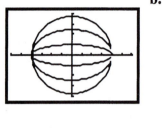

b. The eccentricities are:

$$\dfrac{\sqrt{15}}{4} = .9682$$

$$\dfrac{\sqrt{10}}{4} = .7906$$

$$\dfrac{\sqrt{2}}{4} = .3536$$

c. As the eccentricity approaches 0, the ellipse becomes more circular. An "ellipse" with eccentricity 0 is actually a circle.

21. The vertex of the parabola is $(4, \pi)$. Graph using the window $-5 \le x \le 5, -10 \le y \le 10$.

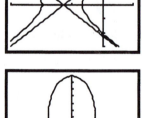

23. The vertices of the hyperbola are $(-2, 0)$ and $\left(\frac{2}{3}, \pi\right)$. Graph using the window $-3 \le x \le 1, -3 \le y \le 3$.

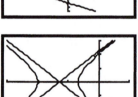

25. The vertices of the ellipse are $\left(10, \frac{\pi}{2}\right)$ and $\left(\frac{10}{7}, \frac{3\pi}{2}\right)$. Graph using the window $-10 \le x \le 10, -2 \le y \le 10$.

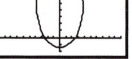

27. The vertices of the ellipse are $(15, 0)$ and $(3, \pi)$. Graph using the window $-5 \le x \le 15, -10 \le y \le 10$.

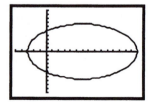

29. The vertex of the parabola is $\left(\frac{3}{2},\frac{\pi}{2}\right)$. Graph using the window $-10 \le x \le 10, -5 \le y \le 2$.

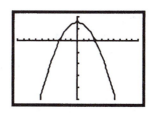

31. The vertices of the hyperbola are $\left(2,\frac{\pi}{2}\right)$ and $\left(-10,\frac{3\pi}{2}\right)$. Graph using the window $-10 \le x \le 10, -5 \le y \le 15$.

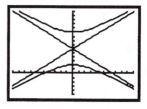

33. The polar equation is of form $r = \dfrac{d}{1-\cos\theta}$. Since $(3,\pi)$ satisfies the equation,

$$3 = \frac{d}{1-\cos\pi}, \quad d = 6, \quad r = \frac{6}{1-\cos\theta}$$

35. The polar equation is of form $r = \dfrac{ed}{1+e\sin\theta}$. Since $\left(2,\dfrac{\pi}{2}\right)$ and $\left(8,\dfrac{3\pi}{2}\right)$ satisfy the

equation, $2 = \dfrac{ed}{1+e\sin\pi/2}$ $\qquad 8 = \dfrac{ed}{1+e\sin 3\pi/2}$

$$2 + 2e = 8 - 8e$$
$$e = 3/5$$
$$ed = 16/5$$
$$r = \frac{16/5}{1+\frac{3}{5}\sin\theta} = \frac{16}{5+3\sin\theta}$$

37. The polar equation is of form $r = \dfrac{ed}{1+e\cos\theta}$. Since $(1,0)$ and $(-3,\pi)$ satisfy the

equation, $1 = \dfrac{ed}{1+e\cos 0}$ $\qquad -3 = \dfrac{ed}{1+e\cos\pi}$

$$1 + e = -3 + 3e$$
$$e = 2$$
$$ed = 3$$
$$r = \frac{3}{1+2\cos\theta}$$

39. Rewrite the equation of the directrix as $r = -2/\cos\theta$, $r\cos\theta = -2$, $x = -2$.
The directrix is a vertical line 2 units to the left of the pole. The equation is of form

$$r = \frac{ed}{1-e\cos\theta} \text{ with } e = 4, d = 2.$$
$$r = \frac{8}{1-4\cos\theta}$$

41. Rewrite the equation of the directrix as $r = -3/\sin\theta$, $r\sin\theta = -3$, $y = -3$.
The directrix is a horizontal line 3 units below the pole. The equation is of form

$$r = \frac{ed}{1 - e\sin\theta} \text{ with } e = 1, d = 3.$$

$$r = \frac{3}{1 - \sin\theta}$$

43. Rewrite the equation of the directrix as $r = 2/\cos\theta$, $r\cos\theta = 2$, $x = 2$.
The directrix is a vertical line 2 units to the right of the pole. The equation is of form

$$r = \frac{ed}{1 + e\cos\theta} \text{ with } e = \frac{1}{2}, d = 2.$$

$$r = \frac{1}{1 + \frac{1}{2}\cos\theta} = \frac{2}{2 + \cos\theta}$$

45. Since the directrix is left of the pole, the equation is of form $r = \dfrac{2d}{1 - 2\cos\theta}$.

Since $(1, 2\pi/3)$ is on the graph,

$$1 = \frac{2d}{1 - 2\cos(2\pi/3)}$$

$$d = 1.$$

$$r = \frac{2}{1 - 2\cos\theta}$$

47. Since $a^2 = \dfrac{e^2 d^2}{\left(1 - e^2\right)^2}$ and $b^2 = \dfrac{e^2 d^2}{1 - e^2}$, $\dfrac{a^2}{b^2} = \dfrac{1}{1 - e^2}$.

Since $e < 1$, $0 < e^2 < 1$, $1 > 1 - e^2 > 0$, $\dfrac{1}{1 - e^2} > 1$.

Therefore $\dfrac{a^2}{b^2} > 1$ and $a^2 > b^2$. $a > b$ since both are positive.

49. Since the orbit is a parabola, $e = 1$.

$$r = \frac{d}{1 - \cos\theta}$$

If $(60, \pi/3)$ is on the graph, then

$$60 = \frac{d}{1 - \cos(\pi/3)}$$

$$d = 30$$

$$r = \frac{30}{1 - \cos\theta} \quad \text{(distance in millions of miles)}$$

Chapter 10 Review Exercises

1. $a^2 = 20, b^2 = 16, c^2 = 20 - 16 = 4$: Ellipse, vertices $\left(0, \pm 2\sqrt{5}\right)$, foci $(0, \pm 2)$.

3. $a^2 = 16, b^2 = 7, c^2 = 16 - 7 = 9$: Ellipse, center $(1,3)$,
   vertices $(1, 3 \pm 4)$, that is , $(1,-1)$ and $(1,7)$, foci $(1, 3 \pm 3)$, that is , $(1,0)$ and $(1,6)$.

5. Rewrite the equation as $x^2 = 10/7\, y$. $4p = 10/7$, $p = 5/14$.
   The axis of the parabola is vertical and it opens upward.
   Focus: $(0, 5/14)$ Directrix: $y = -5/14$

7. Ellipse

9. Ellipse

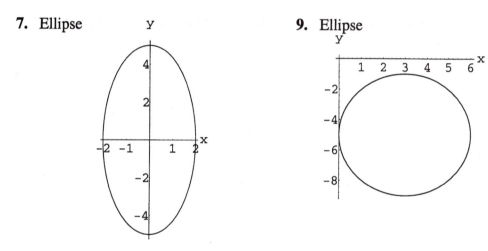

11. $a = 5, b = 2$. Hyperbola
    Asymptotes $y + 4 = \pm\dfrac{5}{2}(x - 1)$

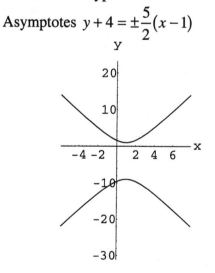

13. Complete the square:

$$x^2 - 10x + 25 + 4y^2 = -9 + 25$$

$$(x-5)^2 + 4y^2 = 16$$

$$\frac{(x-5)^2}{16} + \frac{y^2}{4} = 1$$

Ellipse

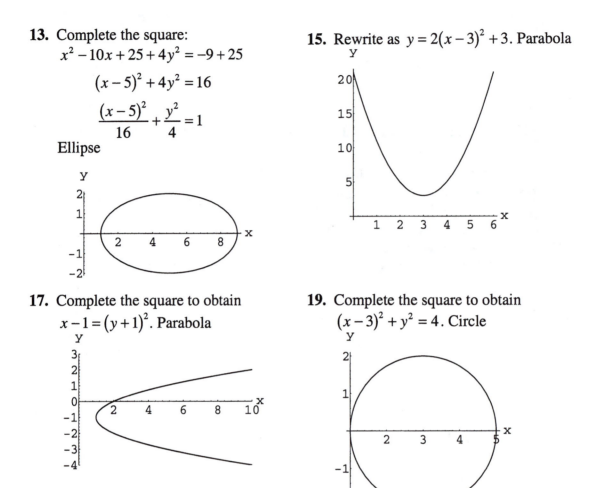

15. Rewrite as $y = 2(x-3)^2 + 3$. Parabola

17. Complete the square to obtain

$x - 1 = (y+1)^2$. Parabola

19. Complete the square to obtain

$(x-3)^2 + y^2 = 4$. Circle

21. Complete the square: $x^2 + 8x + 16 + y^2 + 10y + 25 = -33 + 16 + 25$

$$(x+4)^2 + (y+5)^2 = 8. \text{ Center: } (-4,-5), \text{ radius: } 2\sqrt{2}$$

23. Complete the square: $4(x^2 - 8x + 16) + 3(y^2 + 12y + 36) = -124 + 64 + 108$

$$4(x-4)^2 + 3(y+6)^2 = 48. \text{ Center: } (4,-6)$$

25. Since the distance from center to vertex is 2, $a = 2$. The equation has form

$\dfrac{(x-3)^2}{4} + \dfrac{(y-1)^2}{b^2} = 1$. Since $\left(2, 1 + \sqrt{3/2}\right)$ is on the ellipse,

$$\frac{(2-3)^2}{4} + \frac{\left(1 + \sqrt{3/2} - 1\right)^2}{b^2} = 1$$

$$\frac{1}{4} + \frac{3}{2b^2} = 1$$

$$\frac{1}{b^2} = \frac{1}{2}. \text{ The equation is } \frac{(x-3)^2}{4} + \frac{(y-1)^2}{2} = 1.$$

27. The equation has form $\dfrac{y^2}{a^2} - \dfrac{(x-3)^2}{b^2} = 1$. Since the distance from center to vertex

is 2, write $\dfrac{y^2}{4} - \dfrac{(x-3)^2}{b^2} = 1$. Since $\left(1, \sqrt{5}\right)$ is on the hyperbola,

$$\dfrac{\left(\sqrt{5}\right)^2}{4} - \dfrac{(1-3)^2}{b^2} = 1$$

$$\dfrac{5}{4} - \dfrac{4}{b^2} = 1$$

$$\dfrac{1}{b^2} = \dfrac{1}{16}.\ \text{The equation is } \dfrac{y^2}{4} - \dfrac{(x-3)^2}{16} = 1.$$

29. Since the axis is horizontal and the parabola opens to the left, the equation has form $\left(y-\left(-\tfrac{1}{2}\right)\right)^2 = 4p\left(x - \tfrac{3}{2}\right)$. Since $(-3,1)$ is on the parabola,

$$\left(1 + \tfrac{1}{2}\right)^2 = 4p\left(-3 - \tfrac{3}{2}\right)$$

$$-\tfrac{1}{2} = 4p$$

$$\left(y + \tfrac{1}{2}\right)^2 = -\tfrac{1}{2}\left(x - \tfrac{3}{2}\right).$$

31.

Ellipse.

Parametric equations: $x = 2\cos t$
$\qquad\qquad\qquad\qquad y = 5\sin t$

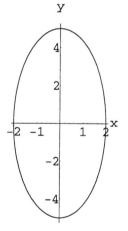

33.

Ellipse.

Parametric equations: $x = 3\cos t + 3$
$\qquad\qquad\qquad\qquad y = 2\sin t - 5$

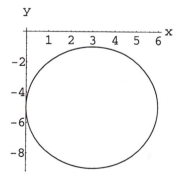

35. Hyperbola.
Parametric equations: $x = 2\tan t + 1$
$$y = 5/\cos t - 4$$

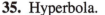

37. Following the hint, write the equation of the ellipse as $\dfrac{x^2}{25} + \dfrac{y^2}{100} = 1$. Then the points

with y-coordinate 8 satisfy $\dfrac{x^2}{25} + \dfrac{8^2}{100} = 1$, $x = \pm 3$. The distance between these points

is $3 - (-3) = 6$ feet.

39. $A = 3,\, B = 2\sqrt{2},\, C = 2,\, B^2 - 4AC = \left(2\sqrt{2}\right)^2 - 4(3)(2) = -8 < 0$. Ellipse

41. $A = -3,\, B = 4,\, C = 0,\, B^2 - 4AC = 4^2 - 4(-3)(0) = 16 > 0$. Hyperbola

43. Solve for y to obtain
$$y = \frac{x \pm \sqrt{24 - 3x^2}}{2}.$$
Graph using $-6 \le x \le 6, -4 \le y \le 4$.

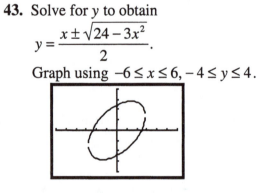

45. Solve for y to obtain $y = \dfrac{2 - x^2}{x}$.
Graph using $-9 \le x \le 9, -6 \le y \le 6$.

47. Solve for y to obtain
$$y = \frac{-2\sqrt{2} - 3x \pm \sqrt{8 + 20x\sqrt{2} + 5x^2}}{2}.$$
Graph using
$-15 \le x \le 10, -10 \le y \le 20$.

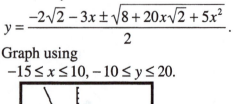

49. $x = u\cos 60° - v\sin 60° = \dfrac{u - v\sqrt{3}}{2}$

$y = u\sin 60° + v\cos 60° = \dfrac{u\sqrt{3} + v}{2}$

51. $A = 1,\, B = 1,\, C = 1$

$\cot 2\theta = \dfrac{A - C}{B} = \dfrac{1 - 1}{1} = 0$

$2\theta = 90°$

$\theta = 45°$

53. $-10 \le x \le 10,\, -10 \le y \le 10$

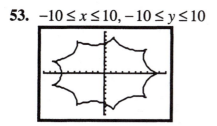

55. $-35 \le x \le 32,\, -2 \le y \le 16$

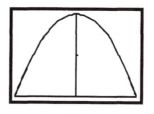

57. Graph using the window
$-10 \le x \le 10,\, -10 \le y \le 10$

$t = 2 - y$

$x = 2(2 - y) - 1$

$x = 3 - 2y$

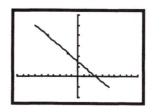

59. Graph using the window
$-1 \le x \le 1,\, 0 \le y \le 2$

$x^2 = \cos^2 t \quad \dfrac{y}{2} = \sin^2 t$

$x^2 + \dfrac{y}{2} = 1$

$y = 2 - 2x^2$

61. **(b)** and **(c)** only yield a portion of the parabola and therefore are not parametrizations of the curve.

63.

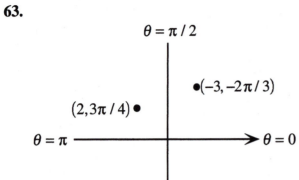

65.

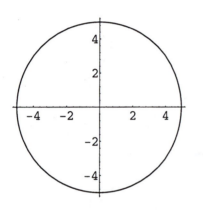

67.

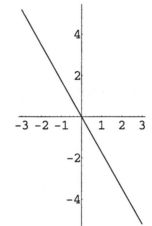

69.

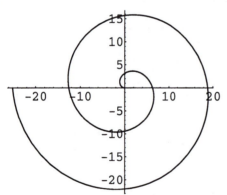

71.

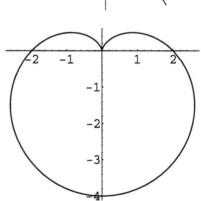

73.

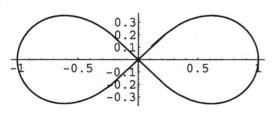

75. $x = r\cos\theta = 3\cos\left(-\dfrac{2\pi}{3}\right) = 3\left(-\dfrac{1}{2}\right) = -\dfrac{3}{2}$

$y = r\sin\theta = 3\sin\left(-\dfrac{2\pi}{3}\right) = 3\left(-\dfrac{\sqrt{3}}{2}\right) = -\dfrac{3\sqrt{3}}{2}$

$\left(-\dfrac{3}{2}, -\dfrac{3\sqrt{3}}{2}\right)$

77. Rewrite the equation as $\dfrac{x^2}{28} + \dfrac{y^2}{84} = 1$.

$a^2 = 84, b^2 = 28, c^2 = 56$

$e = \dfrac{c}{a} = \dfrac{\sqrt{56}}{\sqrt{84}} = \dfrac{\sqrt{6}}{3}$

79. The vertices of the ellipse are $\left(12, \frac{\pi}{2}\right)$ and $\left(4, \frac{3\pi}{2}\right)$.
Graph using the window $-15 \le x \le 15, -5 \le y \le 15$.

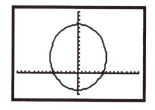

81. The vertices of the hyperbola are $(4,0)$ and $(-2,\pi)$.
Graph using the window $0 \le x \le 6, -8 \le y \le 8$.

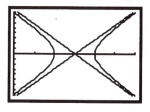

83. The polar equation is of form

$r = \dfrac{ed}{1 + e\cos\theta}$. Since $(4,0)$ and $(6,\pi)$

satisfy the equation

$4 = \dfrac{ed}{1+e} \qquad 6 = \dfrac{ed}{1-e}$.

$4 + 4e = 6 - 6e$

$e = 1/5$

$ed = 24/5$

$r = \dfrac{24/5}{1 + \frac{1}{5}\cos\theta} = \dfrac{24}{5 + \cos\theta}$

85. Rewrite the equation of the directrix as $r = \dfrac{2}{\cos\theta}$, $r\cos\theta = 2$, $x = 2$. The directrix is

a vertical line 2 units to the right of the pole. $d = 2, e = 1$

The polar equation is $r = \dfrac{2}{1 + \cos\theta}$.

Chapter 11
Systems of Equations

11.1 Systems of Linear Equations in Two Variables

1. Substitute:

$2(-1) + 3 = 1?$ $1 = 1$ True

$-3(-1) + 2(3) = 9?$ $9 = 9$ True
The values are a solution.

3. Substitute:

$\frac{1}{3}(2) + \frac{1}{2}(-1) = \frac{1}{6}?$ $\frac{1}{6} = \frac{1}{6}$ True

$\frac{1}{2}(2) + \frac{1}{3}(-1) = \frac{2}{3}?$ $\frac{2}{3} = \frac{2}{3}$ True
The values are a solution.

5. Since the last equation requires $z = 1$, values for which $z = -1$ cannot be a solution.

7.
$$x - 2y = 5$$
$$2x + y = 3$$
$$x = 2y + 5$$
$$2(2y + 5) + y = 3$$
$$y = -7/5$$
$$x = 2(-7/5) + 5$$
$$x = 11/5$$

9.
$$3x - 2y = 4$$
$$2x + y = -1$$
$$y = -2x - 1$$
$$3x - 2(-2x - 1) = 4$$
$$x = 2/7$$
$$y = -2(2/7) - 1$$
$$y = -11/7$$

11.
$$r + s = 0$$
$$r - s = 5$$
$$r = -s$$
$$-s - s = 5$$
$$s = -5/2$$
$$r = 5/2$$

13.
$$x + y = c + d$$
$$x - y = 2c - d$$
$$x = y + 2c - d$$
$$y + 2c - d + y = c + d$$
$$y = d - \tfrac{1}{2}c$$
$$x = d - \tfrac{1}{2}c + 2c - d$$
$$x = \tfrac{3}{2}c$$

15.
$$2x - 2y = 12$$
$$\underline{-2x + 3y = 10}$$
$$y = 22$$
$$2x - 2(22) = 12$$
$$x = 28$$

17.
$$x + 3y = -1$$
$$2x - y = 5$$
Multiply the second
equation by 3.
$$x + 3y = -1$$
$$\underline{6x - 3y = 15}$$
$$7x \quad = 14$$
$$x = 2$$
$$2(2) - y = 5$$
$$y = -1$$

19.
$$2x + 3y = 15$$
$$8x + 12y = 40$$
Multiply the first
equation by -4.
$$-8x - 12y = -60$$
$$\underline{8x + 12y = 40}$$
$$0 = -20$$
No solution.

21.
$$3x - 2y = 4$$
$$6x - 4y = 8$$
Multiply the first
equation by -2.
$$-6x + 4y = -8$$
$$\underline{6x - 4y = 8}$$
$$0 = 0$$
Infinitely many solutions.
Let $x = b$, any real number.
Then $3b - 2y = 4$, $y = \dfrac{3b - 4}{2}$.

318

23.
$$\frac{x}{3}-\frac{y}{2}=-3$$

$$\frac{2x}{5}+\frac{y}{5}=-2$$

Clear of fractions.

$$2x-3y=-18$$

$$2x+y=-10$$

Multiply the first

equation by -1.

$$-2x+3y=18$$

$$\underline{2x+y=-10}$$

$$4y=8$$

$$y=2$$

$$2x+2=-10$$

$$x=-6$$

25.
$$\frac{x+y}{4}-\frac{x-y}{3}=1$$

$$\frac{x+y}{4}+\frac{x-y}{2}=9$$

Clear of fractions
and collect terms to
obtain:

$$-x+7y=12$$

$$3x-y=36$$

Multiply the first

equation by 3.

$$-3x+21y=36$$

$$\underline{3x-y=36}$$

$$20y=72$$

$$y=3.6$$

$$-x+7(3.6)=12$$

$$x=13.2$$

27.
$$\frac{1}{x}-\frac{3}{y}=2$$

$$\frac{2}{x}+\frac{1}{y}=3$$

Multiply the second

equation by 3.

$$\frac{1}{x}-\frac{3}{y}=2$$

$$\frac{6}{x}+\frac{3}{y}=9$$

$$\overline{}$$

$$\frac{7}{x}=11$$

$$\frac{1}{x}=\frac{11}{7}$$

$$x=\frac{7}{11}$$

$$\frac{11}{7}-\frac{3}{y}=2$$

$$y=-7$$

29.
$$\frac{2}{x}+\frac{3}{y}=8$$

$$\frac{3}{x}-\frac{1}{y}=1$$

Multiply the second

equation by 3.

$$\frac{2}{x}+\frac{3}{y}=8$$

$$\frac{9}{x}-\frac{3}{y}=3$$

$$\overline{}$$

$$\frac{11}{x}=11$$

$$x=1$$

$$\frac{2}{1}+\frac{3}{y}=8$$

$$y=\frac{1}{2}$$

31. **a.** Solve each equation for y and observe that a positive
slope represents an increasing population; a negative
slope represents a decreasing population.

b. Write $y=-12.04x+2099.64$, $y=26.47x+660.64$.
Solve graphically to obtain $x=37.3$, which
corresponds to 1987.

33. Write $y=\dfrac{2106x+74{,}148}{3}$, $y=\dfrac{26{,}366+1190x}{2}$.

Graph and observe that since the lines are diverging,
they will never intersect.

35.
$$ax+by=r$$

$$cx+dy=s$$

Multiply the first equation
by d and the second by $-b$.

$$adx+bdy=rd$$

$$\underline{-bcx-bdy=-sb}$$

$$x(ad-bc)=rd-sb$$

$$x=\frac{rd-sb}{ad-bc}$$

$$ax+by=r$$

$$cx+dy=s$$

Multiply the first equation
by $-c$ and the second by a.

$$-acx-bcy=-rc$$

$$\underline{acx+ady=as}$$

$$(ad-bc)y=as-rc$$

$$y=\frac{as-rc}{ad-bc}$$

37. Since the two lines have slopes $-\frac{1}{2}$ and 2, they are perpendicular and intersect in one point, regardless of c. Therefore there is exactly one solution for the system.

39. Substitute: $4d = 2$
$2c + 16d = 2$ Solve to obtain $d = \frac{1}{2}$, $c = -3$.

41. Substitute to obtain $.75x = 16 - 1.25x$. Solve to obtain $x = 8$. Then $p = .75(8) = 6$.
Price: $6; Quantity: 8000.

43. Substitute to obtain $300 - 30x = 80 + 25x$. Solve to obtain $x = 4$.
Then $p = 300 - 30(4) = 180$. Price: $180; Quantity: 4000.

45. Let $x =$ the number of adults and $y =$ the number of children.
$$x + y = 200$$
$$3x + 1.5y = 510$$
Solve to obtain $x = 140$ adults and $y = 60$ children.

47. Let $x =$ the speed of the boat and $y =$ the speed of the current.
Then the speed upstream $= x - y$ and the speed downstream $= x + y$.
$\left(15 \text{ minutes} = \frac{1}{4} \text{ hour, } 12 \text{ minutes} = \frac{1}{5} \text{ hour}\right)$
$$\frac{1}{4}(x - y) = 4$$
$$\frac{1}{5}(x + y) = 4$$
Solve to obtain $x = 18$ mph, $y = 2$ mph.

49. Clearly the cost of x pounds of cashews is $4.4x$ and the cost of y pounds of peanuts is $1.2y$. Hence: $4.4x + 1.2y = 6$
$$x + y = 3$$
Solve to obtain $x = .75 \left(\frac{3}{4} \text{ pounds of cashews}\right)$ and $y = 2.25 \left(2\frac{1}{4} \text{ pounds of peanuts}\right)$.

51. Let x be the amount of 50% alloy and y be the amount of 75% alloy. Then
$$x + y = 40 \quad \text{(amounts of alloy add up)}$$
$$.5x + .75y = .6(40) \quad \text{(amounts of silver add up)}$$
Clear the decimals to obtain: $\quad x + y = 40$
$$50x + 75y = 2400$$
Solve to obtain $x = 24$ grams of 50% alloy, $y = 16$ grams of 75% alloy.

53. Note that $80 per month equals $960 per year, and $9.50 per month is $114 per year.
a. Electric: $y = 2000 + 960x$. Solar: $y = 14{,}000 + 114x$.
b. Electric: $y = 2000 + 960(5) = \$6800$. Solar: $y = 14{,}000 + 114(5) = \$14{,}570$.
c. Solve the system of equations in part **a** by substitution:
$$2000 + 960x = 14{,}000 + 114x$$
$$846x = 12{,}000$$
$$x = 14.2$$
The costs will be the same in the fifteenth year. Electric heating will be cheaper before that: solar afterwards.

55. a. $R = 250,500x$ **b.** $C = 1,295,000 + 206,500x$ **c.** $P = R - C = 44,000x - 1,295,000$
   d. $R = 109,000x$, $C = 440,000 + 82,000x$, $P = R - C = 27,000x - 440,000$
   e. Set the two expressions for profit equal and solve for x,
   $$27,000x - 440,000 = 44,000x - 1,295,000$$
   $$855,000 = 17,000x$$
   $$x = 50.3 \text{ weeks}$$
   $$P = \$918,000$$
   f. It would be better to open off Broadway if a run of 50 weeks or less is expected.

57. a. $R = 60x$
   b. Set $R = C$.
   $$60x = 45x + 6000$$
   $$15x = 6000$$
   $$x = 400 \text{ hedge trimmers}$$

59. Set $R = C$.
   $$95x = 80x + 7500$$
   $$15x = 7500$$
   $$x = 500$$
   Since the break-even point is above the maximum number that can be sold, the product should not be produced.

61. Let x = amount at 9% and y = amount at 11%. Then
   $$y = x - 8000$$
   $$.09x + .11y = 2010$$
   Solve by substitution: $.09x + .11(x - 8000) = 2010$, $x = \$14,450$, $y = \$6,450$.

63. a.   **b.**

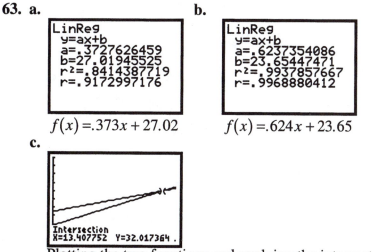

$f(x) = .373x + 27.02$ $f(x) = .624x + 23.65$

   c.

Plotting the two functions and applying the intersection routine yields the year 2003.

65. Let x = how many bowls and y = how many plates. Then
   $$3x + 2y = 480 \quad \text{(time)}$$
   $$.25x + .20y = 44 \quad \text{(costs)}$$
   Solve the system to obtain $x = 80$ bowls and $y = 120$ plates.

11.1A Systems of Non-Linear Equations

1. Solve the first equation for y in terms of x to obtain $y = x^2$. Substitute into the second equation to obtain:

$$-2x + x^2 = 3$$

$$x^2 - 2x - 3 = 0$$
$$x = 3 \quad x = -1$$

If $x = 3$, $y = 9$. If $x = -1$, $y = 1$.

3. Solve the first equation for y in terms of x to obtain $y = x^2$. Substitute into the second equation to obtain:

$$x + 3x^2 = 6$$
$$3x^2 + x - 6 = 0$$

$$x = \frac{-1 \pm \sqrt{73}}{6}$$

If $x = \dfrac{-1 - \sqrt{73}}{6}$, $y = \dfrac{37 + \sqrt{73}}{18}$.

If $x = \dfrac{-1 + \sqrt{73}}{6}$, $y = \dfrac{37 - \sqrt{73}}{18}$.

5. Solve the first equation for y in terms of x to obtain $y = 10 - x$. Substitute into the second equation to obtain:

$$x(10 - x) = 21$$
$$x^2 - 10x + 21 = 0$$
$$x = 7 \quad x = 3$$

If $x = 7$, $y = 3$. If $x = 3$, $y = 7$.

7. Solve the second equation for x in terms of y to obtain $x = 2y + 4$. Substitute into the first equation to obtain: $(2y + 4)y + 2y^2 = 8$

$$y^2 + y - 2 = 0$$
$$y = 1 \quad y = -2$$

If $y = 1$, $x = 6$. If $y = -2$, $x = 0$.

9. Solve the second equation for x in terms of y to obtain $x = y + 2$. Substitute into the first equation to get:

$$(y + 2)^2 + y^2 - 4(y + 2) - 4y = -4$$
$$y^2 + 4y + 4 + y^2 - 4y - 8 - 4y = -4$$
$$2y^2 - 4y = 0$$
$$y = 0 \quad y = 2$$

If $y = 0$, $x = 2$. If $y = 2$, $x = 4$.

11. Solve the second equation for x^2 in terms of y to obtain $x^2 = 19 - y$. Substitute into the first equation to get:

$$19 - y + y^2 = 25$$
$$y^2 - y - 6 = 0$$
$$y = 3 \quad y = -2$$

If $y = 3$, $x = \pm 4$.

If $y = -2$, $x = \pm\sqrt{21}$. (Four solutions.)

13. Graph both equations using the window $-3 \le x \le 5$, $-15 \le y \le 10$, and apply the intersection routine to obtain $(-1.6237, -8.1891)$, $(1.3163, 1.0826)$, and $(2.8073, 2.4814)$. (Only one of the calculator screens is shown.)

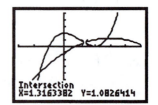

Intersection
X=1.3163382   Y=1.0826414

15. Graph both equations using the window $-3 \le x \le 3$, $-1 \le y \le 5$, and apply the intersection routine to obtain $(-1.9493, .4412)$, $(.3634, .9578)$, and $(1.4184, .5986)$. (Only one of the calculator screens is shown.)

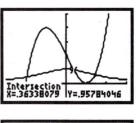

17. Graph both equations using the window $-\pi \le x \le \pi$, $-2 \le y \le 2$, and apply the intersection routine to obtain $(-.9519, -.8145)$.

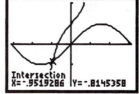

19. $25x^2 - 16y^2 = 400$

$-9x^2 + 4y^2 = -36$

Multiply the second equation by 4.

$25x^2 - 16y^2 = 400$

$\underline{-36x^2 + 16y^2 = -144}$

$-11x^2 \qquad = 256$

This equation has no real solutions, and therefore the system has no solutions.

21. Solve the second equation for x in terms of y to obtain $x = y + 2$. Substitute into the first equation to obtain $5(y+2)^2 + 3y^2 - 20(y+2) + 6y = -8$. Solve this equation using the quadratic formula to obtain $y = \dfrac{-3 \pm \sqrt{105}}{8}$. The solutions are therefore:

$$x = \frac{13 + \sqrt{105}}{8}, \; y = \frac{-3 + \sqrt{105}}{8} \quad \text{and} \quad x = \frac{13 - \sqrt{105}}{8}, \; y = \frac{-3 - \sqrt{105}}{8}$$

23. Solve the second equation for y in terms of x to obtain $y = x^2 + x + 1$. Substitute into the first to get: $x^2 + 4x(x^2 + x + 1) + 4(x^2 + x + 1)^2 - 30x - 90(x^2 + x + 1) + 450 = 0$

Graph the left side of this equation using the window $-7 \le x \le 7$, $-150 \le y \le 500$ and apply the zero (root) routine to obtain four roots: -4.8093, -3.1434, 2.1407 and 2.8120. Since $y = x^2 + x + 1$ the solutions are therefore:

$(-4.8093, 19.3201), (-3.1434, 7.7374), (2.1407, 7.7230)$ and $(2.8120, 11.7195)$.

25. Solve the second equation for y in terms of x to obtain $y = \pm\sqrt{64 - 4x^2}$. Solve the first equation for y in terms of x to obtain: $y = \dfrac{3x - 5 \pm \sqrt{x^2 - 24x + 37}}{2}$.

Graph these four equations using the window $-5 \le x \le 5$, $-10 \le y \le 10$ and apply the intersection routine to obtain $(-3.8371, -2.2596)$, and $(-.9324, -7.7796)$. (Only one of the calculator screens is shown.)

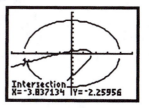

27. Solve the equations for y in terms of x to obtain
$$y = \frac{-3x \pm \sqrt{5x^2 + 8}}{2} \quad \text{and} \quad y = \frac{5x \pm \sqrt{84 - 11x^2}}{6}$$
Graph these four equations using the window $-4 \le x \le 4, -4 \le y \le 4$ and apply the intersection routine to obtain $(\pm 1.4873, \mp .0480)$ and $(\pm .0480, \mp 1.4873)$. (Only one of the calculator screens is shown.)

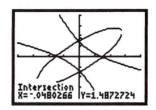

29. The system would become $y = 70 - 2x$, $y^2 = 4\pi\left(100 - x^2/8\right)$.
The calculator shows that the two graphs do not intersect, hence there is no solution.

31. Let x be the length and width of the square ends and y be the third dimension. Then
$x^2 y = 16$ (volume) and $2x^2 + 4xy = 40$ (surface area).
Solve the first equation for y in terms of x to obtain $y = 16/x^2$, then substitute into the second equation to obtain: $2x^2 + 4x\left(16/x^2\right) = 40$ or $2x^3 - 40x + 64 = 0$.
Solve graphically or algebraically to obtain $x = 2$ and $x = 3.1231$ as the two positive roots. If $x = 2$ meters, $y = 4$ meters. If $x = 3.1231$ meters, $y = 1.6404$ meters.

33. Let x and y be the numbers. Then
$$x + y = -16$$
$$xy = 48$$
Solve the first equation for y in terms of x to obtain $y = -16 - x$.
Substitute into the second equation to obtain: $x(-16 - x) = 48$
$$x^2 + 16x + 48 = 0$$
$$x = -4 \quad x = -12$$
The two numbers are $x = -4$ and $y = -12$ (or the other way around).

35. Let x and y be the numbers. Then
$$y = x + 1$$
$$xy = 4.16$$
Substitute the expression for y from the first equation into the second to obtain: $x(x + 1) = 4.16$
$$x^2 + x - 4.16 = 0$$
$$(x - 1.6)(x + 2.6) = 0$$
The only positive solution is $x = 1.6, y = 2.6$.

37. Let x and y be the numbers. Then
$$x + y = 3$$
$$x^2 + y^2 = 369$$
Solve the first equation for y in terms of x to obtain $y = 3 - x$.
Substitute into the second equation to obtain: $x^2 + (3 - x)^2 = 369$
$$x^2 - 3x - 180 = 0$$
$$x = 15 \quad x = -12$$
The two numbers are $x = 15$ and $y = -12$ (or the other way around).

39. Let x be the length and y be the width. Then: $2x + 2y = 58$
$$xy = 204$$
Solve the first equation for y in terms of x to obtain $y = 29 - x$.
Substitute into the second equation to obtain: $x(29 - x) = 204$
$$x^2 - 29x + 204 = 0$$
$$x = 12 \text{ or } x = 17$$
The dimensions are $x = 12$ feet and $y = 17$ feet (or the other way around).

41. Let x be the length and y be the width. Then
$$xy = 120$$
$$x^2 + y^2 = 17^2$$
Solve the first equation for y in terms of x to obtain $y = 120/x$.

Substitute into the second equation to obtain: $x^2 + (120/x)^2 = 289$
$$x^4 - 289x^2 + 14,400 = 0$$
$$(x - 8)(x + 8)(x - 15)(x + 15) = 0$$
The positive solutions are $x = 8$ in and $x = 15$ in. Either yields the same dimensions of 8 in $\times 15$ in for the rectangle.

43. The equation of the line is $y - 9 = m(x - 3)$. It remains to determine m. If the system consisting of this equation and $y = x^2$ is solved, the equation $x^2 - 9 = m(x - 3)$ results.
$$x^2 - mx + 3m - 9 = 0$$
This quadratic equation will have exatly one solution if the discriminant is 0.
$$(-m)^2 - 4(3m - 9) = 0$$
$$m^2 - 12m + 36 = 0$$
$$m = 6 \quad \text{The equation must be } y - 9 = 6(x - 3) \text{ or } y = 6x - 9.$$

11.2 Large Systems of Linear Equations

1.
$$\begin{pmatrix} 2 & -3 & 4 & | & 1 \\ 1 & 2 & -6 & | & 0 \\ 3 & -7 & 4 & | & -3 \end{pmatrix}$$

3.
$$\begin{pmatrix} 1 & -\frac{1}{2} & \frac{7}{4} & | & 0 \\ 2 & -\frac{3}{2} & 5 & | & 0 \\ 0 & -2 & \frac{1}{3} & | & 0 \end{pmatrix}$$

5.
$$2x - 3y = 1$$
$$4x + 7y = 2$$

7.
$$x + \quad z \quad = 1$$
$$x - y + 4z - 2w = 3$$
$$4x + 2y + 5z \quad = 2$$

9. $x = 3/2,\, y = 5,\, z = -2,\, w = 0$

11.
$$x + w = 2$$
$$y + 2w = -3$$
$$z = 4$$
Yields: $x = 2 - w,\, y = -3 - 2w,\, z = 4,\, w$ any real number.

13. Original:

$$-x \quad +3y \quad +2z = 0 \quad [A]$$
$$2x \quad -y \quad -z = 3 \quad [B]$$
$$x \quad +2y \quad +3z = 0 \quad [C]$$

Eliminate x:

$$x \quad +2y \quad +3z = 0 \quad [C]$$
$$5y \quad +5z = 0 \quad [A+C]=[D]$$
$$-5y \quad -7z = 3 \quad [B-2C]=[E]$$

Then eliminate y:

$$x \quad +2y \quad +3z = 0 \quad [C]$$
$$y \quad +z = 0 \quad [D] \div 5$$
$$-2z = 3 \quad [E+D]$$

Then $z = -3/2$. Back substitution
yields $y = 3/2$, $x = 3/2$.

15. Write the augmented matrix

$$\begin{pmatrix} 1 & 1 & 1 & | & 1 \\ 1 & -2 & 2 & | & 4 \\ 2 & -1 & 3 & | & 5 \end{pmatrix}$$

Obtain zeros in Column 1:

$$\begin{pmatrix} 1 & 1 & 1 & | & 1 \\ 0 & -3 & 1 & | & 3 \\ 0 & -3 & 1 & | & 3 \end{pmatrix} \begin{matrix} \\ R_2 - R_1 \\ R_3 - 2R_1 \end{matrix}$$

Get a zero in Col. 2, eliminating Row 3:

$$\begin{pmatrix} 1 & 1 & 1 & | & 1 \\ 0 & -3 & 1 & | & 3 \\ 0 & 0 & 0 & | & 0 \end{pmatrix} \begin{matrix} \\ \\ R_3 - R_2 \end{matrix}$$

This is the augmented matix of:

$$x + \quad y + z = 1$$
$$-3y + z = 3$$

Set $z = t$. Then $y = -1 + \frac{1}{3}t$, $x = 2 - \frac{4}{3}t$.

17. Write the augmented matrix:

$$\begin{pmatrix} 1 & -2 & 4 & | & 6 \\ 1 & 1 & 13 & | & 6 \\ -2 & 6 & -1 & | & -10 \end{pmatrix}$$

First, obtain zeros in Col. 1:

$$\begin{pmatrix} 1 & -2 & 4 & | & 6 \\ 0 & 3 & 9 & | & 0 \\ 0 & 2 & 7 & | & 2 \end{pmatrix} \begin{matrix} \\ R_2 - R_1 \\ R_3 + 2R_1 \end{matrix}$$

Second, obtain a 1 as the leading
coefficient in Row 2, and 0's above
and below:

$$\begin{pmatrix} 1 & 0 & 10 & | & 6 \\ 0 & 1 & 3 & | & 0 \\ 0 & 0 & 1 & | & 2 \end{pmatrix} \begin{matrix} R_1 + \frac{2}{3}R_2 \\ \frac{1}{3}R_2 \\ R_3 - \frac{2}{3}R_2 \end{matrix}$$

Last, obtain zeros in Col. 3:

$$\begin{pmatrix} 1 & 0 & 0 & | & -14 \\ 0 & 1 & 0 & | & -6 \\ 0 & 0 & 1 & | & 2 \end{pmatrix}$$

The solution is read directly:

$$x = -14, y = -6, z = 2.$$

Details of the reduction to reduced row-echelon form are not supplied from here on; the rref routine on the T-83 should be used.

19. Write the augmented matrix:
$$\begin{pmatrix} 1 & 1 & 1 & 200 \\ 1 & -2 & 0 & 0 \\ 2 & 3 & 5 & 600 \\ 2 & -1 & 1 & 200 \end{pmatrix}$$
Reduce to: $\begin{pmatrix} 1 & 0 & 0 & 100 \\ 0 & 1 & 0 & 50 \\ 0 & 0 & 1 & 50 \\ 0 & 0 & 0 & 0 \end{pmatrix}$
The solution is
$$x = 100, \, y = 50, \, z = 50.$$

21. Write the augmented matrix:
$$\begin{pmatrix} 11 & 10 & 9 & 5 \\ 1 & 2 & 3 & 1 \\ 3 & 2 & 1 & 1 \end{pmatrix}$$
Start by exchanging Rows 1 and 2:
$$\begin{pmatrix} 1 & 2 & 3 & 1 \\ 11 & 10 & 9 & 5 \\ 3 & 2 & 1 & 1 \end{pmatrix}$$
In reduced row-echelon form this
becomes: $\begin{pmatrix} 1 & 0 & -1 & 0 \\ 0 & 1 & 2 & \frac{1}{2} \\ 0 & 0 & 0 & 0 \end{pmatrix}$
The solutions are:
$$z = t, \, y = -2t + \tfrac{1}{2}, \, x = t.$$

23. Write the augmented matrix:
$$\begin{pmatrix} 1 & 1 & 3 \\ 5 & -1 & 3 \\ 9 & -4 & 1 \end{pmatrix}$$
In reduced row-echelon form this
becomes: $\begin{pmatrix} 1 & 0 & 1 \\ 0 & 1 & 2 \\ 0 & 0 & 0 \end{pmatrix}$
The solution is $x = 1, \, y = 2$.

25. Write the augmented matrix:
$$\begin{pmatrix} 1 & -4 & -13 & 4 \\ 1 & -2 & -3 & 2 \\ -3 & 5 & 4 & 2 \end{pmatrix}$$
In reduced row-echelon form this
becomes: $\begin{pmatrix} 1 & 0 & 7 & 0 \\ 0 & 1 & 5 & -1 \\ 0 & 0 & 0 & 1 \end{pmatrix}$
The last row represents the equation
$0 = 1$, hence the system has no solution.

27. Write the augmented matrix:
$$\begin{pmatrix} 4 & 1 & 3 & 7 \\ 1 & -1 & 2 & 3 \\ 3 & 2 & 1 & 4 \end{pmatrix}$$
In reduced row-echelon form this
becomes: $\begin{pmatrix} 1 & 0 & 1 & 2 \\ 0 & 1 & -1 & -1 \\ 0 & 0 & 0 & 0 \end{pmatrix}$
The solutions are
$$z = t, \, y = t - 1, \, x = -t + 2.$$

29. Write the augmented matrix:
$$\begin{pmatrix} 1 & 1 & 1 & 0 \\ 3 & -1 & 1 & 0 \\ 5 & -1 & 1 & 0 \end{pmatrix}$$
In reduced row-echelon form this
becomes: $\begin{pmatrix} 1 & 0 & 0 & 0 \\ 0 & 1 & 0 & 0 \\ 0 & 0 & 1 & 0 \end{pmatrix}$
The solution is
$$x = 0, \, y = 0, \, z = 0.$$

31. Write the augmented matrix:

$$\begin{pmatrix} 2 & 1 & 3 & -2 & -6 \\ 4 & 3 & 1 & -1 & -2 \\ 1 & 1 & 1 & 1 & -5 \\ -2 & -2 & 2 & 2 & -10 \end{pmatrix}$$

In reduced row-echelon form this becomes:

$$\begin{pmatrix} 1 & 0 & 0 & 0 & -1 \\ 0 & 1 & 0 & 0 & 1 \\ 0 & 0 & 1 & 0 & -3 \\ 0 & 0 & 0 & 1 & -2 \end{pmatrix}$$

The solution is
$$x = -1, y = 1, z = -3, w = -2.$$

33. Write the augmented matrix:

$$\begin{pmatrix} 1 & -2 & -1 & -3 & -3 \\ -1 & 1 & 1 & 0 & 0 \\ 0 & 4 & 3 & -2 & -1 \\ 2 & -2 & 0 & 1 & 1 \end{pmatrix}$$

In reduced row-echelon form this becomes:

$$\begin{pmatrix} 1 & 0 & 0 & 0 & \frac{7}{31} \\ 0 & 1 & 0 & 0 & \frac{6}{31} \\ 0 & 0 & 1 & 0 & \frac{1}{31} \\ 0 & 0 & 0 & 1 & \frac{29}{31} \end{pmatrix}$$

The solution is $x = 7/31, y = 6/31,$ $z = 1/31, w = 29/31.$

35. Following the hint, write the system as:

$$\begin{aligned} 3u \quad -v \quad +4w &= -13 \\ u \quad +2v \quad -w &= 12 \\ 4u \quad -v \quad +3w &= -7 \end{aligned}$$

Write the augmented matrix:

$$\begin{pmatrix} 3 & -1 & 4 & -13 \\ 1 & 2 & -1 & 12 \\ 4 & -1 & 3 & -7 \end{pmatrix}$$

In reduced row-echelon form this becomes:

$$\begin{pmatrix} 1 & 0 & 0 & 2 \\ 0 & 1 & 0 & 3 \\ 0 & 0 & 1 & -4 \end{pmatrix}$$

Thus $u = 2, v = 3, w = -4$, hence $x = 1/2, y = 1/3, z = -1/4$.

37. Adding the first two equations yields $5x = 3$. Thus $x = 3/5, y = 1/5$ is the only possible solution.
Check that this satisfies all four equations.

39. Write $x + 2y = 3$ as $-2x - 4y = -6$ and add to the second equation to obtain
$-y = -2, y = 2, x = -1,$ as the only possible solution.
Check that this satisfies all four equations.

41. Multiply both sides of the equation by $(x + 1)(x + 2)$ to get
$x = A(x + 2) + B(x + 1) = (A + B)x + (2A + B).$
Equating the coefficients gives the system of equations :
$$A + B = 1$$
$$2A + B = 0$$
which has solution $A = -1, B = 2$.

43. Multiply both sides of the equation by $(x+2)(x-3)^2$ to get
$2x+1 = A(x^2 - 6x + 9) + B(x^2 - x - 6) + C(x+2)$ or
$2x+1 = (A+B)x^2 + (-6A - B + C)x + (9A - 6B + 2C)$. Equate coefficients to get:

$$
\begin{array}{rcl}
A \quad +B \qquad\qquad &=& 0 \\
-6A \quad -B \quad +C &=& 2 \\
9A \quad -6B \quad +2C &=& 1
\end{array}
$$
which has solution $A = -3/25, B = 3/25, C = 7/5$.

45. Multiply both sides of the equation by $(x+1)(x^2 - x + 1)$ to get
$5x^2 + 1 = A(x^2 - x + 1) + (Bx + C)(x+1)$ or
$5x^2 + 1 = (A+B)x^2 + (-A + B + C)x + (A + C)$. Equate coefficients to get:

$$
\begin{array}{rcl}
A + B \qquad &=& 5 \\
-A + B + C &=& 0 \\
A \quad + C &=& 1
\end{array}
$$
which has solution $A = 2, B = 3, C = -1$.

47. Let x = amount borrowed from friend, y = amount borrowed from bank, z = amount borrowed from insurance company.   Then: $x + y + z = 10,000$
$$y = 2x$$
$$.08x + .09y + .05z = 830$$

Form the augmented matrix:

$$
\begin{pmatrix}
1 & 1 & 1 & 10000 \\
-2 & 1 & 0 & 0 \\
.08 & .09 & .05 & 830
\end{pmatrix}
$$

In reduced row-echelon form this becomes:

$$
\begin{pmatrix}
1 & 0 & 0 & 3000 \\
0 & 1 & 0 & 6000 \\
0 & 0 & 1 & 1000
\end{pmatrix}
$$

$3000 from friend, $6000 from bank, $1000 from insurance company.

49. Let x = amount in fund, y = amount in bonds, z = amount in food.
Then: $x + y + z = 70,000$
$$y = 2x$$
$$.02x + .1y + .06z = 4800$$

Form the augmented matrix:

$$
\begin{pmatrix}
1 & 1 & 1 & 70000 \\
-2 & 1 & 0 & 0 \\
.02 & .1 & .06 & 4800
\end{pmatrix}
$$

In reduced row-echelon form this becomes:

$$
\begin{pmatrix}
1 & 0 & 0 & 15000 \\
0 & 1 & 0 & 30000 \\
0 & 0 & 1 & 25000
\end{pmatrix}
$$

$15,000 in fund, $30,000 in bonds, $25,000 in the fast food franchise.

51. Let x = how many cups of Progresso, y = how many cups of H.C., z = how many cups of Campbell. Then

$$100x + 130y + 130z = 2030 \quad \text{(calories)}$$

$$970x + 480y + 880z = 11,900 \quad \text{(sodium)}$$

$$6x + 8y + 8z = 124 \quad \text{(protein)}$$

Form the augmented matrix: In reduced row-echelon form this becomes:

$$\begin{pmatrix} 100 & 130 & 130 & 2030 \\ 970 & 480 & 880 & 11900 \\ 6 & 8 & 8 & 124 \end{pmatrix} \qquad \begin{pmatrix} 1 & 0 & 0 & 6 \\ 0 & 1 & 0 & 9 \\ 0 & 0 & 1 & 2 \end{pmatrix}$$

6 cups of Progresso, 9 cups of Healthy Choice, 2 cups of Campbell's

53. Let x = how many bedroom models, y = how many living room models, z = how many whole-house models. Then: $10x + 20y + 60z = 440$

$$8x + 8y + 28z = 248$$

Form the augmented matrix:

$$\begin{pmatrix} 10 & 20 & 60 & 440 \\ 8 & 8 & 28 & 248 \end{pmatrix}$$

In reduced row-echelon form this becomes:

$$\begin{pmatrix} 1 & 0 & 1 & 18 \\ 0 & 1 & \frac{5}{2} & 13 \end{pmatrix}$$

Thus if $z = t$, $y = 13 - \frac{5}{2}t$, $x = 18 - t$. t must be positive, even, and less than 5. This yields 3 solutions.

$t = 0$: 18 bedroom models, 13 living-room models, 0 whole-house models

$t = 2$: 16 bedroom models,   8 living-room models, 2 whole-house models

$t = 4$: 14 bedroom models,   3 living-room models, 4 whole-house models

55. a. X-ray 1: $a + b$. X-ray 2: $a + c$.
   b. The system is:

$$a + b = .75$$

$$a + c = .60$$

$$b + c = .54$$

Form the augmented matrix:

$$\begin{pmatrix} 1 & 1 & 0 & .75 \\ 1 & 0 & 1 & .60 \\ 0 & 1 & 1 & .54 \end{pmatrix}$$

In reduced row-echelon form this becomes:

$$\begin{pmatrix} 1 & 0 & 0 & .405 \\ 0 & 1 & 0 & .345 \\ 0 & 0 & 1 & .195 \end{pmatrix}$$

$a = .405$, $b = .345$, $c = .195$

   c. a is bone, b is tumorous, c is healthy.

57. Let x = how many chairs, y = how many chests, z = how many tables. Then

$.2x + .5y + .3z = 1950$ (cutting)

$.3x + .4y + .1z = 1490$ (assembly)

$.1x + .6y + .4z = 2160$ (finishing)

Form the augmented matrix: In reduced row-echelon form this becomes:

$$\begin{pmatrix} .2 & .5 & .3 & | & 1950 \\ .3 & .4 & .1 & | & 1490 \\ .1 & .6 & .4 & | & 2160 \end{pmatrix} \qquad \begin{pmatrix} 1 & 0 & 0 & | & 2000 \\ 0 & 1 & 0 & | & 1600 \\ 0 & 0 & 1 & | & 2500 \end{pmatrix}$$

Produce 2000 chairs, 1600 chests, and 2500 tables.

59. a. Following the hint, obtain :

Intersection B: $x + y = 700 + 400$

Intersection C: $y + z = 200 + 500$

Intersection D: $z + t = 300 + 300$

b. Form the augmented matrix: In reduced row-echelon form this becomes:

$$\begin{pmatrix} 1 & 0 & 0 & 1 & | & 1000 \\ 1 & 1 & 0 & 0 & | & 1100 \\ 0 & 1 & 1 & 0 & | & 700 \\ 0 & 0 & 1 & 1 & | & 600 \end{pmatrix} \qquad \begin{pmatrix} 1 & 0 & 0 & 1 & | & 1000 \\ 0 & 1 & 0 & -1 & | & 100 \\ 0 & 0 & 1 & 1 & | & 600 \\ 0 & 0 & 0 & 0 & | & 0 \end{pmatrix}$$

Solution: $z = 600 - t$, $y = 100 + t$, $x = 1000 - t$.

c. The solution requires $0 \le t \le 600$. Thus the smallest number of cars leaving A on 4th avenue is $t = 0$. The largest is 600. Thus the smallest number of cars leaving A on Euclid is $x = 400$. The largest is 1000. Thus the smallest number of cars leaving C on 5th avenue is 100. The largest is 700. Thus the smallest number of cars leaving C on Chester is $z = 0$. The largest is 600.

11.3 Matrix Methods for Square Systems

1. A is a 2×3 matrix. B is a 3×4 matrix. AB is a 2×4 matrix, but BA is not defined.

3. A is a 3×2 matrix. B is a 2×3 matrix. AB is a 3×3 matrix. BA is a 2×2 matrix.

5. A is a 3×2 matrix. B is a 2×2 matrix. AB is a 3×2 matrix, but BA is not defined.

7. AB is a 2×2 matrix.

$$\begin{pmatrix} 3 \cdot 1 + 2 \cdot 0 & 3(-2) + 2 \cdot 3 & 3 \cdot 3 + 2 \cdot 1 \\ 2 \cdot 1 + 4 \cdot 0 & 2(-2) + 4 \cdot 3 & 2 \cdot 3 + 4 \cdot 1 \end{pmatrix} = \begin{pmatrix} 3 & 0 & 11 \\ 2 & 8 & 10 \end{pmatrix}$$

9. AB is a 3×2 matrix.

$$\begin{pmatrix} 1 \cdot 1 + 0 \cdot 1 + (-4) \cdot 0 & 1 \cdot 1 + 0 \cdot 0 + (-4) \cdot 1 \\ 0 \cdot 1 + 2 \cdot 1 + (-1) \cdot 0 & 0 \cdot 1 + 2 \cdot 0 + (-1) \cdot 1 \\ 2 \cdot 1 + 3 \cdot 1 + 4 \cdot 0 & 2 \cdot 1 + 3 \cdot 0 + 4 \cdot 1 \end{pmatrix} = \begin{pmatrix} 1 & -3 \\ 2 & -1 \\ 5 & 6 \end{pmatrix}$$

11. AB is a 4×4 matrix.

$$\begin{pmatrix} 2(1) + 0(1) + (-1)1 & 2(0) + 0(1) + (-1)1 & 2(1) + 0(0) + (-1)1 & 2(1) + 0(1) + (-1)0 \\ 1(1) + 1(1) + 2(1) & 1(0) + 1(1) + 2(1) & 1(1) + 1(0) + 2(1) & 1(1) + 1(1) + 2(0) \\ 0(1) + 2(1) + (-3)1 & 0(0) + 2(1) + (-3)1 & 0(1) + 2(0) + (-3)1 & 0(1) + 2(1) + (-3)0 \\ 2(1) + 3(1) + 0(1) & 2(0) + 3(1) + 0(1) & 2(1) + 3(0) + 0(1) & 2(1) + 3(1) + 0(0) \end{pmatrix}$$

$$= \begin{pmatrix} 1 & -1 & 1 & 2 \\ 4 & 3 & 3 & 2 \\ -1 & -1 & -3 & 2 \\ 5 & 3 & 2 & 5 \end{pmatrix}$$

13. $AB = \begin{pmatrix} 17 & -3 \\ 33 & -19 \end{pmatrix}$ $BA = \begin{pmatrix} -4 & 9 \\ 24 & 2 \end{pmatrix}$ These are not equal.

15. $AB = \begin{pmatrix} 8 & 24 & -8 \\ 2 & -2 & 6 \\ -3 & -21 & 15 \end{pmatrix}$ $BA = \begin{pmatrix} 19 & 9 & 8 \\ -10 & 2 & 0 \\ 0 & 0 & 0 \end{pmatrix}$ These are not equal.

17. Form the augmented matrix:

$$\left(\begin{array}{cc|cc} 1 & 2 & 1 & 0 \\ 3 & 4 & 0 & 1 \end{array} \right)$$

In reduced row-echelon form this is:

$$\left(\begin{array}{cc|cc} 1 & 0 & -2 & 1 \\ 0 & 1 & 3/2 & -1/2 \end{array} \right)$$

Therefore $A^{-1} = \begin{pmatrix} -2 & 1 \\ 3/2 & -1/2 \end{pmatrix}$

19. The matrix is not reducible by row reduction to the identity. Therefore the matrix has no inverse.

21. The matrix is not reducible by row reduction to the identity. Therefore the matrix has no inverse.

23. Form the augmented matrix:

$$\begin{pmatrix} 5 & 0 & 2 & | & 1 & 0 & 0 \\ 2 & 2 & 1 & | & 0 & 1 & 0 \\ -3 & 1 & -1 & | & 0 & 0 & 1 \end{pmatrix}$$

In reduced row-echelon form this becomes:

$$\begin{pmatrix} 1 & 0 & 0 & | & -3 & 2 & -4 \\ 0 & 1 & 0 & | & -1 & 1 & -1 \\ 0 & 0 & 1 & | & 8 & -5 & 10 \end{pmatrix}$$

Therefore the inverse matrix is:

$$\begin{pmatrix} -3 & 2 & -4 \\ -1 & 1 & -1 \\ 8 & -5 & 10 \end{pmatrix}$$

25. Form the matrix equation $AX = B$ with

$$A = \begin{pmatrix} -1 & 1 & 0 \\ -1 & 0 & 1 \\ 6 & -2 & -3 \end{pmatrix}, \; X = \begin{pmatrix} x \\ y \\ z \end{pmatrix}, \; B = \begin{pmatrix} 1 \\ -2 \\ 3 \end{pmatrix}$$

The solution is given by :

$$X = A^{-1}B = \begin{pmatrix} -1 \\ 0 \\ -3 \end{pmatrix}$$

$$x = -1, \, y = 0, \, z = -3$$

27. Form the matrix equation $AX = B$ with

$$A = \begin{pmatrix} 2 & 1 & 0 \\ -4 & -1 & -3 \\ 3 & 1 & 2 \end{pmatrix}, \; X = \begin{pmatrix} x \\ y \\ z \end{pmatrix}, \; B = \begin{pmatrix} 0 \\ 1 \\ 2 \end{pmatrix}$$

The solution is given by :

$$X = A^{-1}B = \begin{pmatrix} -8 \\ 16 \\ 5 \end{pmatrix}$$

$$x = -8, \, y = 16, \, z = 5$$

29. Form the matrix equation $AX = B$ with

$$A = \begin{pmatrix} 1 & 1 & 0 & 2 \\ 2 & -1 & 1 & -1 \\ 3 & 3 & 2 & -2 \\ 1 & 2 & 1 & 0 \end{pmatrix}, \; X = \begin{pmatrix} x \\ y \\ z \\ w \end{pmatrix}, \; B = \begin{pmatrix} 3 \\ 5 \\ 0 \\ 2 \end{pmatrix}$$

The solution is given by :

$$X = A^{-1}B = \begin{pmatrix} -.5 \\ -2.1 \\ 6.7 \\ 2.8 \end{pmatrix}$$

Thus $x = -.5, \, y = -2.1, \, z = 6.7, \, w = 2.8$.

31. Form the matrix equation $AX = B$ with

$$A = \begin{pmatrix} 1 & 1 & 6 & 2 & 0 \\ 1 & 0 & 5 & 2 & -3 \\ 3 & 2 & 17 & 6 & -4 \\ 4 & 3 & 21 & 7 & -2 \\ -6 & -5 & -36 & -12 & 3 \end{pmatrix}, \; X = \begin{pmatrix} x \\ y \\ z \\ v \\ w \end{pmatrix}, \; B = \begin{pmatrix} 1.5 \\ 2 \\ 2.5 \\ 3 \\ 3.5 \end{pmatrix}$$

The solution is given by :

$$X = A^{-1}B = \begin{pmatrix} 10.5 \\ 5 \\ -13 \\ 32 \\ 2.5 \end{pmatrix}$$

Thus $x = 10.5, \, y = 5, \, z = -13, \, v = 32, \, w = 2.5$.

33. Form the matrix equation $AX = B$ with

$$A = \begin{pmatrix} 1 & 2 & 2 & -2 \\ 4 & 4 & -1 & 5 \\ -2 & 5 & 6 & 4 \\ 5 & 13 & 7 & 12 \end{pmatrix}, \ X = \begin{pmatrix} x \\ y \\ z \\ w \end{pmatrix}, \ B = \begin{pmatrix} -23 \\ 7 \\ 0 \\ -7 \end{pmatrix}$$

The solution is (given in exact fractional form):

$$X = A^{-1}B = \begin{pmatrix} -1149/161 \\ 426/161 \\ -1124/161 \\ 579/161 \end{pmatrix}$$

35. Form the matrix equation $AX = B$ as in previous problems. Since A^{-1} exists, $A^{-1}B$ is defined and

$$X = A^{-1}B = \begin{pmatrix} 0 \\ 0 \\ 0 \\ 0 \\ 0 \end{pmatrix}$$

gives the solution
$$x = y = z = v = w = 0.$$

37. Since the matrix A of the system has no inverse, the inverse method fails. Write the augmented matrix:

$$\begin{pmatrix} 1 & 2 & 3 & 0 & | & 1 \\ 3 & 2 & 4 & 0 & | & -1 \\ 2 & 6 & 8 & 1 & | & 3 \\ 2 & 0 & 2 & -2 & | & 3 \end{pmatrix}$$

In reduced row-echelon form this becomes:

$$\begin{pmatrix} 1 & 0 & 0 & 1 & | & 0 \\ 0 & 1 & 0 & 5/2 & | & 0 \\ 0 & 0 & 1 & -1 & | & 0 \\ 0 & 0 & 0 & 0 & | & 1 \end{pmatrix}$$

Since the last row is equivalent to $0 = 1$ the system is inconsistent and has no solution.

39. Since the matrix A of the system has no inverse, the inverse method fails. Write the augmented matrix:

$$\begin{pmatrix} 1 & 0 & 0 & 3 & | & -2 \\ 1 & -4 & -1 & 3 & | & -7 \\ 0 & 4 & 1 & 0 & | & 5 \\ -1 & 12 & 3 & -3 & | & 17 \end{pmatrix}$$

In reduced row-echelon form this becomes:

$$\begin{pmatrix} 1 & 0 & 0 & 3 & | & -2 \\ 0 & 1 & 1/4 & 0 & | & 5/4 \\ 0 & 0 & 0 & 0 & | & 0 \\ 0 & 0 & 0 & 0 & | & 0 \end{pmatrix}$$

The solutions can be written as $w = t$, $z = s$, $y = \frac{5}{4} - \frac{1}{4}s$, $x = -2 - 3t$, where s and t are any real numbers.

41. Write the equation of the parabola as $y = ax^2 + bx + c$. Since the three given points must satisfy the equation, solve the system.

$$\begin{array}{rrrrr} 9a & -3b & +c & = & 15 \\ a & + b & +c & = & -7 \\ 25a & +5b & +c & = & 111 \end{array}$$

Form the matrix equation $AX = B$ with

$$A = \begin{pmatrix} 9 & -3 & 1 \\ 1 & 1 & 1 \\ 25 & 5 & 1 \end{pmatrix}, \ X = \begin{pmatrix} a \\ b \\ c \end{pmatrix}, \ B = \begin{pmatrix} 15 \\ -7 \\ 111 \end{pmatrix}$$

The solution is given by :

$$X = A^{-1}B = \begin{pmatrix} 35/8 \\ 13/4 \\ -117/8 \end{pmatrix}$$

The equation of the parabola is therefore $y = \dfrac{35}{8}x^2 + \dfrac{13}{48}x - \dfrac{117}{8}$.

43. a. The points $(0,315), (21,337)$ and $(43,371)$ must satisfy the equation
$y = ax^2 + bx + c$.
Solve the system:

$$
\begin{array}{rcl}
c &=& 315 \\
441a \;\; +21b \;\; +c &=& 337 \\
1849a \;\; +43b \;\; +c &=& 371
\end{array}
$$

to obtain $a = \dfrac{115}{9933}, b = \dfrac{7991}{9933}, c = 315$.

b. Use the table feature to obtain :

$$
\begin{array}{lll}
1983 & x = 25 & y = 342.35 \text{ ppm} \\
1993 & x = 35 & y = 357.34 \text{ ppm} \\
2003 & x = 45 & y = 374.65 \text{ ppm}
\end{array}
$$

45. a. The points $(2,169), (6,260)$ and $(10,320)$ must satisfy the equation
$y = ax^2 + bx + c$.
Solve the system:

$$
\begin{array}{rcl}
4a \;\;\; +2b \;\;\; c &=& 169 \\
36a \;\;\; +6b \;\; +c &=& 260 \\
100a \;\; +10b \;\; +c &=& 320
\end{array}
$$

to obtain $y = -\dfrac{31}{32}x^2 + \dfrac{61}{2}x + \dfrac{895}{8}$.

b. Use the table feature to obtain :

$$
\begin{array}{lll}
1993 & x = 3 & y = \$194.67 \\
1998 & x = 8 & y = \$293.88 \\
2002 & x = 12 & y = \$338.38
\end{array}
$$

47. Note first that $e^{\ln x} = x$ and $e^{-\ln x} = \dfrac{1}{x}$. Substitute to obtain the system:

$$
\begin{array}{rcl}
a \;\; + b \;\;\;\; c &=& -2 \\
2a \;\; +\tfrac{1}{2}b \;\; +c &=& 1 \\
4a \;\; +\tfrac{1}{4}b \;\; +c &=& 4
\end{array}
$$

Solve the system to obtain $a = 1, b = -4, c = 1, y = e^x - 4e^{-x} + 1$.

49. Let x = price of one pair of jeans, y = price of one jacket , z = price of one sweater,
w = price of one shirt. Solve the system $AX = B$ with

$$
A = \begin{pmatrix} 3000 & 3000 & 2200 & 4200 \\ 2700 & 2500 & 2100 & 4300 \\ 5000 & 2000 & 1400 & 7500 \\ 7000 & 1800 & 600 & 8000 \end{pmatrix}, \; X = \begin{pmatrix} x \\ y \\ z \\ w \end{pmatrix}, \; B = \begin{pmatrix} 507,650 \\ 459,075 \\ 541,225 \\ 571,500 \end{pmatrix}
$$

to obtain $x = \$34.50, \; y = \$72, \; z = \$44, \; w = \$21.75$.

51. Let x = how many of box A, y = how many of box B , z = how many of box C.
Solve the system:

$$
\begin{array}{rcll}
.6x \;\; +.3y \;\; +.5z &=& 41,400 & \text{(chocolates)} \\
.4x \;\; +.4y \;\; +.3z &=& 29,400 & \text{(mints)} \\
.3y \;\; +.2z &=& 16,200 & \text{(caramels)}
\end{array}
$$

to obtain $x = 15,000, \; y = 18,000, \; z = 54,000$ boxes.

Chapter 11 Review Exercises

1. $\begin{array}{rlll} -5x & +3y & = & 4 \quad [A] \\ 2x & -y & = & -3 \quad [B] \\ 6x & -3y & = & -9 \quad 3[B] \\ x & & = & -5 \quad [A]+3[B] \end{array}$

    Substitute $x = -5$ into equation $[B]$
    to obtain $y = -7$.

3. $\begin{array}{rlll} 3x & -5y & = & 10 \quad [A] \\ 4x & -3y & = & 6 \quad [B] \\ 9x & -15y & = & 30 \quad 3[A] \\ -20x & +15y & = & -30 \quad -5[B] \\ -11x & & = & 0 \quad 3[A]-5[B] \\ & & x = 0. \end{array}$

    Substitute into equation $[B]$
    to obtain $y = -2$.

5. Multiply the first equation by 2 and add the two equations.
    $$21.78x + 2y = 369.84$$
    $$\underline{16.06x - 2y = 32.04}$$
    $$37.84x = 401.88$$
    $$x = 10.62$$
    $$y = 184.92 - 10.89(10.62) = 69.26$$
    10.62 corresponds to 2001. 69 to 70 days were found.

7. Let $x =$ amount of 40% alloy,
    $y =$ amount of 70% alloy. Then
    $$\begin{array}{rll} x + y = & 50 \quad [A] \\ .4x + .7y = & .6(50) \quad [B] \end{array}$$
    $$-4x - 4y = -200 \quad -4[A]$$
    $$\underline{4x + 7y = 300 \quad 10[B]}$$
    $$3y = 100$$
    $$y = 33\tfrac{1}{3} \text{ lbs of 70\% alloy}$$
    $$x = 16\tfrac{2}{3} \text{ lbs of 40\% alloy}$$

9. Solve the second equation for y in
    terms of x to obtain $y = 2x + 3$.
    Substitute into the first equation
    to obtain: $x^2 - (2x + 3) = 0$
    $$x^2 - 2x - 3 = 0$$
    $$x = 3 \text{ or } x = -1$$
    If $x = 3$, $y = 9$. If $x = -1$, $y = 1$.

11. Solve the second equation for y in terms of x to obtain $y = 2 - x$.
    Substitute into the first equation to obtain:
    $$x^2 + (2 - x)^2 = 16$$
    $$x^2 + 4 - 4x + x^2 = 16$$
    $$x^2 - 2x - 6 = 0$$
    $$x = 1 \pm \sqrt{7}$$
    If $x = 1 + \sqrt{7}$, $y = 1 - \sqrt{7}$. If $x = 1 - \sqrt{7}$, $y = 1 + \sqrt{7}$.

13. The augmented matrix is:

$$\begin{pmatrix} 1 & -2 & 3 & | & 4 \\ 2 & 1 & -4 & | & 3 \\ -3 & 4 & -1 & | & -2 \end{pmatrix}$$

15. The matrix of the system is:

$$\begin{pmatrix} 2 & -1 & -2 & 2 \\ 1 & 3 & -2 & 1 \\ -1 & 4 & 2 & -3 \end{pmatrix}$$

17. Form the matrix equation $AX = B$ with

$$A = \begin{pmatrix} 3 & 1 & -1 \\ 1 & 0 & 2 \\ -3 & -1 & 2 \end{pmatrix}, \ X = \begin{pmatrix} x \\ y \\ z \end{pmatrix}, \ B = \begin{pmatrix} 13 \\ 9 \\ 9 \end{pmatrix}$$

The solution is given by :

$$X = A^{-1}B = \begin{pmatrix} -35 \\ 140 \\ 22 \end{pmatrix}$$

$x = -35, \ y = 140, \ z = 22$

19. Form the matrix equation $AX = B$ with

$$A = \begin{pmatrix} 4 & 3 & -3 \\ 5 & -3 & 2 \\ 2 & -2 & 3 \end{pmatrix}, \ X = \begin{pmatrix} x \\ y \\ z \end{pmatrix}, \ B = \begin{pmatrix} 2 \\ 10 \\ 14 \end{pmatrix}$$

The solution is given by :

$$X = A^{-1}B = \begin{pmatrix} 2 \\ 4 \\ 6 \end{pmatrix}$$

$x = 2, \ y = 4, \ z = 6$

21. The augmented matrix is:

$$\begin{pmatrix} 1 & -2 & -3 & | & 1 \\ 0 & 5 & 10 & | & 0 \\ 8 & -6 & -4 & | & 8 \end{pmatrix}$$

In reduced row-echelon form this is:

$$\begin{pmatrix} 1 & 0 & 1 & | & 1 \\ 0 & 1 & 2 & | & 0 \\ 0 & 0 & 0 & | & 0 \end{pmatrix}$$

The solution is $z = t, \ y = -2t,$
$x = 1 - t, \ t$ any real number.

23. Since each equation is a multiple of $2x - y = 3,$ only (**c**) is true.

25. Multiply both sides of the equation by $(x - 3)(x + 2)$ to get
$4x - 7 = A(x + 2) + B(x - 3), \ 4x - 7 = (A + B)x + (2A - 3B).$ Equate coefficients to obtain the system of equations: $A + B = 4$
$$2A - 3B = -7$$
Multiply the first equation by –2 and add to get $B = 3.$ Use this value to find $A = 1.$

27. $\begin{pmatrix} -1(2) + 0(4) & -1(-3) + 0(1) \\ 0(2) + (-1)(4) & 0(-3) + (-1)(1) \end{pmatrix} = \begin{pmatrix} -2 & 3 \\ -4 & -1 \end{pmatrix}$

29. Since Since A is 2×2 and E is 3×2 the product AE does not exist.

31. Form the augmented matrix:

$$\begin{pmatrix} 3 & -7 & 1 & 0 \\ 4 & -9 & 0 & 1 \end{pmatrix}$$

In reduced row-echelon form this is:

$$\begin{pmatrix} 1 & 0 & -9 & 7 \\ 0 & 1 & -4 & 3 \end{pmatrix}$$

The inverse matrix is: $\begin{pmatrix} -9 & 7 \\ -4 & 3 \end{pmatrix}$

33. Form the augmented matrix:

$$\begin{pmatrix} 3 & 2 & 6 & 1 & 0 & 0 \\ 1 & 1 & 2 & 0 & 1 & 0 \\ 2 & 2 & 5 & 0 & 0 & 1 \end{pmatrix}$$

In reduced row-echelon form this is:

$$\begin{pmatrix} 1 & 0 & 0 & 1 & 2 & -2 \\ 0 & 1 & 0 & -1 & 3 & 0 \\ 0 & 0 & 1 & 0 & -2 & 1 \end{pmatrix}$$

The inverse matrix is:

$$\begin{pmatrix} 1 & 2 & -2 \\ -1 & 3 & 0 \\ 0 & -2 & 1 \end{pmatrix}$$

35. Form the matrix equation $AX = B$ with

$$A = \begin{pmatrix} 1 & 0 & 2 & 6 \\ 3 & 4 & -2 & -1 \\ 5 & 0 & 2 & -5 \\ 4 & -4 & 2 & 3 \end{pmatrix}, \ X = \begin{pmatrix} x \\ y \\ z \\ w \end{pmatrix}, \ B = \begin{pmatrix} 2 \\ 0 \\ -4 \\ 1 \end{pmatrix}$$

The solution is given by :

$$X = A^{-1}B = \begin{pmatrix} -1/85 \\ -14/85 \\ -21/34 \\ 46/85 \end{pmatrix}$$

$x = -1/85, \ y = -14/85, \ z = -21/34, \ w = 46/85$

37. Write the equation of the parabola as $y = ax^2 + bx + c$. Since the three given points must satisfy the equation, solve the system.

$$\begin{array}{rrrcr} 9a & -3b & +c & = & 52 \\ 4a & +2b & +c & = & 17 \\ 64a & +8b & +c & = & 305 \end{array}$$

Form the matrix equation $AX = B$ with

$$A = \begin{pmatrix} 9 & -3 & 1 \\ 4 & 2 & 1 \\ 64 & 8 & 1 \end{pmatrix}, \ X = \begin{pmatrix} a \\ b \\ c \end{pmatrix}, \ B = \begin{pmatrix} 52 \\ 17 \\ 305 \end{pmatrix}$$

The solution is given by :

$$X = A^{-1}B = \begin{pmatrix} 5 \\ -2 \\ 1 \end{pmatrix}$$

The equation of the parabola is therefore $y = 5x^2 - 2x + 1$.

39. a. The points $(6,25)$, $(10,76)$ and $(14,161)$ must satisfy the equation $y = ax^2 + bx + c$.

Solve the system:
$$
\begin{aligned}
36a &+6b &c &= 25 \\
100a &+10b &+c &= 76 \\
196a &+14b &+c &= 161
\end{aligned}
$$

to obtain $y = \dfrac{17}{16}x^2 - \dfrac{17}{4}x + \dfrac{49}{4}$.

b. Use the table feature to obtain:

1998 — $x = 8$	$y = 46.25$ hours
2002 — $x = 12$	$y = 114.25$ hours
2006 — $x = 16$	$y = 216.25$ hours

41. Let x = number of students, y = number of faculty, z = number of others.

Solve the system:
$$
\begin{aligned}
x &+y &+z &= 460 \\
x &-3y & &= 0 \\
x &+1.5y &+2z &= 570
\end{aligned}
$$

to obtain (among other information) $y = 100$ faculty.

43. Let x = how many pounds of corn, y = how many pounds of soybeans, z = how many pounds of by-products.

Solve the system:
$$
\begin{aligned}
10x &+20y &+30z &= 1800 &\text{(fiber)} \\
30x &+20y &+40z &= 2800 &\text{(fat)} \\
20x &+40y &+25z &= 2200 &\text{(protein)}
\end{aligned}
$$

to obtain $x = 30$ pounds of corn, $y = 15$ pounds of soybeans, $z = 40$ pounds of by-products.

Chapter 12
Discrete Algebra

12.1 Sequences and Series

1. $a_1 = 2 \cdot 1 + 6 = 8, a_2 = 2 \cdot 2 + 6 = 10, a_3 = 2 \cdot 3 + 6 = 12, a_4 = 2 \cdot 4 + 6 = 14,$
$a_5 = 2 \cdot 5 + 6 = 16$

3. $a_1 = \dfrac{1}{1^3} = 1, a_2 = \dfrac{1}{2^3} = \dfrac{1}{8}, a_3 = \dfrac{1}{3^3} = \dfrac{1}{27}, a_4 = \dfrac{1}{4^3} = \dfrac{1}{64}, a_5 = \dfrac{1}{5^3} = \dfrac{1}{125}$

5. $a_1 = (-1)^1 \sqrt{1+2} = -\sqrt{3}$

$a_2 = (-1)^2 \sqrt{2+2} = 2$

$a_3 = (-1)^3 \sqrt{3+2} = -\sqrt{5}$

$a_4 = (-1)^4 \sqrt{4+2} = \sqrt{6}$

$a_5 = (-1)^5 \sqrt{5+2} = -\sqrt{7}$

7. $a_1 = 4 + (-.1)^1 = 3.9$

$a_2 = 4 + (-.1)^2 = 4.01$

$a_3 = 4 + (-.1)^3 = 3.999$

$a_4 = 4 + (-.1)^4 = 4.0001$

$a_5 = 4 + (-.1)^5 = 3.99999$

9. $a_1 = (-1)^1 + 3 \cdot 1 = 2$

$a_2 = (-1)^2 + 3 \cdot 2 = 7$

$a_3 = (-1)^3 + 3 \cdot 3 = 8$

$a_4 = (-1)^4 + 3 \cdot 4 = 13$

$a_5 = (-1)^5 + 3 \cdot 5 = 14$

11. $\displaystyle\sum_{i=1}^{11} i$

13. $\displaystyle\sum_{i=1}^{7} \frac{1}{2^{i+6}}$ or $\displaystyle\sum_{i=7}^{13} \frac{1}{2^i}$

15. $\displaystyle\sum_{i=1}^{5} 3i = 3 + 6 + 9 + 12 + 15 = 45$

17. $\displaystyle\sum_{n=1}^{16} (2n - 3) = (-1) + (1) + (3) + \ldots + (29) = 224$

19. $\displaystyle\sum_{n=15}^{36} (n^2 - 8) = (15^2 - 8) + (16^2 - 8) + (17^2 - 8) + \ldots + (36^2 - 8) = 15{,}015$

21. $a_n = (-1)^n$

23. $a_n = \dfrac{n}{n+1}$

25. The terms differ by 5, so $a_n = 5n + k$. Since $a_1 = 2, 2 = 5 + k, k = -3; a_n = 5n - 3$.

27. $a_1 = 4$

$a_2 = 2a_1 + 3 = 2 \cdot 4 + 3 = 11$

$a_3 = 2a_2 + 3 = 2 \cdot 11 + 3 = 25$

$a_4 = 2a_3 + 3 = 2 \cdot 25 + 3 = 53$

$a_5 = 2a_4 + 3 = 2 \cdot 53 + 3 = 109$

29. $a_1 = 1$

$a_2 = -2$

$a_3 = 3$

$a_4 = a_3 + a_2 + a_1 = 3 + (-2) + 1 = 2$

$a_5 = a_4 + a_3 + a_2 = 2 + 3 + (-2) = 3$

31. $a_0 = 2$

$a_1 = 3$

$a_2 = a_1\left(\frac{1}{2}a_0\right) = 3\left(\frac{1}{2}\cdot 2\right) = 3$

$a_3 = a_2\left(\frac{1}{2}a_1\right) = 3\left(\frac{1}{2}\cdot 3\right) = \frac{9}{2}$

$a_4 = a_3\left(\frac{1}{2}a_2\right) = \frac{9}{2}\left(\frac{1}{2}\cdot 3\right) = \frac{27}{4}$

33. $a_1 = 3$

$a_2 = 1$

$a_3 = 4$

$a_4 = 1$

$a_5 = 5$

35. $a_1 = 1^2 - 5\cdot 1 + 2 = -2$

$a_2 = 2^2 - 5\cdot 2 + 2 = -4$

$a_3 = 3^2 - 5\cdot 3 + 2 = -4$

$a_4 = 4^2 - 5\cdot 4 + 2 = -2$

$a_5 = 5^2 - 5\cdot 5 + 2 = 2$

$a_6 = 6^2 - 5\cdot 6 + 2 = 8$

$a_1 + a_2 + a_3 = -10$

$a_1 + a_2 + a_3 + a_4 + a_5 + a_6 = -2$

37. $a_1 = (-1)^{1+1}5 = 5$

$a_2 = (-1)^{2+1}5 = -5$

$a_3 = (-1)^{3+1}5 = 5$

$a_4 = (-1)^{4+1}5 = -5$

$a_5 = (-1)^{5+1}5 = 5$

$a_6 = (-1)^{6+1}5 = -5$

$a_1 + a_2 + a_3 = 5$

$a_1 + a_2 + a_3 + a_4 + a_5 + a_6 = 0$

39. $\displaystyle\sum_{n=1}^{6} \frac{1}{2n+1}$

41. $\displaystyle\sum_{n=1}^{5} \frac{(-1)^{n+1}n}{n+7}$

43. $a_{12} = \left(1+\frac{1}{12}\right)^{12} = 2.613035$

45. $a_{102} = \dfrac{102^3 - 102^2 + 5(102)}{3(102)^2 + 2(102) - 1} = 33.465$

47. $\displaystyle\sum_{k=1}^{14} \frac{1}{k^2} = \frac{1}{1^2} + \frac{1}{2^2} + \frac{1}{3^2} + \ldots + \frac{1}{14^2} = 1.5759958$

49. After 1 month the interest is $\frac{.05}{12}(9500)$. After paying \$218.78, the principal is
$9500 + \frac{.05}{12}(9500) - 218.78 = 9500\left(1+\frac{.05}{12}\right) - 218.78$. Each month, then, the previous month's principal is multiplied by $\left(1+\frac{.05}{12}\right)$ and 218.78 is subtracted. Thus

$a_0 = 9500$

$a_n = a_{n-1}\left(1 + \dfrac{.05}{12}\right) - 218.78$

Using the table feature of a calculator in sequence mode,

$a_{12} = \$7299.70$

$a_n = \$2555.50$

51. **a.** 2000 — $n = 3$, $a_3 = 937.84 + 37.55\ln 3 = 979.1$ hours

2003 — $n = 6$, $a_6 = 937.84 + 37.55\ln 6 = 1005.1$ hours

b. $\displaystyle\sum_{i=1}^{6}(937.84 + 37.55\ln i) = 5874.1$ hours

53. a. $2002 — n = 7, c_7 = .42(7)^2 + 1.528(7) + 10.396 = \$41.67$

$2004 — n = 9, c_9 = .42(9)^2 + 1.528(9) + 10.396 = \$58.17$

b. $\sum_{i=0}^{8} \left(.42i^2 + 1.528i + 10.396\right) = \$234.25$

55. b. $a_{17} = 59, a_{18} = 61, a_{19} = 67, a_{20} = 71, a_{21} = 73$

57. $a_1 = 2^2 = 4, a_2 = 3^2 = 9, a_3 = 5^2 = 25, a_4 = 7^2 = 49, a_5 = 11^2 = 121$

59. a_1 is the largest prime integer less than 5. $a_1 = 3$
a_2 is the largest prime integer less than 10. $a_2 = 7$
a_3 is the largest prime integer less than 15. $a_3 = 13$
a_4 is the largest prime integer less than 20. $a_4 = 19$
a_5 is the largest prime integer less than 25. $a_5 = 23$

61. a. 1, 1, 2, 3, 5, 8, 13, 21, 34, 55
b. 1, 2, 4, 7, 12, 20, 33, 54, 88, 143
c. The nth partial sum is 1 less than the (n + 2)nd term.

63.

$5(1)^2 + 4(-1)^1 = 1 = 1^2$ $5(8)^2 + 4(-1)^6 = 324 = 18^2$

$5(1)^2 + 4(-1)^2 = 9 = 3^2$ $5(13)^2 + 4(-1)^7 = 841 = 29^2$

$5(2)^2 + 4(-1)^3 = 16 = 4^2$ $5(21)^2 + 4(-1)^8 = 2209 = 47^2$

$5(3)^2 + 4(-1)^4 = 49 = 7^2$ $5(34)^2 + 4(-1)^9 = 5776 = 76^2$

$5(5)^2 + 4(-1)^5 = 121 = 11^2$ $5(55)^2 + 4(-1)^{10} = 15129 = 123^2$

65. Since $a_n = a_{n-1} + a_{n-2}$, $a_n - a_{n-1} = a_{n-2}$.

Thus $a_3 - a_2 = a_1$

$a_4 - a_3 = a_2$

$a_5 - a_4 = a_3$

$\cdots\cdots\cdots$

$\underline{a_{k+2} - a_{k+1} = a_k}$

Adding: $a_{k+2} - a_2 = \sum_{n=1}^{k} a_n$

Since $a_2 = 1$, $a_{k+2} - 1 = \sum_{n=1}^{k} a_n$.

12.2 Arithmetic Sequences

1. $3-1=5-3=7-5=9-7=2$. Therefore, the sequence is arithmetic, with common difference 2.

3. $2-1=1; 4-2=2$. Therefore, the sequence is not arithmetic.

5. $\log 6 - \log 3 = \log 2; \log 9 - \log 6 = \log 1.5$. Therefore, the sequence is not arithmetic.

7. $a_1 = 5 + 4 \cdot 1 = 9$
    $a_2 = 5 + 4 \cdot 2 = 13$
    $a_3 = 5 + 4 \cdot 3 = 17$
    $a_4 = 5 + 4 \cdot 4 = 21$
    $a_5 = 5 + 4 \cdot 5 = 25$
    The sequence is arithmetic, with common difference $13 - 9 = 4$.

9. $c_1 = (-1)^1 = -1$
    $c_2 = (-1)^2 = 1$
    $c_3 = (-1)^3 = -1$
    $c_4 = (-1)^4 = 1$
    $c_5 = (-1)^5 = -1$
    Since $c_2 - c_1 = 2$ but $c_3 - c_2 = -2$, the sequence is not arithmetic.

11. Note: $e^{\ln n} = n$. Thus $a_1 = 1, a_2 = 2, a_3 = 3, a_4 = 4, a_5 = 5$, the sequence is arithmetic, with common difference 1.

13. $a_n = 3 - 2n, \quad a_{n+1} = 3 - 2(n+1)$
    $a_{n+1} - a_n = [3 - 2(n+1)] - (3 - 2n) = -2$
    Thus the sequence is arithmetic, with common difference -2.

15. $a_n = \dfrac{5 + 3n}{2}, \quad a_{n+1} = \dfrac{5 + 3(n+1)}{2}$
    $a_{n+1} - a_n = \dfrac{5 + 3(n+1)}{2} - \dfrac{5 + 3n}{2} = \dfrac{5 + 3n + 3 - 5 - 3n}{2} = \dfrac{3}{2}$
    Thus the sequence is arithmetic, with common difference $\frac{3}{2}$.

17. $a_n = c + 2n \quad a_{n+1} = c + 2(n+1)$
    $a_{n+1} - a_n = [c + 2(n+1)] - (c + 2n) = 2$
    Thus the sequence is arithmetic, with common difference 2.

19. $a_5 = a_1 + (5-1)d = 5 + (5-1)2 = 13$
    $a_n = a_1 + (n-1)d = 5 + (n-1)2 = 2n + 3$

21. $a_5 = a_1 + (5-1)d = 4 + (5-1)\frac{1}{4} = 5$
    $a_n = a_1 + (n-1)d = 4 + (n-1)\frac{1}{4}$
    $= \dfrac{n + 15}{4}$

23. $a_5 = a_1 + (5-1)d = 10 + (5-1)\left(-\frac{1}{2}\right) = 8$
    $a_n = a_1 + (n-1)d = 10 + (n-1)\left(-\frac{1}{2}\right)$
    $= \dfrac{21 - n}{2}$

25. $a_4 = a_1 + (4-1)2$
$12 = a_1 + (4-1)2$
$a_1 = 6$
$a_n = a_1 + (n-1)2$
$a_n = 6 + (n-1)2$
$a_n = 2n + 4$

27. $a_2 = a_1 + d = 4$
$a_6 = a_1 + 5d = 32$
Eliminating a_1 yields
$d = 7$, hence $a_1 = -3$.
$a_n = a_1 + (n-1)d$
$a_n = 7n - 10$

29. $a_5 = a_1 + 4d = 0$
$a_9 = a_1 + 8d = 6$
Eliminating a_1 yields
$d = \frac{3}{2}$, hence $a_1 = -6$.
$a_n = a_1 + (n-1)(\frac{3}{2})$
$a_n = \dfrac{3n-15}{2}$

31. $\displaystyle\sum_{n=1}^{k} a_n = ka_1 + \frac{k(k-1)}{2}d = 6\cdot 2 + \frac{6(6-1)}{2}5 = 87$

33. $\displaystyle\sum_{n=1}^{k} a_n = ka_1 + \frac{k(k-1)}{2}d = 7\left(\frac{3}{4}\right) + \frac{7(7-1)}{2}\left(-\frac{1}{2}\right) = -\frac{21}{4}$

35. $\displaystyle\sum_{n=1}^{k} a_n = \frac{k}{2}(a_1 + a_k) = \frac{6}{2}((-4)+14) = 30$

37. $\displaystyle\sum_{n=1}^{k} a_n = \frac{k}{2}(a_1 + a_k) = \frac{20}{2}((3\cdot 1 + 4) + (3\cdot 20 + 4)) = 10(7 + 64) = 710$

39. $\displaystyle\sum_{n=1}^{k} a_n = \frac{k}{2}(a_1 + a_k) = \frac{40}{2}\left(\frac{1+3}{6} + \frac{40+3}{6}\right) = 20\left(\frac{47}{6}\right) = \frac{470}{3}$

41. These form an arithmetic sequence with $a_1 = 2$, $d = 2$. Find k, the number of terms: $100 = a_1 + (k-1)d$

$100 = 2 + (k-1)2$. Thus $k = 50$.

$\displaystyle\sum_{n=1}^{k} a_n = \frac{k}{2}(a_1 + a_k) = \frac{50}{2}(2 + 100) = 2550$

43. These form an arithmetic sequence with $a_1 = 1$, $d = 1$. Clearly, $k = 200$.

$\displaystyle\sum_{n=1}^{k} a_n = \frac{k}{2}(a_1 + a_k)$

$= \dfrac{200}{2}(1 + 200) = 20{,}100$

45. These form an arithmetic sequence with $a_1 = 10{,}000$, $d = 7500$. In the tenth year $a_{10} = a_1 + (10-1)\,d = 10{,}000 + 9\cdot 75000 = \$77{,}500$. The total profit is given by

$\displaystyle\sum_{n=1}^{k} a_n = \frac{k}{2}(a_1 + a_k) = \frac{10}{2}(10{,}000 + 77{,}500) = \$437{,}500$

47. a. $a_1 = 1223.50$, $a_{10} = 2786.80$

$a_{10} = a_1 + (10-1)d$

$d = \dfrac{a_{10} - a_1}{9} = \dfrac{2786.80 - 1223.50}{9} = \$173.7$

$a_n = 1223.50 + (n-1)173.7$

b. Use the above formula, 2000 — $n = 20$ — $a_{19} = \$4523.80$

2003 — $n = 23$ — $a_{23} = \$5044.90$

49. a. $c_1 = 15{,}828.80$, $c_{19} = 28{,}932.80$

$$c_{19} = c_1 + (19 - 1)d$$

$$d = \frac{c_{19} - c_1}{18} = \frac{28{,}932.80 - 15{,}828.80}{18} = \$728$$

$$c_n = 15{,}828.80 + (n - 1)728$$

b. Use the above formula, 2002 — $n = 22$

$$\sum_{n=1}^{22} a_n = \frac{22}{2}(c_1 + c_{22})$$

$$= \frac{22}{2}(15828.80 + 15{,}828.80 + (22 - 1)728) = \$516{,}401.60$$

51. The first 12 rows form an arithmetic sequence with $a_1 = 6$, $d = 2$.

The twelfth row has $a_{12} = a_1 + (12 - 1)d = 6 + 11 \cdot 2 = 28$ seats.

The first twelve rows have $\sum_{n=1}^{n} a_n = \frac{12}{2}(6 + 28) = 204$ seats.

Then there are 8 more rows of 28 seats each — 224 more seats for a total of 428 seats.

53. The lengths form an arithmetic sequence with $a_1 = 24$ and $a_9 = 18$.

$$a_9 = a_1 + (9 - 1)d$$

$$d = \frac{a_9 - a_1}{8} = \frac{18 - 24}{8} = -\frac{3}{4}$$

Thus the rungs decrease by $\frac{3}{4}$ of an inch: $23\frac{1}{4}, 22\frac{1}{2}, 21\frac{3}{4}, 21, 20\frac{1}{4}, 19\frac{1}{2}, 18\frac{3}{4}$ inches are the seven intermediate lengths.

12.3 Geometric Sequences

1. $a_2 - a_1 = a_3 - a_2 = a_4 - a_3 = a_5 - a_4 = 5$. Arithmetic

3. $a_2 \div a_1 = a_3 \div a_2 = a_4 \div a_3 = a_5 \div a_4 = 1/2$. Geometric

5. $a_2 - a_1 = a_3 - a_2 = a_4 - a_3 = -2$. Arithmetic

7. $a_2 \div a_1 = a_3 \div a_2 = a_4 \div a_3 = a_5 \div a_4 = -1/2$. Geometric

9. $a_2 \div a_1 = a_3 \div a_2 = a_4 \div a_3 = a_5 \div a_4 = \sqrt{2}$. Geometric

11. Since this sequence is constant it is both arithmetic ($d = 0$) and geometric ($r = 1$).

13. $a_6 = a_1 r^5 = 5 \cdot 2^5 = 160$

$a_n = a_1 r^{n-1} = 5 \cdot 2^{n-1}$

15. $a_6 = a_1 r^5 = 4\left(\frac{1}{4}\right)^5 = 1/256$

$a_n = a_1 r^{n-1} = 4\left(\frac{1}{4}\right)^{n-1} = 4^{2-n}$

17. $a_6 = a_1 r^5 = 10\left(-\frac{1}{2}\right)^5 = -\frac{5}{16}$

$a_n = a_1 r^{n-1} = 10\left(-\frac{1}{2}\right)^{n-1} = \frac{5(-1)^{n-1}}{2^{n-2}}$

19. $a_2 = a_1 r$

$12 = a_1\left(\frac{1}{3}\right)$

$a_1 = 36$

$a_6 = a_1 r^5 = 36\left(\frac{1}{3}\right)^5 = \frac{4}{27}$

$a_n = a_1 r^{n-1} = 36\left(\frac{1}{3}\right)^{n-1}$

21. $a_4 = a_1 r^3$

$-\frac{4}{5} = a_1\left(\frac{2}{5}\right)^3$

$a_1 = -\frac{25}{2}$

$a_6 = a_1 r^5 = \left(-\frac{25}{2}\right)\left(\frac{2}{5}\right)^5 = -\frac{16}{125}$

$a_n = a_1 r^{n-1} = \left(-\frac{25}{2}\right)\left(\frac{2}{5}\right)^{n-1} = -\frac{2^{n-2}}{5^{n-3}}$

23. $a_n \div a_{n-1} = \left(-\frac{1}{2}\right)^n \div \left(-\frac{1}{2}\right)^{n-1} = -\frac{1}{2}$. The sequence is geometric, with common ratio $-\frac{1}{2}$.

25. $a_n \div a_{n-1} = 5^{n+2} \div 5^{n+2-1} = 5$. Thus the sequence is geometric, with common ratio 5.

27. $a_n \div a_{n-1} = \left(\sqrt{5}\right)^n \div \left(\sqrt{5}\right)^{n-1} = \sqrt{5}$. Geometric sequence, with common ratio $\sqrt{5}$.

29. $a_2 \div a_1 = -64 \div 256 = -\frac{1}{4} = r$

$a_5 = a_1 r^4 = 256\left(-\frac{1}{4}\right)^4 = 1$

$a_n = a_1 r^{n-1} = 256\left(-\frac{1}{4}\right)^{n-1} = \left(-\frac{1}{4}\right)^{n-5}$

31. $a_2 \div a_1 = 5 \div \frac{1}{2} = 10 = r$

$a_5 = a_1 r^4 = \frac{1}{2}(10)^4 = 5000$

$a_n = a_1 r^{n-1} = \frac{1}{2}(10)^{n-1}$

33. $a_5 = a_1 r^4 = \frac{1}{16}$

$a_2 = a_1 r = 4$

$a_5 \div a_2 = r^3 = \frac{1}{16} \div 4 = \frac{1}{64}$,

hence $r = \frac{1}{4}$, $a_1 = a_2 \div r = 4 \div \frac{1}{4} = 16$

$a_5 = \frac{1}{16}$

$a_n = a_1 r^{n-1} = 16\left(\frac{1}{4}\right)^{n-1} = 4^{3-n}$

35. $a_5 = a_1 r^4$

$20 = 5 \cdot r^4$

$r^4 = 4$

$r = \sqrt[4]{4} = \sqrt{2}$ (since $r > 0$)

$a_5 = 20$

$a_n = a_1 r^{n-1} = 5\left(\sqrt{2}\right)^{n-1}$

37. $\sum_{n=1}^{k} a_n = a_1\left(\frac{1-r^k}{1-r}\right)$

$\sum_{n=1}^{6} a_n = 5\left(\frac{1-\left(\frac{1}{2}\right)^6}{1-\frac{1}{2}}\right) = \frac{315}{32}$

39. $\sum_{n=1}^{k} a_n = a_1\left(\frac{1-r^k}{1-r}\right)$

$a_2 = a_1 r, 6 = a_1 \cdot 2, a_1 = 3$

$\sum_{n=1}^{7} a_n = 3\left(\frac{1-2^7}{1-2}\right) = 381$

41. This is a geometric sequence with $a_1 = 2, r = 2$.

$$\sum_{n=1}^{k} a_n = a_1\left(\frac{1-r^k}{1-r}\right)$$

$$\sum_{n=1}^{7} 2^n = 2\left(\frac{1-2^7}{1-2}\right) = 254$$

43. This is a geometric sequence with $a_1 = -\frac{1}{3}, r = -\frac{1}{3}$.

$$\sum_{n=1}^{k} a_n = a_1\left(\frac{1-r^k}{1-r}\right)$$

$$\sum_{n=1}^{9}\left(-\frac{1}{3}\right)^n = -\frac{1}{3}\left(\frac{1-\left(-\frac{1}{3}\right)^9}{1-\left(-\frac{1}{3}\right)}\right) = -\frac{4921}{19683}$$

45. This is a geometric sequence with $a_1 = 4, r = \frac{3}{2}$.

$$\sum_{j=1}^{6}\left(\frac{3}{2}\right)^{j-1} = 4\left(\frac{1-\left(\frac{3}{2}\right)^6}{1-\left(\frac{3}{2}\right)}\right) = \frac{665}{8}$$

47. a. $r = a_2 \div a_1 = 38.0154 \div 44.4313 = .8556$, $a_n = a_1 r^{n-1} = 44.4313(.8556)^{n-1}$

   b. 2003 corresponds to $n = 19$.

   $a_{19} = 44.4313(.8556)^{18} = 2.6826$ students per computer

   c. Solve $1 = 44.4313(.8556)^n$

   $$n = \frac{\ln(1/44.4313)}{\ln(.8556)} = 24.3.$$ This corresponds to 2009 or early 2010.

49. a. $r = a_2 \div a_1 = 4.1672 \div 3.9631 = 1.0515$, $a_n = a_1 r^{n-1} = 3.9631(1.0515)^{n-1}$

   b. 1998 — $n = 8$ $a_8 = 3.9631(1.0515)^7 = \$5.632$ billion

   2002 — $n = 12$ $a_{12} = 3.9631(1.0515)^{11} = \$6.886$ billion

   2004 — $n = 14$ $a_{14} = 3.9631(1.0515)^{13} = \$7.613$ billion

51. a. $b_3 = 151.39$, $b_1 = 130.76$

   $b_3 = b_1 r^2$

   $r = \sqrt{b_3 \div b_1} = \sqrt{151.39 \div 130.76} = 1.076$

   $b_n = b_1 r^{n-1} = 130.76(1.076)^{n-1}$

   b. $\sum_{n=1}^{k} b_n = b_1\left(\frac{1-r^k}{1-r}\right)$; $k = 9$ in 2003

   $$\sum_{n=1}^{9} b_n = 130.76\left(\frac{1-(1.076)^9}{1-1.076}\right) = \$1605.85$$

53. Following Example **7** of the text, during the first bounce the ball travels $4 + 4 = 8$ ft, during the second, half of this, and so on. This is a geometric sequence with $a_1 = 8, r = \frac{1}{2}$.

$$\text{Total distance} = \text{initial drop} + \sum_{n=1}^{7} a_1 r^n = 8 + a_1\left(\frac{1-r^7}{1-r}\right) = 8 + 8\left(\frac{1-\left(\frac{1}{2}\right)^7}{1-\frac{1}{2}}\right) = 23.875 \text{ ft.}$$

55. This forms a geometric sequence with $a_1 = .01$, $r = 2$.

$$\sum_{n=1}^{31} a_n = .01\left(\frac{1-2^{31}}{1-2}\right) = \$21,474,836.47$$

57. This forms a geometric sequence with $a_1 = 8000(.75)$ (value after 1 yr), $r = .75$.

$$a_5 = a_1 r^4 = 8000(.75)(.75)^4 = \$1898.44$$

59. Since $a_n = a_1 r^{n-1}$, $\log a_n = \log a_1 + (n-1)\log r$.

Therefore $\log a_n - \log a_{n-1} = \left[\log a_1 + (n-1)\log r\right] - \left[\log a_1 + (n-2)\log r\right] = \log r$.

This shows that the $\{\log a_n\}$ sequence is arithmetic with common difference $\log r$.

61. Each term $a_k = a_1 r^{k-1} = 2^{k-1}$. The sum of the preceding terms is

$$\sum_{n=1}^{k-1} a_n = a_1\left(\frac{1-r^{k-1}}{1-r}\right) = 1\left(\frac{1-2^{k-1}}{1-2}\right) = 2^{k-1} - 1$$

Thus each term 2^{k-1} equals 1 plus the sum of the preceding terms.

63. First determine how many payments are necessary before 1/25 of the balance falls below \$5. Form a geometric sequence with $a_1 = 200$ and $r = 24/25$. Then $a_n = a_1 r^{n-1} = 200\left(\frac{24}{25}\right)^{n-1}$. Use the Table feature of the calculator to observe that $a_{11} = 127.65$, $a_{12} = 122.54$ and $a_{13} = 117.64$.

Thus after 12 payments, with a balance of \$122.54, 1/25 of the outstanding balance is less than \$5. Then it would take \$122.54 ÷ \$5 or 25 more payments of \$5 to complete the payment, for a total of 3 years and 1 month.

12.3A Infinite Series

1. This is a convergent geometric series with $a_1 = \frac{1}{2}$ and $r = \frac{1}{2}$. Its sum is

$$\frac{a_1}{1-r} = \frac{\frac{1}{2}}{1-\frac{1}{2}} = 1.$$

3. This is a convergent geometric series with $a_1 = .06$ and $r = .06$. Its sum is

$$\frac{a_1}{1-r} = \frac{.06}{1-.06} = \frac{3}{47}.$$

5. This is a convergent geometric series with $a_1 = 500$ and $r = \frac{200}{500} = \frac{2}{5}$. Its sum is

$$\frac{a_1}{1-r} = \frac{500}{1-\frac{2}{5}} = \frac{2500}{3} = 833\frac{1}{3}.$$

7. This is a convergent geometric series with $a_1 = 2$ and $r = \sqrt{2}/2$. Its sum is

$$\frac{a_1}{1-r} = \frac{2}{1-\left(\sqrt{2}/2\right)} = \frac{4}{2-\sqrt{2}} = 4 + 2\sqrt{2}.$$

9. $.2+.02+.002+\dots$ is a convergent infinite series with $a_1 = .2$ and $r = .1$. Its sum is

$$\frac{a_1}{1-r} = \frac{.2}{1-.1} = \frac{2}{9}.$$

11. $5.4+.027+.00027+.0000027+\dots$, after the first term, is a convergent infinite series with $a_1 = .027$ and $r = .01$. Its sum is $5.4 + \dfrac{a_1}{1-r} = 5.4 + \dfrac{.027}{1-.01} = \dfrac{54}{10} + \dfrac{27}{990} = \dfrac{597}{110}.$

13. $2.1+.0425+.0000425+\dots$, after the first term, is a convergent infinite series with $a_1 = .0425$ and $r = .001$. Its sum is $2.1 + \dfrac{a_1}{1-r} = 2.1 + \dfrac{.0425}{1-.001} = \dfrac{21}{10} + \dfrac{425}{9990} = \dfrac{10702}{4995}.$

15. $1.74+.00241+.00000241+\dots$, after the first term, is a convergent infinite series with $a_1 = .00241$ and $r = .001$.

Its sum is $1.74 + \dfrac{a_1}{1-r} = 1.74 + \dfrac{.00241}{1-.001} = \dfrac{174}{100} + \dfrac{241}{99900} = \dfrac{174,067}{99,900}.$

17. a. Since $\displaystyle\sum_{n=1}^{\infty} 2(1.5)^n = 2(1.5)^1 + 2(1.5)^2 + 2(1.5)^3 + \dots = 3 + 3(1.5)^1 + 3(1.5)^2 + \dots$, this is a geometric series with $a_1 = 3$, $r = 1.5$.

b. $\displaystyle\sum_{n=1}^{k} a_n = a_1\left(\frac{1-r^k}{1-r}\right) = 3\left(\frac{1-1.5^k}{1-1.5}\right) = 6(1.5^k - 1)$. Hence, let $f(x) = 6(1.5^k - 1)$.

c. Graph the function of part **b** using the window $0 \le x \le 30, 0 \le y \le 10{,}000$ and observe that the function increases faster and faster as x gets large; the graph does not approach a horizontal line, and the series does not converge.

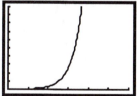

19. $s_1 = \dfrac{\pi}{2} = 1.570796327$

$$s_2 = \frac{\pi}{2} - \frac{(\pi/2)^3}{3!} = .9248322293$$

$$s_3 = \frac{\pi}{2} - \frac{(\pi/2)^3}{3!} + \frac{(\pi/2)^5}{5!} = 1.004524856$$

$$s_4 = \frac{\pi}{2} - \frac{(\pi/2)^3}{3!} + \frac{(\pi/2)^5}{5!} - \frac{(\pi/2)^7}{7!} = .9998431014$$

$$s_5 = \frac{\pi}{2} - \frac{(\pi/2)^3}{3!} + \frac{(\pi/2)^5}{5!} - \frac{(\pi/2)^7}{7!} + \frac{(\pi/2)^9}{9!} = 1.000003543$$

At this point, we may conjecture that the sum is 1.

12.4 The Binomial Theorem

1. $6! = 6 \cdot 5 \cdot 4 \cdot 3 \cdot 2 \cdot 1 = 720$

3. $\dfrac{12!}{9!3!} = \dfrac{12 \cdot 11 \cdot 10 \cdot 9!}{9!3 \cdot 2 \cdot 1} = \dfrac{1320}{6} = 220$

5. $\dbinom{5}{3} + \dbinom{5}{2} - \dbinom{6}{3} = \dfrac{5!}{3!2!} + \dfrac{5!}{2!3!} - \dfrac{6!}{3!3!} = \dfrac{5 \cdot 4 \cdot 3!}{3!2 \cdot 1} + \dfrac{5!}{2!3!} - \dfrac{6 \cdot 5 \cdot 4 \cdot 3!}{3!3 \cdot 2 \cdot 1} = 10 + 10 - 20 = 0$

7. Display the coefficients on the calculator to show

$$\dbinom{6}{0} + \dbinom{6}{1} + \dbinom{6}{2} + \dbinom{6}{3} + \dbinom{6}{4} + \dbinom{6}{5} + \dbinom{6}{6} = 1 + 6 + 15 + 20 + 15 + 6 + 1 = 64$$

9. $\dbinom{100}{96} = \dfrac{100!}{96!4!} = \dfrac{100 \cdot 99 \cdot 98 \cdot 97 \cdot 96!}{96!4 \cdot 3 \cdot 2 \cdot 1} = 3{,}921{,}225$

11. $(x+y)^5 = \dbinom{5}{0}x^5 + \dbinom{5}{1}x^4y + \dbinom{5}{2}x^3y^2 + \dbinom{5}{3}x^2y^3 + \dbinom{5}{4}xy^4 + \dbinom{5}{5}y^5$

$\qquad = x^5 + 5x^4y + 10x^3y^2 + 10x^2y^3 + 5xy^4 + y^5$

13. $(a-b)^5 = (a+(-b))^5$

$\qquad = \dbinom{5}{0}a^5 + \dbinom{5}{1}a^4(-b) + \dbinom{5}{2}a^3(-b)^2 + \dbinom{5}{3}a^2(-b)^3 + \dbinom{5}{4}a(-b)^4 + \dbinom{5}{5}(-b)^5$

$\qquad = a^5 - 5a^4b + 10a^3b^2 - 10a^2b^3 + 5ab^4 - b^5$

15. $\left(2x+y^2\right)^5 = \dbinom{5}{0}(2x)^5 + \dbinom{5}{1}(2x)^4\left(y^2\right) + \dbinom{5}{2}(2x)^3\left(y^2\right)^2 + \dbinom{5}{3}(2x)^2\left(y^2\right)^3$

$$+ \dbinom{5}{4}(2x)\left(y^2\right)^4 + \dbinom{5}{5}\left(y^2\right)^5$$

$\qquad = 32x^5 + 80x^4y^2 + 80x^3y^4 + 40x^2y^6 + 10xy^8 + y^{10}$

17. $\left(\sqrt{x}+1\right)^6 = \dbinom{6}{0}\left(\sqrt{x}\right)^6 + \dbinom{6}{1}\left(\sqrt{x}\right)^5 1 + \dbinom{6}{2}\left(\sqrt{x}\right)^4 1^2 + \dbinom{6}{3}\left(\sqrt{x}\right)^3 1^3$

$$+ \dbinom{6}{4}\left(\sqrt{x}\right)^2 1^4 + \dbinom{6}{5}\left(\sqrt{x}\right)1^5 + \dbinom{6}{6}1^6$$

$\qquad = x^3 + 6x^2\sqrt{x} + 15x^2 + 20x\sqrt{x} + 15x + 6\sqrt{x} + 1$

19. $(1-c)^{10} = \binom{10}{0}1^{10} + \binom{10}{1}1^9(-c) + \binom{10}{2}1^8(-c)^2 + \ldots + \binom{10}{9}1(-c)^9 + \binom{10}{10}(-c)^{10}$

$= 1 - 10c + 45c^2 - 120c^3 + 210c^4 - 252c^5 + 210c^6 - 120c^7 + 45c^8 - 10c^9 + c^{10}$

21. $\left(x^{-3} + x\right)^4 = \binom{4}{0}\left(x^{-3}\right)^4 + \binom{4}{1}\left(x^{-3}\right)^3 x + \binom{4}{2}\left(x^{-3}\right)^2 x^2 + \binom{4}{3}\left(x^{-3}\right)x^3 + \binom{4}{4}x^4$

$= x^{-12} + 4x^{-8} + 6x^{-4} + 4 + x^4$

23. $\left(1 + \sqrt{3}\right)^4 + \left(1 - \sqrt{3}\right)^4 = \binom{4}{0}1^4 + \binom{4}{1}1^3\left(\sqrt{3}\right) + \binom{4}{2}1^2\left(\sqrt{3}\right)^2 + \binom{4}{3}1\left(\sqrt{3}\right)^3 + \binom{4}{4}\left(\sqrt{3}\right)^4$

$+ \binom{4}{0}1^4 + \binom{4}{1}1^3\left(-\sqrt{3}\right) + \binom{4}{2}1^2\left(-\sqrt{3}\right)^2 + \binom{4}{3}1\left(-\sqrt{3}\right)^3 + \binom{4}{4}\left(-\sqrt{3}\right)^4$

$= 1 + 4\sqrt{3} + 18 + 12\sqrt{3} + 9 + 1 - 4\sqrt{3} + 18 - 12\sqrt{3} + 9 = 56$

25. $(1+i)^6 = \binom{6}{0}1^6 + \binom{6}{1}1^5 i + \binom{6}{2}1^4 i^2 + \binom{6}{3}1^3 i^3 + \binom{6}{4}1^2 i^4 + \binom{6}{5}1 i^5 + \binom{6}{6}i^6$

$= 1 + 6i + 15i^2 + 20i^3 + 15i^4 + 6i^5 + i^6$

$= 1 + 6i - 15 - 20i + 15 + 6i - 1 = -8i$

27. In this term, y has exponent 2, the binomial coefficient is $\binom{5}{2}$, and x has exponent 3.

$\binom{5}{2}x^3 y^2 = 10x^3 y^2$

29. In this term, $-d$ has exponent 4, the binomial coefficient is $\binom{7}{4}$, and c has

exponent 3. $\binom{7}{4}c^3(-d)^4 = 35c^3 d^4$

31. In this term, $\dfrac{u}{2}$ has exponent 3, the binomial coefficient is $\binom{7}{3}$, and u^{-2} has

exponent 4. $\binom{7}{3}\left(\dfrac{u}{2}\right)^3\left(u^{-2}\right)^4 = 35\dfrac{u^3}{8}\left(u^{-8}\right) = \dfrac{35}{8u^5}$

33. This is the term with exponent 5 for x, 4 for $-y^2$, and binomial coefficient $\binom{9}{4}$.

$\binom{9}{4}(2x)^5\left(-y^2\right)^4 = 126\left(32x^5\right)y^8 = 4032x^5 y^8$.   Coefficient: 4032

35. This term results from $(2x)^3$ and $\left(\dfrac{1}{x^2}\right)^3$ and thus has binomial coefficient $\dbinom{6}{3}$.

$\dbinom{6}{3}(2x)^3\left(\dfrac{1}{x^2}\right)^3 = 20(8x)^3\left(\dfrac{1}{x^6}\right) = \dfrac{160}{x^3}$. Coefficient: 160

37. a. $\dbinom{9}{1} = \dfrac{9!}{1!8!} = \dfrac{9 \cdot 8!}{8!} = 9$ $\dbinom{9}{8} = \dfrac{9!}{8!1!} = 9$

b. $\dbinom{n}{1} = \dfrac{n!}{1!(n-1)!} = \dfrac{n(n-1)!}{(n-1)!} = n$ $\dbinom{n}{n-1} = \dfrac{n!}{(n-1)!1!} = n$

39. Since $2 = 1 + 1$,

$$2^n = (1+1)^n = \dbinom{n}{0}1^n + \dbinom{n}{1}1^{n-1} \cdot 1 + \dbinom{n}{2}1^{n-2} \cdot 1^2 + \ldots + \dbinom{n}{n}1^n$$

$$= \dbinom{n}{0} + \dbinom{n}{1} + \dbinom{n}{2} + \ldots + \dbinom{n}{n}$$

41. $(\cos\theta + i\sin\theta)^4 = \dbinom{4}{0}(\cos\theta)^4 + \dbinom{4}{1}(\cos\theta)^3(i\sin\theta) + \dbinom{4}{2}(\cos\theta)^2(i\sin\theta)^2$

$$+ \dbinom{4}{3}(\cos\theta)(i\sin\theta)^3 + \dbinom{4}{4}(i\sin\theta)^4$$

$$= \cos^4\theta + 4\cos^3\theta(i\sin\theta) + 6\cos^2\theta\left(-\sin^2\theta\right)$$

$$+ 4\cos\theta\left(-i\sin^3\theta\right) + \sin^4\theta$$

$$= \cos^4\theta - 6\cos^2\theta\sin^2\theta + \sin^4\theta + i\left(4\cos^3\theta\sin\theta - 4\cos\theta\sin^3\theta\right)$$

43. a. $f(x+h) - f(x) = (x+h)^5 - x^5$

$$= \dbinom{5}{0}x^5 + \dbinom{5}{1}x^4h + \dbinom{5}{2}x^3h^2 + \dbinom{5}{3}x^2h^3 + \dbinom{5}{4}xh^4 + \dbinom{5}{5}h^5 - x^5$$

$$= \dbinom{5}{1}x^4h + \dbinom{5}{2}x^3h^2 + \dbinom{5}{3}x^2h^3 + \dbinom{5}{4}xh^4 + \dbinom{5}{5}h^5$$

b. $\dfrac{f(x+h) - f(x)}{h} = \dfrac{h\left[\dbinom{5}{1}x^4 + \dbinom{5}{2}x^3h + \dbinom{5}{3}x^2h^2 + \dbinom{5}{4}xh^3 + \dbinom{5}{5}h^4\right]}{h}$

$$= \dbinom{5}{1}x^4 + \dbinom{5}{2}x^3h + \dbinom{5}{3}x^2h^2 + \dbinom{5}{4}xh^3 + \dbinom{5}{5}h^4$$

c. If h is very close to 0, the above quantity is very close to $\dbinom{5}{1}x^4$ or $5x^4$.

45. a. $f(x+h) - f(x) = (x+h)^{12} - x^{12}$

$$= \binom{12}{0}x^{12} + \binom{12}{1}x^{11}h + \binom{12}{2}x^{10}h^2 + \ldots + \binom{12}{12}h^{12} - x^{12}$$

$$= \binom{12}{1}x^{11}h + \binom{12}{2}x^{10}h^2 + \ldots + \binom{12}{12}h^{12}$$

b. $\dfrac{f(x+h) - f(x)}{h} = \dfrac{h\left[\binom{12}{1}x^{11} + \binom{12}{2}x^{10}h + \ldots + \binom{12}{12}h^{11}\right]}{h}$

$$= \binom{12}{1}x^{11} + \binom{12}{2}x^{10}h + \ldots + \binom{12}{12}h^{11}$$

c. If h is very close to 0, the above quantity is very close to $\binom{12}{1}x^{11}$ or $12x^{11}$.

47. In general, $a! = a(a-1)(a-2)\ldots 1 = a(a-1)!$

a. Thus, $(n-r)! = (n-r)[(n-r)-1]! = (n-r)[n-(r+1)]!$

b. $(n-r)! = (n+1-r-1)! = [(n+1)-(r+1)]!$

c. $\binom{n}{r+1} + \binom{n}{r} = \dfrac{n!}{(r+1)![n-(r+1)]!} + \dfrac{n!}{r!(n-r)!}$

$$= \dfrac{(n-r)n!}{(r+1)!(n-r)[n-(r+1)]!} + \dfrac{(r+1)n!}{(r+1)r!(n-r)!}$$

$$= \dfrac{(n-r)n!}{(r+1)!(n-r)!} + \dfrac{(r+1)n!}{(r+1)!(n-r)!}$$

$$= \dfrac{[(n-r)+(r+1)]n!}{(r+1)!(n-r)!}$$

$$= \dfrac{(n+1)n!}{(r+1)![(n+1)-(r+1)]!}$$

$$= \dfrac{(n+1)!}{(r+1)![(n+1)-(r+1)]!} = \binom{n+1}{r+1}$$

d. Since each entry in row n has form $\binom{n}{k}$, the sum of two adjacent entries is

$\binom{n}{r+1} + \binom{n}{r}$. These are the entries above and to the right and left of $\binom{n+1}{r+1}$. This explains the assertion.

12.5 Mathematical Induction

1. When $n = 1$, both sides of the equation are equal to 1, thus the statement is true. Assume that the statement is true for $n = k$. Then
    $$1 + 2 + 2^2 + 2^3 + \ldots + 2^{k-1} = 2^k - 1.$$ Add 2^k to both sides of the equation to get:
    $$1 + 2 + 2^2 + 2^3 + \ldots + 2^{k-1} + 2^k = 2^k - 1 + 2^k = 2(2^k) - 1 = 2^{k+1} - 1$$
    Thus the statement is true for $n = k + 1$. Therefore, by the Principle of Mathematical Induction the statement is true for all positive integers n.

3. When $n = 1$, both sides of the equation are equal to 1, thus the statement is true. Assume that the statement is true for $n = k$. Then
    $$1 + 3 + 5 + 7 + \ldots + (2k - 1) = k^2$$
    Add the next odd integer, $(2k + 1)$ to both sides:
    $$1 + 3 + 5 + 7 + \ldots + (2k - 1) + (2k + 1) = k^2 + (2k + 1) = (k + 1)^2$$
    Thus the statement is true for $n = k + 1$. Therefore, by the Principle of Mathematical Induction the statement is true for all positive integers n. That is, the sum of the first n odd inters is n^2.

5. When $n = 1$, both sides of the equation are equal to 1, thus the statement is true. Assume that the statement is true for $n = k$. Then
    $$1^2 + 2^2 + 3^2 + \ldots + k^2 = \frac{k(k+1)(2k+1)}{6}$$
    Add the next square $(k + 1)^2$ to both sides of the equation:
    $$1^2 + 2^2 + 3^2 + \ldots + k^2 + (k+1)^2 = \frac{k(k+1)(2k+1)}{6} + (k+1)^2$$
    $$= (k+1)\left[\frac{k(2k+1)}{6} + (k+1)\right] = (k+1)\left[\frac{2k^2 + 7k + 6}{6}\right]$$
    $$= (k+1)\frac{(k+2)(2k+3)}{6} = \frac{(k+1)[(k+1)+1][2(k+1)+1]}{6}$$
    Thus the statement is true for $n = k + 1$. Therefore the statement is true for all positive integers n.

7. When $n = 1$, both sides of the equation are equal to $\frac{1}{2}$, thus the statement is true. Assume that the statement is true for $n = k$. Then
    $$\frac{1}{1 \cdot 2} + \frac{1}{2 \cdot 3} + \frac{1}{3 \cdot 4} + \ldots + \frac{1}{k(k+1)} = \frac{k}{k+1}$$
    Add the next term $\dfrac{1}{(k+1)(k+2)}$ to both sides of the equation:

$$\frac{1}{1\cdot 2}+\frac{1}{2\cdot 3}+\frac{1}{3\cdot 4}+\ldots+\frac{1}{k(k+1)}+\frac{1}{(k+1)(k+2)}=\frac{k}{k+1}+\frac{1}{(k+1)(k+2)}$$

$$=\frac{(k+1)^2}{(k+1)(k+2)}=\frac{k+1}{k+2}$$

Thus the statement is true for $n = k + 1$. Therefore the statement is true for all positive integers n.

9. When $n = 1$, we have $1+2>1$, which is true. Assume that the statement is true for $n = k$.. Then $k+2>k$. Add 1 to both sides of the inequality to get $k+2+1>k+1$, $(k+1)+2>(k+1)$ and the statement is true for $n = k + 1$. Therefore, the statement $n+2>n$ is true for all positive integers n.

11. When $n = 1$, we have $3^1 \geq 3(1)$, which is true. Assume that the statement is true for $n = k$. Then $3^k \geq 3(k)$. Multiply both sides of the inequality by 3 to get $3(3^k)\geq 3(3k), 3^{k+1}\geq 9k, 3^{k+1}\geq 3k+6k$. Since k is a positive integer, $6k>3$ and the statement becomes $3^{k+1}\geq 3k+3, 3^{k+1}\geq 3(k+1)$. Thus the statement is true for $n = k + 1$. Therefore, $3^n \geq 3n$ is true for all positive integers n.

13. When $n = 1$, we have $3(1)>1+1$, which is true. Assume that the statement is true for $n = k$. Then $3k>k+1$. Add 3 to both sides of the inequality to get $3k+3>k+4$, and since $k+4>k+2$ we have $3k+3>k+2, 3(k+1)>(k+1)+1$ and the statement is true for $n = k + 1$. Therefore, the statement $3n>n+1$ is true for all positive integers n.

15. When $n=1$, $2^{2n+1}+1=2^3+1=9$ and the statement is true since 3 is a factor of 9. Assume that the statement is true for $n = k$. Then 3 is a factor of $2^{2k+1}+1$. Thus there is an integer p we can multiply by 3 to get $2^{2k+1}+1$. That is $3p=2^{2k+1}+1$ or $3p-1=2^{2k+1}$. Now consider:
$$2^{2(k+1)+1}+1=2^{2k+3}+1=2^{2k+1}2^2+1=(3p-1)2^2+1=12p-3=3(4p-1).$$
Therefore, 3 is a factor of $2^{2(k+1)+1}+1$ and the statement is true for $n = k + 1$. Thus, 3 is a factor of $2^{2n+1}+1$ for all positive integers n.

17. When $n=1$, $3^{2n+2}-8n-9=64$ and the statement is true since 64 is a factor of 64. Assume that the statement is true for $n = k$. Then 64 is a factor of $3^{2k+2}-8k-9$. Thus there is an integer p we can multiply by 64 to get $3^{2k+2}-8k-9$. That is $64p=3^{2k+2}-8k-9$ or $3^{2k+2}=64p+8k+9$. Now consider:
$$3^{2(k+1)+2}-8(k+1)-9=3^{2k+4}-8k-17=9(3^{2k+2})-8k-17$$

$$=9(64p+8k+9)-8k-17$$

$$=64(9p+k+1)$$

Therefore, 64 is a factor of $3^{2(k+1)+2}-8(k+1)-9$ and the statement is true for $n = k + 1$. Thus, 64 is a factor of $3^{2n+2}-8n-9$ for all positive integers n.

19. When $n = 1$, both sides of the equation are equal to c and the statement is true. Assume that the statement is true for $n = k$. Then

$$c + (c + d) + (c + 2d) + (c + 3d) + \ldots + + (c + (k-1)d) = \frac{k(2c + (k-1)d)}{2}$$

Add $c + kd$ to both sides of the equation to obtain:

$$c + (c + d) + (c + 2d) + \ldots + (c + (k-1)d) + (c + kd) = \frac{k(2c + (k-1)d)}{2} + (c + kd)$$

$$= \frac{2kc + k(k-1)d + 2c + 2kd}{2} = \frac{2c(k+1) + (k+1)kd}{2}$$

$$= \frac{(k+1)(2c + kd)}{2}$$

and the statement is true for $n = k + 1$. Thus it is true for all positive integers n.

21. a. $x^2 - y^2 = (x - y)(x + y)$

$$x^3 - y^3 = (x - y)(x^2 + xy + y^2)$$

$$x^4 - y^4 = (x - y)(x^3 + x^2y + xy^2 + y^3)$$

b. $x^n - y^n = (x - y)(x^{n-1} + x^{n-2}y + x^{n-3}y^2 + \ldots + y^{n-1})$

From part **a** we see the statement is true for $n = 2$. Assume that the statement is true for $n = k$.. Then $x^k - y^k = (x - y)(x^{k-1} + x^{k-2}y + x^{k-3}y^2 + \ldots + y^{k-1})$.

Consider $x^{k+1} - y^{k+1} = xx^k - yy^k = xx^k - xy^k + xy^k - yy^k$

$$= x(x^k - y^k) + (x - y)y^k$$

$$= x(x - y)(x^{k-1} + x^{k-2}y + x^{k-3}y^2 + \ldots + y^{k-1}) + (x - y)y^k$$

$$= (x - y)[x(x^{k-1} + x^{k-2}y + x^{k-3}y^2 + \ldots + y^{k-1}) + y^k]$$

$$= (x - y)(x^k + x^{k-1}y + x^{k-2}y^2 + \ldots + xy^{k-1} + y^k)$$

and the statement is true for $n = k + 1$. Thus it is true for all positive integers n.

23. This statement is false. For example 9 is an odd positive integer and it is not a prime.

25. When $n = 1$ the statement is true. Assume that the statement is true for $n = k$. Then $(k+1)^2 > k^2 + 1$. Then

$$(k+2)^2 = k^2 + 4k + 4 = k^2 + 2k + 1 + 2k + 3 = (k+1)^2 + (2k+3)$$

$$> k^2 + 1 + (2k+3) = k^2 + 2k + 1 + 3 = (k+1)^2 + 1 + 3$$

$$> (k+1)^2 + 1$$

and the statement is true for $n = k + 1$. Thus $(n+1)^2 > n^2 + 1$ is true for all positive integers n.

27. This statement is false, for example when $n = 2$, $n^4 - n + 4 = 18$ and 4 is not a factor of 18.

29. When $n = 5$, we have $2(5) - 4 > 5$, which is true. Assume that the statement is true for $n = k$ where $k \geq 5$. Then $2k - 4 > k$.
And $2(k+1) - 4 = 2k - 4 + 2 > k + 2 > k + 1$. Thus the statement is true for $n = k + 1$.
Therefore by induction the statement is true for all $n \geq 5$.

31. When $n = 2$, we have $2^2 > 2$, which is true. Assume that the statement is true for $n = k$ where $k \geq 2$. Then $k^2 > k$. Then $(k+1)^2 = k^2 + 2k + 1 > k + 2k + 1 > k + 1$.
Then the statement is true for $n = k + 1$. Thus $n^2 > n$ for all $n \geq 2$.

33. When $n = 4$, we have $3^4 > 2^4 + 10(4)$, which is true. Assume that the statement is true for $n = k$ where $k \geq 4$. Then $3^k > 2^k + 10k$. And
$$3^{k+1} = 3 \cdot 3^k > 3(2^k + 10k) = 3 \cdot 2^k + 30k > 2 \cdot 2^k + 30k > 2^{k+1} + 10k + 10$$
$$= 2^{k+1} + 10(k+1).$$
Therefore the statement is true for $n = k + 1$. Thus $3^n > 2^n + 10n$ for all $n \geq 4$.

35. a. When $n = 2$, you can move the stack in 3 moves. If you start on peg 1, you move the the top ring to peg 2, then the bottom ring to peg 3 and the the top ring on top of the bottom ring.
When $n = 3$, you can move the stack in 7 moves. It takes 3 moves to put the top 2 rings onto peg 2, then the fourth move puts the bottom ring to peg 3 and then 3 more moves to move the top two on top of the bottom. That is $3 + 1 + 3 = 7$.
When $n = 4$, it takes 15 moves, 7 to move the top three rings to peg 2, 1 to move the bottom ring to peg 3 and 7 more to put the top three on the bottom. That is $7 + 1 + 7 = 15$.
b. You can move 1 ring in one move, 2 rings in three moves, 3 rings in seven moves, 4 rings in fifteen moves. Thus we conjecture that it takes $2^n - 1$ moves to move n rings. Clearly this holds true for $n = 1$. Assume that the statement is true for $n = k$. That is, it takes $2^k - 1$ moves to move k rings. Now consider $k + 1$ rings. You can move the top k rings in $2^k - 1$ moves, one move to move the bottom ring, and $2^k - 1$ moves to move the top k rings on the bottom. This gives a total of $2^k - 1 + 1 + 2^k - 1 = 2(2^k) - 1 = 2^{k+1} - 1$ moves. Thus the conjecture is true for $n = k + 1$. It will take $2^n - 1$ moves to move n rings for all positive integers n.

37. The statement is true for $n = 1$. Assume that the statement is true for $n = k$. Then $z^k = r^k[\cos(k\theta) + i\sin(k\theta)]$. And
$$z^{k+1} = z \cdot z^k = z[r^k(\cos(k\theta) + i\sin(k\theta))] = r(\cos\theta + i\sin\theta)r^k(\cos(k\theta) + i\sin(k\theta))$$
$$= r^{k+1}[\cos\theta\cos(k\theta) + i^2\sin\theta\sin(k\theta) + i\cos\theta\sin(k\theta) + i\sin\theta\cos(k\theta)]$$
$$= r^{k+1}[\cos\theta\cos(k\theta) - \sin\theta\sin(k\theta) + i(\cos\theta\sin(k\theta) + \sin\theta\cos(k\theta))]$$
$$= r^{k+1}[\cos(\theta + k\theta) + i\sin(\theta + k\theta)] = r^{k+1}[\cos((k+1)\theta) + i\sin((k+1)\theta)]$$
Therefore the statement is true for $n = k + 1$. Hence the statement is true for all positive integers n.

Chapter 12 Review Exercises

1. $a_1 = 2 \cdot 1 - 5 = -3$

$a_2 = 2 \cdot 2 - 5 = -1$

$a_3 = 2 \cdot 3 - 5 = 1$

$a_4 = 2 \cdot 4 - 5 = 3$

3. $a_1 = \left(\frac{-1}{1}\right)^2 = 1$

$a_2 = \left(\frac{-1}{2}\right)^2 = \frac{1}{4}$

$a_3 = \left(\frac{-1}{3}\right)^2 = \frac{1}{9}$

$a_4 = \left(\frac{-1}{4}\right)^2 = \frac{1}{16}$

5.   a. September 1999 — $n = 1$; $a_1 = 152,600$

July 2001 — $n = 23$; $a_{23} = 86,634$

August 2002 — $n = 36$; $a_{36} = 48,620$

   b. Use the Table feature to obtain $a_{30} = 60,814$, $a_{31} = 58,155$. $n = 30$ corresponds to February 2002, $n = 31$ to March 2002.

7. This is a geometric sequence with

$a_1 = \frac{1}{9}, r = 3$

$$\sum_{n=1}^{4} a_n = a_1 \left(\frac{1 - r^4}{1 - r}\right)$$

$$= \frac{1}{9}\left(\frac{1 - 3^4}{1 - 3}\right) = \frac{40}{9}$$

9. $\displaystyle\sum_{n=2}^{4} \left(3n^2 - n + 1\right) = \left(3 \cdot 2^2 - 2 + 1\right) + \left(3 \cdot 3^2 - 3 + 1\right) + \left(3 \cdot 4^2 - 4 + 1\right) = 11 + 25 + 45 = 81$

11. $a_n = a_1 + (n - 1)d$

$a_n = 3 + (n - 1)(-6)$

$a_n = 9 - 6n$

13. $a_3 = a_1 + 2d$

$d = \dfrac{a_3 - a_1}{2} = \dfrac{7 - (-5)}{2} = 6$

$a_n = a_1 + (n - 1)d$

$a_n = -5 + (n - 1)6$

$a_n = 6n - 11$

15. $\displaystyle\sum_{n=1}^{k} a_n = ka_1 + \frac{k(k - 1)}{2}d$

$\displaystyle\sum_{n=1}^{11} a_n = 11 \cdot 5 + \frac{11 \cdot 10}{2}(-2) = -55$

17. $a_n = a_1 r^{n-1}$

$a_n = 2 \cdot 3^{n-1}$

19. $a_2 = a_1 r$

$a_7 = a_1 r^6$

$r^5 = \dfrac{a_7}{a_2} = \dfrac{6}{192} = \dfrac{1}{32}$

$r = \dfrac{1}{2},\ a_1 = \dfrac{a_2}{r} = \dfrac{192}{1/2} = 384$

$a_n = a_1 r^{n-1}$

$a_n = 384\left(\tfrac{1}{2}\right)^{n-1} = \dfrac{3}{2^{n-8}}$

21. $\displaystyle\sum_{n=1}^{5} a_n = a_1\left(\dfrac{1-r^5}{1-r}\right) = \dfrac{1}{4}\left(\dfrac{1-3^5}{1-3}\right) = \dfrac{121}{4}$

23. Since $a_1 = 4,\ a_5 = 23,\ a_5 = a_1 + 4D,\ D = \dfrac{a_5 - a_1}{4} = \dfrac{19}{4}.$ (using D = common difference)

Then $b = a_2 = 4 + \dfrac{19}{4} = \dfrac{35}{4},\ c = a_3 = \dfrac{35}{4} + \dfrac{19}{4} = \dfrac{27}{2},\ d = a_4 = \dfrac{27}{2} + \dfrac{19}{4} = \dfrac{73}{4}.$

25. a. Since $a_1 = 5.123,\ a_2 = 5.5067.$

Hence $r = \dfrac{a_2}{a_1} = \dfrac{5.5067}{5.123} = 1.0749$ (approximated to 4 decimal places)

$a_n = a_1 r^{n-1} = 5.123(1.0749)^{n-1}$

b. $2002 \longrightarrow n = 9;\ a_9 = 5.123(1.0749)^{9-1} = \$9.1298$ billion

c. $\displaystyle\sum_{n=1}^{9} a_n = a_1\left(\dfrac{1-r^9}{1-r}\right) = 5.123\left(\dfrac{1-(1.0749)^9}{1-1.0749}\right) = \$62.6274$ billion

27. This is a geometric series with $a_1 = 1$ and $r = \tfrac{1}{2}.$ $\dfrac{a_1}{1-r} = \dfrac{1}{1-\frac{1}{2}} = 2$

29. $(1.02)^{51} = 1^{51} + 51(1)^{50}(.02) + \dfrac{51 \cdot 50}{2 \cdot 1}(1)^{49}(.02)^2 + \ldots$

The first three terms add to 2.53 and the rest are positive, hence $(1.02)^{51} > 2.5.$

31. $\dbinom{15}{12} = \dfrac{15!}{12!\,3!} = \dfrac{15 \cdot 14 \cdot 13 \cdot 12!}{12!\,3 \cdot 2 \cdot 1} = \dfrac{15 \cdot 14 \cdot 13}{3 \cdot 2 \cdot 1} = 455$

33. $\dbinom{n+1}{n} = \dfrac{(n+1)!}{n!(n+1-n)!} = \dfrac{(n+1)!}{n!\,1!} = \dfrac{(n+1)n!}{n!\cdot 1} = n+1$

35. $\dfrac{20!5!}{6!17!} = \dfrac{20 \cdot 19 \cdot 18 \cdot 17!5!}{6 \cdot 5!17!} = \dfrac{20 \cdot 19 \cdot 18}{6} = 1140$

37. This term results from $(2y)^4$ and $(x^2)^1$ and thus has binomial coefficient $\binom{5}{1}$.

$$\binom{5}{1}(2y)^4(x^2)^1 = 5 \cdot 16y^4x^2 = 80x^2y^4. \text{ Coefficient: } 80$$

39. When $n = 1$, both sides of the equation are equal to 1, thus the statement is true. Assume that the statement is true for $n = k$. Then

$$1 + 5^1 + 5^2 + \ldots + 5^{k-1} = \frac{5^k - 1}{4}. \text{ Add the } 5^k \text{ to both sides of the equation:}$$

$$1 + 5^1 + 5^2 + \ldots + 5^{k-1} + 5^k = \frac{5^k - 1}{4} + 5^k = \frac{5^k - 1 + 4 \cdot 5^k}{4} = \frac{5 \cdot 5^k - 1}{4} = \frac{5^{k+1} - 1}{4}$$

Thus the statement is true for $n = k + 1$. Therefore, by induction the statement is true for all positive integers n.

41. Clearly the statement is true for $x = 0$, hence assume $x \neq 0$. Then the statement is true for $n = 1$. Assume that the statement is true for $n = k$, then $|x^k| < 1$. Multiply both sides of this inequality by $|x|$ to obtain $|x^k| \cdot |x| < 1 \cdot |x|$. Thus $|x^{k+1}| = |x^k| \cdot |x| < |x| < 1$. Thus the statement is true for $n = k + 1$. Therefore, by induction the statement is true for all positive integers n.

43. When $n = 1$, both sides of the equation are equal to 1, thus the statement is true. Assume that the statement is true for $n = k$. Then

$$1 + 4^1 + 4^2 + \ldots + 4^{k-1} = \frac{4^k - 1}{3}. \text{ Add the } 4^k \text{ to both sides of the equation:}$$

$$1 + 4^1 + 4^2 + \ldots + 4^{k-1} + 4^k = \frac{4^k - 1^k}{3} + 4^k = \frac{4^k - 1 + 3 \cdot 4^k}{3} = \frac{4^{k+1} - 1}{3}$$

Thus the statement is true for $n = k + 1$. Therefore, by induction the statement is true for all positive integers n.

45. When $n = 1$, $9^n - 8n - 1 = 0$ and the the statement is true since 8 is a factor of 0. Assume that the statement is true for $n = k$. Then 8 is a factor of $9^k - 8k - 1$. That is, there is an integer p that we can multiply by 8 to get $9^k - 8k - 1$. Thus $8p = 9^k - 8k - 1$ or $8p + 8k + 1 = 9^k$. Consider $9^{k+1} - 8(k+1) - 1$. Rearrange this expression to get $9(9^k) - 8k - 9 = 9(8p + 8k + 1) - 8k - 9 = 72p + 64k = 8(9p + 8k)$ and so 8 is a factor of $9^{k+1} - 8(k+1) - 1$. Thus the statement is true for $n = k + 1$. Therefore, by induction the statement is true for all positive integers n.

Chapter 13
Limits and Continuity

13.1 Limits of Functions

1. For the function $f(x) = \dfrac{x^6 - 1}{x^4 - 1}$, $f(-1.01) = 1.515$, $f(-1.0001) = 1.50015$,

$f(-1.000001) = 1.5000015$. Also $f(-.99) = 1.485$, $f(-.9999) = 1.49985$. The limit seems to be 1.5.

3. For the function $f(x) = \dfrac{\tan x - x}{x^3}$, $f(.01) = .3333$, $f(.001) = .333333$,

$f(.0001) = .3333333$. Also $f(-.01) = .3333$, $f(-.001) = .333333$. The limit seems to be $\frac{1}{3}$.

5. For the function $f(x) = \dfrac{x - \tan x}{x - \sin x}$, $f(.01) = -2.001$, $f(.001) = -2.000001$,

$f(.0001) = -1.999988$. Also $f(-.01) = -2.0001$, $f(-.001) = -2.000001$. The limit seems to be -2.

7. For the function $f(x) = \dfrac{x}{\ln|x|}$, $f(.01) = -.002$, $f(.001) = -.0001$,

$f(.0001) = -.00001$. Also $f(-.01) = .002$, $f(-.001) = .0001$. The limit seems to be 0.

9. For the function $f(x) = \dfrac{e^x - 1}{\sin x}$, $f(.01) = 1.005$, $f(.001) = 1.0005$, $f(.0001) = 1.00005$.

Also $f(-.01) = .995$, $f(-.001) = .9995$. The limit seems to be 1.

11. For the function $f(x) = \dfrac{\sin x}{1 - \cos x}$, choose values approaching π from the left.

$f(3.1) = .0208$, $f(3.14) = .0008$, $f(3.141) = .0003$, $f(3.1415) = .00005$. The limit seems to be 0.

13. For the function $f(x) = \sqrt{x}\ln x$, choose values approaching 0 from the right.

$f(.0001) = -.0921$, $f(.00001) = -.0364$, $f(.000001) = -.0138$,

$f(.00000001) = -.0018$. The limit seems to be 0.

15. For the function $f(x) = \dfrac{\sin 6x}{x}$, choose values approaching 0 from the left.

$f(-.1) = 5.6464$, $f(-.01) = 5.9964$, $f(-.001) = 5.999964$,

$f(-.0001) = 5.99999964$. The limit seems to be 6.

17. $\lim\limits_{x \to -3} f(x) = -1$, $\lim\limits_{x \to 0} f(x) = 1$, $\lim\limits_{x \to 2} f(x)$ does not exist.

19. $\lim_{x \to -3} f(x) = 2$, $\lim_{x \to 0} f(x)$ does not exist, $\lim_{x \to 2} f(x) = 0$.

21. $\lim_{x \to -3} f(x) = 2$, $\lim_{x \to 0} f(x) = 1$, $\lim_{x \to 2} f(x) = 1$.

23. a. $\lim_{x \to -2^-} f(x) = 0$. **b.** $\lim_{x \to 0^+} f(x) = 1$. **c.** $\lim_{x \to 3^-} f(x) = -1$. **d.** $\lim_{x \to 3^+} f(x) = -1$.

25. a. $\lim_{x \to -2^-} f(x) = 1$.   **b.** $\lim_{x \to 0^+} f(x) = 0$.

    c. $\lim_{x \to 3^-} f(x)$ is not defined.   **d.** $\lim_{x \to 3^+} f(x)$ is not defined.

27. $\lim_{x \to 3} f(x) = \dfrac{11}{5} = f(3)$.

29. $\lim_{x \to 2} \dfrac{x^2 - 5x + 6}{x^2 - 6x + 8} = \lim_{x \to 2} \dfrac{(x-2)(x-3)}{(x-2)(x-4)} = \lim_{x \to 2} \dfrac{(x-3)}{(x-4)} = \dfrac{1}{2}$. $f(2)$ is not defined.

31. The limit does not exist.   **33.** $\lim_{x \to 2^+} \sqrt{x^2 - 4} = 0 = f(2)$.

35. The limit does not exist.

37. For the function $f(x) = \dfrac{x}{e^x - 1}$, $f(.01) = .995$, $f(.001) = .9995$, $f(.0001) = .99995$.
Also $f(-.01) = 1.005$, $f(-.001) = 1.0005$. The limit seems to be 1. $f(0)$ is not defined.

39. For the function $f(x) = \dfrac{\ln x}{x - 1}$, $f(1.01) = .995$, $f(1.001) = .9995$, $f(1.0001) = .99995$.
Also $f(.99) = 1.005$, $f(.9999) = 1.00005$. The limit seems to be 1. $f(1)$ is not defined.

41. For the function $f(x) = x \ln|x|$, $f(.01) = -.046$, $f(.001) = -.0069$,
$f(.0001) = -.00092$. Also $f(-.01) = .046$, $f(-.0001) = .00092$. The limit seems to be 0. $f(0)$ is not defined.

43. $\lim_{x \to \pi/2} x \cos x = \lim_{x \to \pi/2} x \lim_{x \to \pi/2} \cos x = \frac{\pi}{2} \cdot 0 = 0 = f\left(\frac{\pi}{2}\right)$.

45. $\lim_{x \to 1} e^x \sin \dfrac{\pi x}{2} = \lim_{x \to 1} e^x \lim_{x \to 1} \sin \dfrac{\pi x}{2} = e \cdot 1 = e = f(1)$.

47. a.

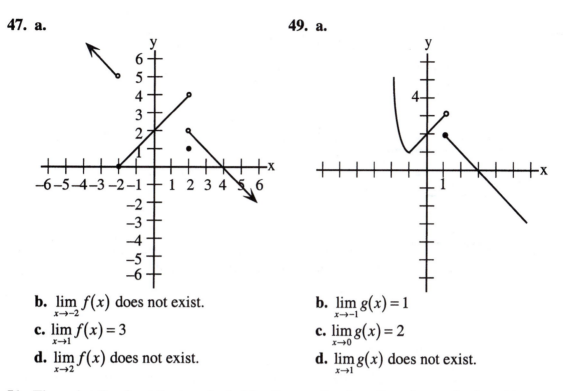

49. a.

b. $\lim_{x \to -2} f(x)$ does not exist.

c. $\lim_{x \to 1} f(x) = 3$

d. $\lim_{x \to 2} f(x)$ does not exist.

b. $\lim_{x \to -1} g(x) = 1$

c. $\lim_{x \to 0} g(x) = 2$

d. $\lim_{x \to 1} g(x)$ does not exist.

51. The point P, where the terminal side of an angle of t radians in standard position meets the unit circle, has coordinates $(\cos t, \sin t)$. The picture shows that as t gets

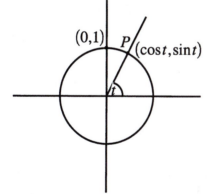

closer and closer to $\pi / 2$, the terminal side gets very close to the y-axis and the point P moves along the unit circle toward the point $(0,1)$. Hence the second coordinate of P, $\sin t$, gets closer to 1.

53. For the function $h(x) = [x] + [-x]$, when x is close to 2, but slightly greater than 2, $[x] = 2$ and $[-x] = -3$. When x is close to 2, but slightly less than 2, $[x] = 1$ and $[-x] = -2$. In either case, for all values of x except 2 itself, $h(x) = -1$, so the limit of the function is -1.

55. When x is close to 2, but slightly greater than 2, $[x] = 2$, hence $\lim_{x \to 2^+} [x] = 2$. When x is close to 2, but slightly less than 2, $[x] = 1$, hence $\lim_{x \to 2^-} [x] = 1$.

57. Write $g(x) = \dfrac{h(x)}{x}$, where $h(x)$ is the function of Exercise **53**. Then:

$$\lim_{x\to 3} g(x) = \frac{\lim_{x\to 3} h(x)}{\lim_{x\to 3} x} = -\frac{1}{3}$$

59. The graph of the function is shown here using the window $-.2 \le x \le .2, -1 \le y \le 1$.

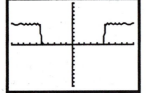

The trace function shows that all values near 0 are mapped to zero, which would suggest a value of the limit as 0 rather than the actual value of 0.5.

13.2 Properties of Limits

1. $\lim_{x\to 4}\big(f(x) + g(x)\big) = \lim_{x\to 4} f(x) + \lim_{x\to 4} g(x) = 5 + 0 = 5$

3. $\lim_{x\to 4} \dfrac{f(x)}{g(x)}$ does not exist because $\lim_{x\to 4} g(x) = 0$

5. $\lim_{x\to 4} f(x)g(x) = \lim_{x\to 4} f(x) \lim_{x\to 4} g(x) = 5 \cdot 0 = 0$

7. $\lim_{x\to 4} \dfrac{3h(x)}{2f(x) + g(x)} = \dfrac{3\lim_{x\to 4} h(x)}{2\lim_{x\to 4} f(x) + \lim_{x\to 4} g(x)} = \dfrac{3(-2)}{2\cdot 5 + 0} = -\dfrac{3}{5}$

9. This is a polynomial function, hence $\lim_{x\to 2} f(x) = f(2) = 6\cdot 2^3 - 2\cdot 2^2 + 5\cdot 2 - 3 = 47$

11. This is a rational function, $f(-2)$ is defined, hence

$$\lim_{x\to -2} f(x) = f(-2) = \frac{3(-2)-1}{2(-2)+3} = 7$$

13. $\lim_{x\to 3} \dfrac{x^2 - x - 6}{x^2 - 2x - 3} = \lim_{x\to 3} \dfrac{(x-3)(x+2)}{(x-3)(x+1)} = \dfrac{3+2}{3+1} = \dfrac{5}{4}$

15. $\lim\limits_{x \to 1} \dfrac{x^3 - 1}{x^2 - 1} = \lim\limits_{x \to 1} \dfrac{(x-1)(x^2 + x + 1)}{(x-1)(x+1)} = \lim\limits_{x \to 1} \dfrac{x^2 + x + 1}{x+1} = \dfrac{1+1+1}{1+1} = \dfrac{3}{2}$

17. $\lim\limits_{x \to 4^-} \dfrac{x - 4}{x^2 - 16} = \lim\limits_{x \to 4^-} \dfrac{x - 4}{(x-4)(x+4)} = \lim\limits_{x \to 4^-} \dfrac{1}{x+4} = \dfrac{1}{4+4} = \dfrac{1}{8}$

19. This rational function is not defined at $x = 3$, and as $x \to 3^+$, values of the function exceed all bounds. The limit does not exist.

21. $\lim\limits_{x \to 1} \sqrt{x^3 + 6x^2 + 2x + 5} = \lim\limits_{x \to 1} \sqrt{1^3 + 6 \cdot 1^2 + 2 \cdot 1 + 5} = \sqrt{14}$

23. $\lim\limits_{x \to 1^+} \left(\sqrt{x-1} + 3 \right) = \sqrt{1-1} + 3 = 3$

25. $\lim\limits_{x \to -2.5^+} \left(\sqrt{5 + 2x} + x \right) = \sqrt{5 + 2(-2.5)} + (-2.5) = -2.5$

27. $\lim\limits_{x \to 3} \dfrac{\sqrt{x} - \sqrt{3}}{x - 3} = \lim\limits_{x \to 3} \dfrac{\left(\sqrt{x} - \sqrt{3} \right)}{\left(\sqrt{x} - \sqrt{3} \right)\left(\sqrt{x} + \sqrt{3} \right)} = \lim\limits_{x \to 3} \dfrac{1}{\sqrt{x} + \sqrt{3}} = \dfrac{1}{2\sqrt{3}}$ or $\dfrac{\sqrt{3}}{6}$

29. $\lim\limits_{x \to 0} \dfrac{\dfrac{1}{x+5} - \dfrac{1}{5}}{x} = \lim\limits_{x \to 0} \dfrac{5 - (x+5)}{5x(x+5)} = \lim\limits_{x \to 0} \dfrac{-1}{5(x+5)} = -\dfrac{1}{25}$

31. $\lim\limits_{x \to 1} \left[\dfrac{1}{x-1} - \dfrac{2}{x^2 - 1} \right] = \lim\limits_{x \to 1} \dfrac{x+1-2}{x^2 - 1} = \lim\limits_{x \to 1} \dfrac{x-1}{(x-1)(x+1)} = \lim\limits_{x \to 1} \dfrac{1}{x+1} = \dfrac{1}{1+1} = \dfrac{1}{2}$

33. $\lim\limits_{x \to 0} \dfrac{x^2}{|x|} = \lim\limits_{x \to 0} \dfrac{|x|^2}{|x|} = \lim\limits_{x \to 0} |x| = |0| = 0$

35. $\lim\limits_{x \to -3^+} \left[\dfrac{|x+3|}{x+3} + \sqrt{x+3} + 1 \right] = \lim\limits_{x \to -3^+} \left[\dfrac{x+3}{x+3} + \sqrt{x+3} + 1 \right] = \lim\limits_{x \to -3^+} \left[1 + \sqrt{x+3} + 1 \right]$

$= 1 + \sqrt{-3+3} + 1 = 2$

37. $\lim\limits_{x \to 0} \dfrac{\sqrt{2-x} - \sqrt{2}}{x} = \lim\limits_{x \to 0} \left[\dfrac{\sqrt{2-x} - \sqrt{2}}{x} \cdot \dfrac{\sqrt{2-x} + \sqrt{2}}{\sqrt{2-x} + \sqrt{2}} \right] = \lim\limits_{x \to 0} \dfrac{2 - x - 2}{x\left(\sqrt{2-x} + \sqrt{2} \right)}$

$= \lim\limits_{x \to 0} \dfrac{-1}{\sqrt{2-x} + \sqrt{2}} = \dfrac{-1}{2\sqrt{2}}$ or $\dfrac{-\sqrt{2}}{4}$

39. Note: $\lim\limits_{x\to 0^+}\left[\dfrac{|x|}{x}-\dfrac{x}{|x|}\right] = \lim\limits_{x\to 0^+}\left[\dfrac{x}{x}-\dfrac{x}{x}\right] = \lim\limits_{x\to 0^+}0 = 0.$

$\lim\limits_{x\to 0^-}\left[\dfrac{|x|}{x}-\dfrac{x}{|x|}\right] = \lim\limits_{x\to 0^-}\left[\dfrac{-x}{x}-\dfrac{x}{-x}\right] = \lim\limits_{x\to 0^-}0 = 0.$

Since the left and right limits exist and are equal to 0, $\lim\limits_{x\to 0}\left[\dfrac{|x|}{x}-\dfrac{x}{|x|}\right] = 0.$

41. $\lim\limits_{h\to 0}\dfrac{f(2+h)-f(2)}{h} = \lim\limits_{h\to 0}\dfrac{(2+h)^2-2^2}{h} = \lim\limits_{h\to 0}\dfrac{4h+h^2}{h} = \lim\limits_{h\to 0}(4+h) = 4$

43. $\lim\limits_{h\to 0}\dfrac{f(2+h)-f(2)}{h} = \lim\limits_{h\to 0}\dfrac{\left[(2+h)^2+(2+h)+1\right]-\left(2^2+2+1\right)}{h}$

$= \lim\limits_{h\to 0}\dfrac{4+4h+h^2+2+h+1-7}{h} = \lim\limits_{h\to 0}\dfrac{5h+h^2}{h} = \lim\limits_{h\to 0}(5+h) = 5$

45. $\lim\limits_{h\to 0}\dfrac{f(0+h)-f(0)}{h} = \lim\limits_{h\to 0}\dfrac{|h|-|0|}{h} = \lim\limits_{h\to 0}\dfrac{|h|}{h}$ does not exist, because the left hand limit is -1 and the right hand limit is 1.

47. $\lim\limits_{h\to 0}\dfrac{f(x+h)-f(x)}{h} = \lim\limits_{h\to 0}\dfrac{\left[2(x+h)+3\right]-(2x+3)}{h} = \lim\limits_{h\to 0}\dfrac{2x+2h+3-2x-3}{h}$

$= \lim\limits_{h\to 0}\dfrac{2h}{h} = \lim\limits_{h\to 0}2 = 2$

49. $\lim\limits_{h\to 0}\dfrac{f(x+h)-f(x)}{h} = \lim\limits_{h\to 0}\dfrac{\left[(x+h)^2+(x+h)\right]-\left(x^2+x\right)}{h}$

$= \lim\limits_{h\to 0}\dfrac{x^2+2xh+h^2+x+h-x^2-x}{h}$

$= \lim\limits_{h\to 0}\dfrac{2xh+h^2+h}{h} = \lim\limits_{h\to 0}(2x+h+1) = 2x+1$

51. $\lim\limits_{h\to 0}\dfrac{f(x+h)-f(x)}{h} = \lim\limits_{h\to 0}\dfrac{\sqrt{x+h+1}-\sqrt{x+1}}{h}$

$= \lim\limits_{h\to 0}\dfrac{1}{\sqrt{x+h+1}+\sqrt{x+1}} = \dfrac{1}{2\sqrt{x+1}}$

53. $\lim\limits_{h\to 0}\dfrac{f(x+h)-f(x)}{h} = \lim\limits_{h\to 0}\dfrac{\sqrt{(x+h)^2+1}-\sqrt{x^2+1}}{h}$

$= \lim\limits_{h\to 0}\dfrac{2x+h}{\sqrt{(x+h)^2+1}+\sqrt{x^2+1}} = \dfrac{2x}{2\sqrt{x^2+1}} = \dfrac{x}{\sqrt{x^2+1}}$

55. One of many possible examples: let $f(x) = 1 + \dfrac{1}{x}$, $g(x) = -\dfrac{1}{x}$, $c = 0$. Then $\lim\limits_{x \to 0} f(x)$

and $\lim\limits_{x \to 0} g(x)$ do not exist, but $\lim\limits_{x \to 0}\big(f(x) + g(x)\big) = \lim\limits_{x \to 0}\left(1 + \dfrac{1}{x} - \dfrac{1}{x}\right) = \lim\limits_{x \to 0} 1 = 1$.

13.2A The Formal Definition of Limit

1. Given $\varepsilon > 0$, we require a $\delta > 0$ such that if $0 < |x - 3| < \delta$, $|(3x - 2) - 7| < \varepsilon$.
Since $|(3x - 2) - 7| = |3x - 9| = 3|(x - 3)| < \varepsilon$ if $|x - 3| < \varepsilon/3$, choose $\delta = \varepsilon/3$.
Then if $0 < |x - 3| < \delta$, $0 < |x - 3| < \varepsilon/3$
$$0 < 3|x - 3| < \varepsilon$$
$$|3x - 9| < \varepsilon$$
$$|3x - 2 - 7| < \varepsilon \text{ as required.}$$

3. We require a $\delta > 0$ such that if $0 < |x - 5| < \delta$, $|x - 5| < \varepsilon$. Clearly $\delta = \varepsilon$ will do.
Then if $0 < |x - 5| < \delta$, $0 < |x - 5| < \varepsilon$ as required.

5. We require a $\delta > 0$ such that if $0 < |x - 2| < \delta$, $|x - 2| < \varepsilon$, then $|(6x + 3) - 15| < \varepsilon$.
Since $|(6x + 3) - 15| = |6x - 12| = 6|x - 2| < \varepsilon$ if $|x - 2| < \varepsilon/6$, choose $\delta = \varepsilon/6$.
Then if $0 < |x - 2| < \delta$, $0 < |x - 2| < \varepsilon/6$
$$0 < 6|x - 2| < \varepsilon$$
$$|6x - 12| < \varepsilon$$
$$|(6x + 3) - 15| < \varepsilon \text{ as required.}$$

7. We require a $\delta > 0$ such that if $0 < |x - 1| < \delta$, $|4 - 4| < \varepsilon$.
Any positive δ, whether $\delta = 2000$ or 6 or 0.005 will do, since $|4 - 4| = 0 < \varepsilon$
for any positive ε whatsoever.

9. We require a $\delta > 0$ such that if $0 < |x - 4| < \delta$, $|x - 6 + 2| < \varepsilon$. Clearly $\delta = \varepsilon$ will do.
Then if $0 < |x - 4| < \delta$, $0 < |x - 6 + 2| < \varepsilon$ as required.

11. Given $\varepsilon > 0$, we require a $\delta > 0$ such that if $0 < |x + 2| < \delta$, $|2x + 5 - 1| < \varepsilon$.
Since $0 < |2x + 5 - 1| = |2x + 4| = 2|x + 2| < \varepsilon$ if $|x + 2| < \varepsilon/2$, choose $\delta = \varepsilon/2$.
Then if $0 < |x + 2| < \delta$, $0 < |x + 2| < \varepsilon/2$
$$0 < 2|x + 2| < \varepsilon$$
$$|2 + 4x| < \varepsilon$$
$$|3x + 5 - 1| < \varepsilon \text{ as required.}$$

13. Given $\varepsilon > 0$, we require a $\delta > 0$ such that if $0 < |x - 0| < \delta, |x^2 - 0| < \varepsilon$.

Since $|x^2| < \varepsilon$ exactly when $|x| < \sqrt{\varepsilon}$, choose $\delta = \sqrt{\varepsilon}$.

Then if $0 < |x - 0| < \delta, \ 0 < |x| < \sqrt{\varepsilon}$

$$|x^2| < \varepsilon \text{ as required.}$$

15. We require a $\delta > 0$ such that if $0 < |x - c| < \delta,$ then $0 < |(f(x) - g(x)) - (L - M)| < \varepsilon$.

By the triangle inequality,

$$|(f(x) - g(x)) - (L - M)| = |(f(x) - L) + (M - g(x))|$$
$$\leq |f(x) - L| + |M - g(x)|$$
$$\leq |f(x) - L| + |g(x) - M|$$

Now let ε be any positive number. There are positive numbers δ_1 and δ_2 such that
if $0 < |x - c| < \delta_1$ then $|f(x) - L| < \varepsilon/2$,

if $0 < |x - c| < \delta_2$ then $|g(x) - M| < \varepsilon/2$.

So choose δ to be the smaller of δ_1, δ_2. Then if $0 < |x - c| < \delta$,

$$|(f(x) - g(x)) - (L - M)| \leq |f(x) - L| + |g(x) - M| \leq \frac{\varepsilon}{2} + \frac{\varepsilon}{2} \leq \varepsilon.$$

This proves that if $\lim_{x \to c} f(x) = L$ and $\lim_{x \to c} g(x) = M$, then $\lim_{x \to c} (f(x) - g(x)) = L - M$.

13.3 Continuity

1. The function is undefined at 3 and 6, and therefore discontinuous at these points.

3. The function is continuous at –2 and 3, but undefined and thus discontinuous at 0.

5. The function has no limit as x approaches –2, hence it is discontinuous there. It is continuous at 0 and 3.

7. $f(3) = 3^2 + 5(3 - 2)^7 = 14, \ \lim_{x \to 3} f(x) = \lim_{x \to 3} x^2 + 5\left[\lim_{x \to 3}(x - 2)\right]^7 = 14$

Since $\lim_{x \to 3} f(x) = f(3)$, the function is continuous at $x = 3$.

9. $f(2) = \dfrac{2^2 - 9}{(2^2 - 2 - 6)(2^2 + 6 \cdot 2 + 9)} = \dfrac{1}{20}$

$$\lim_{x \to 2} f(x) = \frac{\lim_{x \to 2}(x^2 - 9)}{\lim_{x \to 2}(x^2 - x - 6)\lim_{x \to 2}(x^2 + 6x + 9)} = \frac{1}{20}$$

Since $\lim_{x \to 2} f(x) = f(2)$, the function is continuous at $x = 2$.

11. $f(36) = \dfrac{36\sqrt{36}}{(36-6)^2} = \dfrac{6}{25}$, $\displaystyle\lim_{x\to36} f(x) = \dfrac{\displaystyle\lim_{x\to36} x \displaystyle\lim_{x\to36} \sqrt{x}}{\displaystyle\lim_{x\to36}(x-6)^2} = \dfrac{6}{25}$

Since $\displaystyle\lim_{x\to36} f(x) = f(36)$, the function is continuous at $x = 36$.

13. The function is undefined at $x = 3$, and therefore discontinuous there.

15. The function is undefined at $x = -1$, and therefore discontinuous there.
$$\left[\text{Note: } \lim_{x\to-1} f(x) = -3/2, \text{ but this is irrelevant.}\right]$$

17. $\displaystyle\lim_{x\to0} f(x) = \lim_{x\to0} x^2 = 0$. However, $f(0) = 1 \neq \displaystyle\lim_{x\to0} f(x)$, hence the function is discontinuous.

19. $\displaystyle\lim_{x\to2^+} f(x) = \lim_{x\to2^+}(2x-4) = 0 = f(2)$. $\displaystyle\lim_{x\to2^-} f(x) = \lim_{x\to2^-}(-2x+4) = 0 = f(2)$.
Hence $\displaystyle\lim_{x\to2} f(x) = f(2)$; the function is continuous at $x = 2$.

21. $\displaystyle\lim_{x\to0^+} f(x) = \lim_{x\to0^+} 2x^2 = 0 = f(0)$. $\displaystyle\lim_{x\to0^-} f(x) = \lim_{x\to0^-}\left(x^2 - x\right) = 0 = f(0)$.
Hence $\displaystyle\lim_{x\to0} f(x) = f(0)$; the function is continuous at $x = 0$.

23. $\displaystyle\lim_{x\to3^+} f(x) = \lim_{x\to3^+}(x-3) = 0 = f(3)$. $\displaystyle\lim_{x\to3^-} f(x) = \lim_{x\to3^-}(3-x) = 0 = f(3)$.
Hence $\displaystyle\lim_{x\to3} f(x) = f(3)$; the function is continuous at $x = 3$.

25. The function is continuous everywhere with the possible exception of $x = 1$ and $x = 3$, the zeros of the denominator.

At $x = 1$, $\displaystyle\lim_{x\to1} f(x) = \lim_{x\to1}\dfrac{x^2+x-2}{x^2-4x+3} = \lim_{x\to1}\dfrac{(x+2)(x-1)}{(x-3)(x-1)} = \lim_{x\to1}\dfrac{x+2}{x-3} = -\dfrac{3}{2} = f(1)$.

Hence the function is continuous at $x = 1$.
At $x = 3$, however, the function is undefined and is therefore not continuous.

27. The function is continuous everywhere with the possible exception of $x = 0$ and $x = 2$. In fact, since $\displaystyle\lim_{x\to0^-} f(x) = 1$ and $\displaystyle\lim_{x\to0^+} f(x) = 0$, it is discontinuous at $x = 0$.

Similarly, since $\displaystyle\lim_{x\to2^-} f(x) = 2$ and $\displaystyle\lim_{x\to2^+} f(x) = -1$, it is also discontinuous at $x = 2$.

29. The function is discontinuous on the interval because it is discontinuous at every multiple of $\frac{1}{5}$.

31. The function is continuous everywhere on the interval.

33. Solve $x^2 - 7 = 0$ by choosing $x = 2$ and $x = 3$, noting that there is a sign change between 2 and 3. Then choose $x = 2.1$, $x = 2.2$, etc. until a sign change is observed between 2.6 and 2.7. Then choose $x = 2.61$, $x = 2.62$, etc. until a sign change is observed. The process can be continued.

35. The function will be discontinuous on $[-3,3]$ because it is undefined at $x = 1$. Note that $f(-3) = -10$ and $f(3) = 20$. We must show that f satisfies the conclusion of the Intermediate Value Theorem: that is, for any number k between -10 and 20, there is a number c between -3 and 3 such that $f(c) = k$.

Let k be a number between -10 and 20. First, assume that $-10 \le k \le 0$ and note that $f(-2) = 0$. Since f is continous on $[-3,-2]$, the Intermediate Value Theorem applies on that interval. Hence, there is a number c between -3 and -2 such that $f(c) = k$. Now suppose that $0 < k \le 20$ and note that $f(2) = 0$. Since f is continous on $[2,3]$, the Intermediate Value Theorem applies on that interval. Hence, there is a number c between 2 and 3 such that $f(c) = k$. Thus in all cases, there is a number c between -3 and 3 such that $f(c) = k$.

37. The function will be continuous if $f(3) = 3b + 4 = \lim_{x \to 3^+} f(x) = bx^2 - 2 = 9b - 2$.
Solving, $b = 1$ is the only possibility.

39. Since $\lim_{x \to 1} f(x) = \lim_{x \to 1} \dfrac{1}{x+1} = \dfrac{1}{2}$, and $f(x) = \dfrac{1}{x+1}$ if $x \ne 1$,
choose $g(x) = \dfrac{1}{x+1}$.

41. Since $\lim_{x \to 4} f(x) = \lim_{x \to 4} \dfrac{1}{2+\sqrt{x}} = \dfrac{1}{4}$, and $f(x) = \dfrac{1}{2+\sqrt{x}}$ if $x \ne 4$,
choose $g(x) = \dfrac{1}{2+\sqrt{x}}$.

43. Since $\lim_{x \to 0^+} f(x) = 1$ and $\lim_{x \to 0^-} f(x) = -1$, no choice of $f(0)$ could possibly reconcile these two facts to provide a continuous function.

13.4 Limits Involving Infinity

1. For the function $f(x) = \sqrt{x^2 + 1} - (x + 1)$, $f(100) = -.995$, $f(1000) = -.9995$, $f(10,000) = -.99995$, $f(100,000) = -.999995$. The limit appears to be -1.

3. For the function $f(x) = \dfrac{x^{2/3} - x^{4/3}}{x^3}$, $f(-100) = .0004$, $f(-1000) = .00001$, $f(-10,000) = .0000002$, $f(-100,000) = .000000005$. The limit appears to be 0.

5. For the function $f(x) = \sin\left(\dfrac{1}{x}\right)$ some values are as shown:

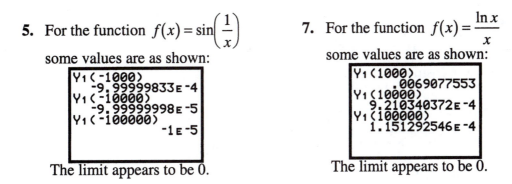

```
Y1(-1000)
     -9.99999833E-4
Y1(-10000)
     -9.99999998E-5
Y1(-100000)
             -1E-5
```

The limit appears to be 0.

7. For the function $f(x) = \dfrac{\ln x}{x}$ some values are as shown:

```
Y1(1000)
         .0069077553
Y1(10000)
     9.210340372E-4
Y1(100000)
     1.151292546E-4
```

The limit appears to be 0.

9. No vertical asymptotes. $\displaystyle\lim_{x\to\infty} f(x) = \infty.\ \lim_{x\to-\infty} f(x) = 0.$

11. No vertical asymptotes. $\displaystyle\lim_{x\to\infty} f(x) = 2.\ \lim_{x\to-\infty} f(x) = -1.$

13. Vertical asymptotes: $x = -10.\ \displaystyle\lim_{x\to\infty} f(x) = \infty.\ \lim_{x\to-\infty} f(x) = -\infty.$

15. $\displaystyle\lim_{x\to\pm\infty} \frac{3x^2 + 5}{4x^2 - 6x + 2} = \lim_{x\to\pm\infty} \frac{3 + 5/x^2}{4 - 6/x + 2/x^2} = \frac{3}{4}.\ y = \frac{3}{4}$ is the horizontal asymptote.

17. $\displaystyle\lim_{x\to\pm\infty} \frac{2x^2 - 6x + 1}{2 + x - x^2} = \lim_{x\to\pm\infty} \frac{2 - 6/x + 1/x^2}{2/x^2 + 1/x - 1} = -2.\ y = -2$ is the horizontal asymptote.

19. Since the numerator is of greater degree than the denominator, the limits as $x \to \infty$ and as $x \to -\infty$ are infinite and there is no horizontal asymptote.

21. $\displaystyle\lim_{x\to-\infty} \frac{(x-3)(x+2)}{2x^2 + x + 1} = \lim_{x\to-\infty} \frac{(1 - 3/x)(1 + 2/x)}{2 + 1/x + 1/x^2} = \frac{1}{2}.$

23. $\displaystyle\lim_{x\to\infty}\left(3x - 1/x^2\right) = \lim_{x\to\infty} 3x - \lim_{x\to\infty} 1/x^2 = \infty - 0 = \infty.$

25. $\displaystyle\lim_{x\to-\infty}\left(\frac{3x}{x+2} + \frac{2x}{x-1}\right) = \lim_{x\to-\infty} \frac{3x}{x+2} + \lim_{x\to-\infty} \frac{2x}{x-1} = 3 + 2 = 5.$

27. $\displaystyle\lim_{x\to\infty} \frac{2x}{\sqrt{x^2 - 2x}} = \lim_{x\to\infty} \frac{2}{\sqrt{1 - 2/x}} = 2.$

29. $\displaystyle\lim_{x\to-\infty} \frac{3x - 2}{\sqrt{2x^2 + 1}} = \lim_{x\to-\infty} \frac{3x - 2}{\sqrt{2x^2 + 1}} \div \frac{(-x)}{\sqrt{x^2}} = \lim_{x\to-\infty} \frac{-3 + 2/x}{\sqrt{2 + 1/x^2}} = \frac{-3}{\sqrt{2}}.$

$\left(\text{since when } x \text{ is negative, } \sqrt{x^2} = -x\right)$

31. $\lim\limits_{x\to\infty}\dfrac{\sqrt{2x^2+1}}{3x-5}=\lim\limits_{x\to\infty}\dfrac{\sqrt{2x^2+1}}{3x-5}\div\dfrac{\sqrt{x^2}}{x}=\lim\limits_{x\to\infty}\dfrac{\sqrt{2+1/x^2}}{3-5/x}=\dfrac{\sqrt{2}}{3}.$

$\left(\text{since when }x\text{ is positive, }\sqrt{x^2}=x\right)$

33. $\lim\limits_{x\to-\infty}\dfrac{\sqrt{3x^2+3}}{x+3}=\lim\limits_{x\to-\infty}\dfrac{\sqrt{3x^2+3}}{x+3}\div\dfrac{\sqrt{x^2}}{(-x)}=\lim\limits_{x\to-\infty}\dfrac{\sqrt{3+3/x^2}}{-1-3/x}=-\sqrt{3}.$

$\left(\text{since when }x\text{ is negative, }\sqrt{x^2}=-x\right)$

35. $\lim\limits_{x\to\infty}\dfrac{x^2+2x+1}{\sqrt{x^4+2x}}=\lim\limits_{x\to\infty}\dfrac{\left(x^2+2x+1\right)\div x^2}{\sqrt{x^4+2x}\div\sqrt{x^4}}=\lim\limits_{x\to\infty}\dfrac{1+2/x+1/x^2}{\sqrt{1+2/x^3}}=1.$

37. $\lim\limits_{x\to\infty}\dfrac{1-\sqrt{x}}{1+\sqrt{x}}=\lim\limits_{x\to\infty}\dfrac{\left(1-\sqrt{x}\right)\left(1-\sqrt{x}\right)}{\left(1+\sqrt{x}\right)\left(1-\sqrt{x}\right)}=\lim\limits_{x\to\infty}\dfrac{1/x-2/\sqrt{x}+1}{1/x-1}=-1.$

39. $\lim\limits_{x\to\infty}\left(\sqrt{x^2+1}-x\right)=\lim\limits_{x\to\infty}\dfrac{\left(\sqrt{x^2+1}-x\right)\left(\sqrt{x^2+1}+x\right)}{1}=\lim\limits_{x\to\infty}\dfrac{1}{\sqrt{x^2+1}+x}=0.$

41. $\lim\limits_{x\to\infty}\left(\sqrt{x^2-1}-\sqrt{x^2+1}\right)=\lim\limits_{x\to\infty}\dfrac{\left(\sqrt{x^2-1}-\sqrt{x^2+1}\right)}{1}\dfrac{\left(\sqrt{x^2-1}+\sqrt{x^2+1}\right)}{\left(\sqrt{x^2-1}+\sqrt{x^2+1}\right)}$

$$=\lim\limits_{x\to\infty}\dfrac{-2}{\left(\sqrt{x^2-1}+\sqrt{x^2+1}\right)}=0.$$

43. $\lim\limits_{x\to\infty}\dfrac{x}{|x|}=\lim\limits_{x\to\infty}\dfrac{x}{x}=\lim\limits_{x\to\infty}1=1.$

45. Since if $|x|$ is very large, $[x]$ differs from x by at most 1,

a. $\lim\limits_{x\to\infty}\dfrac{[x]}{x}=\lim\limits_{x\to\infty}\dfrac{x-\alpha}{x}=\lim\limits_{x\to\infty}\dfrac{1-\alpha/x}{1}=1.$

b. $\lim\limits_{x\to-\infty}\dfrac{[x]}{x}=\lim\limits_{x\to-\infty}\dfrac{x-\beta}{x}=\lim\limits_{x\to-\infty}\dfrac{1-\beta/x}{1}=1.$

Here α and β are variable, but always less than 1.

47. $\lim\limits_{x\to\infty}\dfrac{\sqrt{x+\sqrt{x+\sqrt{x}}}}{\sqrt{x+1}}=\lim\limits_{x\to\infty}\dfrac{\sqrt{x+\sqrt{x+\sqrt{x}}}\div\sqrt{x}}{\sqrt{x+1}\div\sqrt{x}}=\lim\limits_{x\to\infty}\dfrac{\sqrt{1+\sqrt{x+\sqrt{x}}/x}}{\sqrt{1+1/x}}$

$$=\lim\limits_{x\to\infty}\dfrac{\sqrt{1+\sqrt{1/x+\sqrt{x}/x^2}}}{\sqrt{1+1/x}}=\lim\limits_{x\to\infty}\dfrac{\sqrt{1+\sqrt{1/x+\sqrt{1/x^3}}}}{\sqrt{1+1/x}}=1.$$

49. The first part of the informal definition is included in the second part, which states: the value of $f(x)$ can be made arbitrarily close to L by taking large enough values of x. This means that whenever you specify how close $f(x)$ should be to L, we can tell you how large x must be to guarantee this. In other words, you specify how close you want $f(x)$ to be to L by giving a positive number ε and we tell you how large x must be to guarantee that $f(x)$ is within ε of L, that is, to guarantee that $|f(x) - L| < \varepsilon$. We do this by giving a positive number k such that $|f(x) - L| < \varepsilon$ whenever $x > k$. In other words, for every positive number ε, there is a positive number k (depending on ε) such that: If $x > k$, then $|f(x) - L| < \varepsilon$.

51. a. For the function $f(x) = \left(1 + \dfrac{1}{x}\right)^x$ some values are as shown:

```
Y₁(10000000)
        2.718281693
Y₁(1000000000)
        2.718281827
Y₁(100000000000)
        2.718281828
■
```

The last two figures agree to seven decimal places: 2.7182818.

b. $e \approx 2.718281828$ **c.** e **d.** The limits are the same.

Chapter 13 Review Exercises

1. For the function $f(x) = \dfrac{3x - \sin x}{x}$ some values are as shown:

```
Y₁(.01)
        2.000016667
Y₁(.0001)
        2.000000002
Y₁(.000001)
                  2
```

```
Y₁(-.01)
        2.000016667
Y₁(-.0001)
        2.000000002
Y₁(-.000001)
                  2
```

The limit appears to be 2.

3. For the function $f(x) = \dfrac{10^x - .1}{x + 1}$, some values are as shown:

```
Y₁(-1.01)
        .2276277904
Y₁(-1.0001)
        .2302320018
Y₁(-1.0000001)
        .23025848
```

```
Y₁(-.99)
        .2329299228
Y₁(-.9999)
        .2302850208
Y₁(-.9999999)
        .2302585
```

The limit appears to be approximately .2302585. (This is actually an approximation of $(.1)\ln 10$.)

5. Since $\lim_{x\to 2^+} f(x) = 2$, $\lim_{x\to 2^-} f(x) = 2$, $\lim_{x\to 2} f(x) = 2$.

7. $\lim_{x\to 3}[3f(x) - 15] = 3\lim_{x\to 3} f(x) - 15 = 3\cdot 5 - 15 = 0$.

9. $\lim_{x\to 3} \dfrac{f(x)g(x) - 2f(x)}{[g(x)]^2} = \dfrac{\lim_{x\to 3} f(x)\lim_{x\to 3} g(x) - 2\lim_{x\to 3} f(x)}{\left[\lim_{x\to 3} g(x)\right]^2} = \dfrac{5(-2) - 2(5)}{(-2)^2} = -5$.

11. $\lim_{x\to -2}(x^3 - 3x + 1) = (-2)^3 - 3(-2) + 1 = -1$.

13. $\lim_{x\to 1} \dfrac{x^2 - 1}{x^2 - 3x + 2} = \lim_{x\to 1} \dfrac{x + 1}{x - 2} = \dfrac{1 + 1}{1 - 2} = -2$.

15. $\lim_{x\to 0} \dfrac{\sqrt{1+x} - 1}{x} = \lim_{x\to 0} \dfrac{\left(\sqrt{1+x} - 1\right)\left(\sqrt{1+x} + 1\right)}{x\left(\sqrt{1+x} + 1\right)} = \lim_{x\to 0} \dfrac{x}{x\left(\sqrt{1+x} + 1\right)} = \dfrac{1}{2}$.

17. $\lim_{x\to -1^-} \sqrt{9 + 8x - x^2} = \sqrt{9 + 8(-1) - (-1)^2} = 0$.

19. $\lim_{x\to -5^+} \dfrac{|x + 5|}{x + 5} = \lim_{x\to -5^+} \dfrac{x + 5}{x + 5} = \lim_{x\to -5^+} 1 = 1$.

21. $\lim_{h\to 0} \dfrac{f(2 + h) - f(2)}{h} = \lim_{h\to 0} \dfrac{\left[(2+h)^2 + 1\right] - (2^2 + 1)}{h}$

$\qquad = \lim_{h\to 0} \dfrac{5 + 4h + h^2 - 5}{h} = \lim_{h\to 0}(4 + h) = 4$.

23. $\lim_{h\to 0} \dfrac{f(x + h) - f(x)}{h} = \lim_{h\to 0} \dfrac{[4(x + h) - 2] - (4x - 2)}{h}$

$\qquad = \lim_{h\to 0} \dfrac{4x + 4h - 2 - 4x + 2}{h} = \lim_{h\to 0} 4 = 4$.

25. We require a $\delta > 0$ such that if $0 < |x - 3| < \delta$, then $|(2x + 1) - 7| < \varepsilon$.
Since $|(2x + 1) - 7| = |2x - 6| = 2|x - 3| < \varepsilon$ if $|x - 3| < \varepsilon/2$, choose $\delta = \varepsilon/2$.
Then if $0 < |x - 3| < \delta$, $0 < |x - 3| < \varepsilon/2$
$$0 < 2|x - 3| < \varepsilon$$
$$|2x - 6| < \varepsilon$$
$$|(2x + 1) - 7| < \varepsilon \text{ as required.}$$

27. Since $\lim_{x \to -3} f(x)$ appears to equal $f(-3)$, the function is continuous at $x = -3$. Since $f(2)$ is not defined, the function is not continuous at $x = 2$.

29. $\lim_{x \to 4} \sqrt{x^2 - 8} = \sqrt{4^2 - 8} = \sqrt{8}$, hence the function is continuous at $x = 4$.

Since $\lim_{x \to \sqrt{8}^-} f(x)$ does not exist, the function is not continuous at $x = \sqrt{8}$, however it is right-continuous there.

31. Since $h(x)$ is not defined at $x = -4$, the function is not continuous at $x = -4$. Since $\lim_{x \to 0} h(x) = h(0)$, the function is continuous at $x = 0$.

33. $f(x) = \dfrac{x^2 - x - 6}{x^2 - 9} = \dfrac{(x-3)(x+2)}{(x-3)(x+3)} = \dfrac{x+2}{x+3}$ if $x \neq 3$.

a. Since $\lim_{x \to 2} f(x) = 4/5 = f(2)$, the function is continuous at $x = 2$.

b. Since $f(x)$ is not defined at $x = 3$, the function is not continuous at $x = 3$.

35. $f(x) = \dfrac{x^2 + 1}{(x - 2)(x + 1)}$. Asymptotes: $x = -1$, $x = 2$. Since if $x < -1$, $f(x) > 0$, the graph

moves upward as $x = -1$ is approached from the left. Since if $-1 < x < 2$, $f(x) < 0$, the graph moves downward as $x = -1$ is approached from the right or as $x = 2$ is approached from the left. Since if $x > 2$, $f(x) > 0$, the graph moves upward as $x = 2$ is approached from the right.

37. The limit does not exist. One can write $\lim_{x \to 3} \dfrac{1}{(x - 3)^2} = \infty$, however.

39. Since $\lim_{x \to \infty} \dfrac{1}{x^2} = \lim_{x \to \infty} \left(-\dfrac{1}{x^2} \right) = 0$, and $-\dfrac{1}{x^2} \leq \dfrac{\sin x}{x^2} \leq \dfrac{1}{x^2}$ for all x, $\lim_{x \to \infty} \dfrac{\sin x}{x^2} = 0$.

41. $\lim_{x \to \infty} \dfrac{2x^3 - 3x^2 + 5x - 1}{4x^3 + 2x^2 - x + 10} = \lim_{x \to \infty} \dfrac{2 - 3/x + 5/x^2 - 1/x^3}{4 + 2/x - 1/x^2 + 10/x^3} = \dfrac{1}{2}$.

43. $\lim_{x \to -\infty} \left(\dfrac{2x + 1}{x - 3} + \dfrac{4x - 1}{3x} \right) = \lim_{x \to -\infty} \dfrac{2 + 1/x}{1 - 3/x} + \lim_{x \to -\infty} \dfrac{4 - 1/x}{3} = 2 + \dfrac{4}{3} = \dfrac{10}{3}$.

45. $\lim_{x \to -\infty} f(x) = \lim_{x \to -\infty} \dfrac{1 - 1/x + 7/x^2}{2 + 5/x + 7/x^2} = \dfrac{1}{2}$. The horizontal asymptote is $y = \dfrac{1}{2}$. Graph using the window $-20 \leq x \leq 20, 0 \leq y \leq 3$.

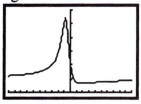

Appendix 1
Algebra Review

1.A Integral Exponents

1. 36   **3.** $5 + 4(3^2 + 2^3) = 5 + 4(9 + 8) = 5 + 4 \cdot 17 = 73$

5. $\dfrac{(-3)^2 + (-2)^4}{-2^2 - 1} = \dfrac{9 + 16}{-4 - 1} = \dfrac{25}{-5} = -5$ **7.** $\left(-\dfrac{5}{4}\right)^3 = -\dfrac{5^3}{4^3} = -\dfrac{125}{64}$

9. $\left(\dfrac{1}{3}\right)^3 + \left(\dfrac{2}{3}\right)^3 = \dfrac{1}{27} + \dfrac{8}{27} = \dfrac{9}{27} = \dfrac{1}{3}$ **11.** $2^4 - 2^7 = 16 - 128 = -112$

13. $\left(2^{-2} + 2\right)^2 = \left(\dfrac{1}{4} + 2\right)^2 = \left(\dfrac{9}{4}\right)^2 = \dfrac{81}{16}$ **15.** $2^2 \cdot 3^{-3} - 3^2 \cdot 2^{-3} = 4 \cdot \dfrac{1}{27} - 9 \cdot \dfrac{1}{8} = -\dfrac{211}{216}$

17. $\dfrac{1}{2^3} + \dfrac{1}{2^{-4}} = \dfrac{1}{8} + 2^4 = \dfrac{129}{8}$ **19.** $x^2 \cdot x^3 \cdot x^5 = x^{2+3+5} = x^{10}$

21. $(.03)y^2 \cdot y^7 = .03y^{2+7} = .03y^9$ **23.** $\left(2x^2\right)^3 3x = 2^3 x^6 \cdot 3x = 24x^7$

25. $\left(3x^2 y\right)^2 = 3^2 \left(x^2\right)^2 y^2 = 9x^4 y^2$ **27.** $\left(a^2\right)(7a)\left(-3a^3\right) = -21a^{2+1+3} = -21a^6$

29. $(2w)^3 (3w)(4w)^2 = 8w^3 \cdot 3w \cdot 16w^2 = 384w^6$ **31.** ab^3

33. $(2x)^{-2}(2y)^3(4x) = 2^{-2} x^{-2} \cdot 2^3 y^3 \cdot 4x = 8x^{-1} y^3 = 8y^3/x$

35. $\left(-3a^4\right)^2 \left(9x^3\right)^{-1} = 9a^8 \cdot \dfrac{1}{9x^3} = \dfrac{a^8}{x^3}$ **37.** $\left(2x^2 y\right)^0 (3xy) = 1 \cdot 3xy = 3xy$

39. $(64)^2 = \left(2^6\right)^2 = 2^{12}$ **41.** $\left(2^4 \cdot 16^{-2}\right)^3 = \left[2^4 \cdot \left(2^4\right)^{-2}\right]^3 = \left(2^4 \cdot 2^{-8}\right)^3 = \left(2^{-4}\right)^3 = 2^{-12}$

43. $\dfrac{x^4 \left(x^2\right)^3}{x^3} = \dfrac{x^4 x^6}{x^3} = \dfrac{x^{10}}{x^3} = x^7$ **45.** $\left(\dfrac{e^6}{c^4}\right)^2 \cdot \left(\dfrac{c^3}{e}\right)^3 = \dfrac{e^{12}}{c^8} \cdot \dfrac{c^9}{e^3} = e^9 c$

47. $\left(\dfrac{ab^2 c^3 d^4}{abc^2 d}\right)^2 = \left(bcd^3\right)^2 = b^2 c^2 d^6$ **49.** $\left(\dfrac{a^6}{b^{-4}}\right)^2 = \dfrac{a^{12}}{b^{-8}} = a^{12} b^8$

51. $\left(\dfrac{c^5}{d^{-3}}\right)^{-2} = \dfrac{c^{-10}}{d^6} = \dfrac{1}{c^{10} d^6}$

53. $\left(\dfrac{3x}{y^2}\right)^{-3}\left(\dfrac{-x}{2y^3}\right)^2 = \left(\dfrac{y^2}{3x}\right)^3\left(\dfrac{-x}{2y^3}\right)^2 = \dfrac{y^6}{27x^3}\cdot\dfrac{x^2}{2^2y^6} = \dfrac{1}{108x}$

55. $\dfrac{\left(a^{-3}b^2c\right)^{-2}}{\left(ab^{-2}c^3\right)^{-1}} = \dfrac{a^6b^{-4}c^{-2}}{a^{-1}b^2c^{-3}} = \dfrac{a^7c}{b^6}$ **57.** c^3d^6

59. $a^2\left(a^{-1}+a^{-3}\right) = a+a^{-1} = a+\dfrac{1}{a}$

61. Since the first factor is negative and the second is positive, the product is negative.

63. Since the first factor is negative and the second is positive, the product is negative.

65. Since the first factor is negative and the second and third factors are positive, the product is negative.

67. $\dfrac{3^{-r}}{3^{-s-r}} = 3^{-r-(-s-r)} = 3^s$ **69.** $\left(\dfrac{a^6}{b^{-4}}\right)^t = \left(a^6b^4\right)^t = a^{6t}b^{4t}$ **71.** $\dfrac{\left(c^{-r}b^s\right)^t}{\left(c^tb^{-s}\right)^r} = \dfrac{c^{-rt}b^{st}}{c^{rt}b^{-sr}} = \dfrac{b^{st+rs}}{c^{2rt}}$

73. Let $a=1, b=1, r=2.$
$a^r + b^r = 1^2 + 1^2 = 2$
$(a+b)^r = (1+1)^2 = 2^2 = 4$

75. Let $a=2, b=3, r=1, s=2.$
$a^rb^s = 2^1\cdot 3^2 = 18$
$(ab)^{rs} = (2\cdot 3)^3 = 6^3 = 216$

77. Let $c=2, r=6, s=3.$
$\dfrac{c^r}{c^s} = \dfrac{2^6}{2^3} = 2^3 = 8$
$c^{\frac{r}{s}} = 2^{\frac{6}{3}} = 2^2 = 4$

79. Let $a=1.$
$(-a)^2 = (-1)^2 = 1$
$-a^2 = -1^2 = -1$

1.B Arithmetic of Algebraic Expressions

1. $8x$ **3.** $6a^2b + (-8b)a^2 = 6a^2b - 8a^2b = -2a^2b$

5. $\left(x^2+2x+1\right)-\left(x^3-3x^2+4\right) = x^2+2x+1-x^3+3x^2-4 = -x^3+4x^2+2x-3$

7. $\left[u^4-(-3)u^3+u/2+1\right]-\left(u^4-2u^3+5-u/2\right) = u^4-3u^3+u/2+1-u^4+2u^3-5+u/2$
$= 5u^3+u-4$

9. $\left[4z-6z^2w-(-2)z^3w^2\right]+\left(8-6z^2w-zw^3+4z^3w^2\right)$
$= 4z-6z^2w+2z^3w^2+8-6z^2w-zw^3+4z^3w^2$
$= 6z^3w^3-12z^2w-zw^3+4z+8$

11. $\left(9x - x^3 + 1\right) - \left[2x^3 + (-6)x + (-7)\right] = 9x - x^3 + 1 - 2x^3 + 6x + 7 = -3x^3 + 15x + 8$

13. $\left(x^2 - 3xy\right) - (x + xy) - \left(x^2 + xy\right) = x^2 - 3xy - x - xy - x^2 - xy = -5xy - x$

15. $15y^3 - 5y$ **17.** $12a^2x^2 - 6a^3xy + 6a^2xy$ **19.** $12z^4 + 30z^3$

21. $12a^2b - 18ab^2 + 6a^3b^2$ **23.** $(x + 1)(x - 2) = x^2 - 2x + 1x - 2 = x^2 - x - 2$

25. $(-2x + 4)(-x - 3) = 2x^2 + 6x - 4x - 12 = 2x^2 + 2x - 12$

27. $(y + 3)(y + 4) = y^2 + 4y + 3y + 12 = y^2 + 7y + 12$

29. $(3x + 7)(-2x + 5) = -6x^2 + 15x - 14x + 35 = -6x^2 + x + 35$

31. $3y^3 - 9y^2 + 4y - 12$ **33.** $(x + 4)(x - 4) = x^2 - 4x + 4x - 16 = x^2 - 16$

35. $(4a + 5b)(4a - 5b) = 16a^2 - 20ab + 20ab - 25b^2 = 16a^2 - 25b^2$

37. $(y - 11)^2 = (y - 11)(y - 11) = y^2 - 11y - 11y + 121 = y^2 - 22y + 121$

39. $(5x - b)^2 = (5x - b)(5x - b) = 25x^2 - 5bx - 5bx + b^2 = 25x^2 - 10bx + b^2$

41. $\left(4x^3 - y^4\right)^2 = \left(4x^3 - y^4\right)\left(4x^3 - y^4\right) = 16x^6 - 4x^3y^4 - 4x^3y^4 + y^8 = 16x^6 - 8x^3y^4 + y^8$

43. $\left(-3x^2 + 2y^4\right)^2 = \left(-3x^2 + 2y^4\right)\left(-3x^2 + 2y^4\right) = 9x^4 - 6x^2y^4 - 6x^2y^4 + 4y^8$
$$= 9x^4 - 12x^2y^4 + 4y^8$$

45. $(2y + 3)\left(y^2 + 3y - 1\right) = 2y\left(y^2 + 3y - 1\right) + 3\left(y^2 + 3y - 1\right)$
$$= 2y^3 + 6y^2 - 2y + 3y^2 + 9y - 3 = 2y^3 + 9y^2 + 7y - 3$$

47. $(5w + 6)\left(-3w^2 + 4w - 3\right) = 5w\left(-3w^2 + 4w - 3\right) + 6\left(-3w^2 + 4w - 3\right)$
$$= -15w^3 + 20w^2 - 15w - 18w^2 + 24w - 18 = -15w^3 + 2w^2 + 9w - 18$$

49. $2x(3x + 1)(4x - 2) = 2x\left(12x^2 - 6x + 4x - 2\right) = 2x\left(12x^2 - 2x - 2\right) = 24x^3 - 4x^2 - 4x$

51. $(x - 1)(x - 2)(x - 3) = (x - 1)\left(x^2 - 3x - 2x + 6\right) = (x - 1)\left(x^2 - 5x + 6\right)$
$$= x\left(x^2 - 5x + 6\right) - 1\left(x^2 - 5x + 6\right) = x^3 - 5x^2 + 6x - x^2 + 5x - 6 = x^3 - 6x^2 + 11x - 6$$

53. $(x + 4y)(2y - x)(3x - y) = (x + 4y)\left(6xy - 2y^2 - 3x^2 + xy\right)$
$$= (x + 4y)\left(-3x^2 + 7xy - 2y^2\right) = x\left(-3x^2 + 7xy - 2y^2\right) + 4y\left(-3x^2 + 7xy - 2y^2\right)$$
$$= -3x^3 + 7x^2y - 2xy^2 - 12x^2y + 28xy^2 - 8y^3 = -3x^3 - 5x^2y + 26xy^2 - 8y^3$$

55. The terms involving x^2 will be $x^2(-3) + (3x)(2x) = -3x^2 + 6x^2 = 3x^2$.   3

57. The term involving x^2 will be $-6(x^2)$. -6

59. $(x+2)^3 = (x+2)(x+2)(x+2) = (x+2)(x^2 + 4x + 4)$

The terms involving x^2 will be $2x^2 + x(4x) = 2x^2 + 4x^2 = 6x^2$.   6

61. The terms involving x^2 will be $x^2(1) + x(-x) + 1(x^2) = x^2 - x^2 + x^2 = x^2$.   1

63. The terms involving x^2 will be $2x(3x) - 1(x^2) = 6x^2 - x^2 = 5x^2$.   5

65. $(\sqrt{x} + 5)(\sqrt{x} - 5) = \sqrt{x}\sqrt{x} - 5\sqrt{x} + 5\sqrt{x} - 25 = x - 25$

67. $(3 + \sqrt{y})^2 = (3 + \sqrt{y})(3 + \sqrt{y}) = 9 + 3\sqrt{y} + 3\sqrt{y} + \sqrt{y}\sqrt{y} = 9 + 6\sqrt{y} + y$

69. $(1 + \sqrt{3}x)(x + \sqrt{3}) = x + \sqrt{3} + \sqrt{3}x^2 + \sqrt{3}x\sqrt{3} = \sqrt{3}x^2 + 4x + \sqrt{3}$

71. $(ax + b)(3x + 2) = 3ax^2 + 2ax + 3bx + 2b = 3ax^2 + (2a + 3b)x + 2b$

73. $(ax + b)(bx + a) = abx^2 + a^2x + b^2x + ab = abx^2 + (a^2 + b^2)x + ab$

75. $(x - a)(x - b)(x - c) = (x - a)(x^2 - cx - bx + bc)$

$$= x(x^2 - cx - bx + bc) - a(x^2 - cx - bx + bc)$$

$$= x^3 - cx^2 - bx^2 + bcx - ax^2 + acx + abx - abc$$

$$= x^3 - (a + b + c)x^2 + (bc + ac + ab)x - abc$$

77. $16x^{n+k}$ **79.** $(y^r + 1)(y^s - 4) = y^r y^s - 4y^r + y^s - 4 = y^{r+s} - 4y^r + y^s - 4$

81. The outer rectangle has dimensions $10 + 2x$ and $6 + 2x$ and area $(10 + 2x)(6 + 2x)$.
The inner rectangle has area $6 \cdot 10 = 60$.
The shaded area is $(10 + 2x)(6 + 2x) - 60 = 60 + 20x + 12x + 4x^2 - 60 = 4x^2 + 32x$.

83. Let $y = 0$. $3(y + 2) = 3(0 + 2) = 6$ but $3y + 2 = 3 \cdot 0 + 2 = 2$.
The correct statement is $3(y + 2) = 3y + 6$.

85. Let $x = y = 1$. $(x + y)^2 = (1 + 1)^2 = 4$ but $x + y^2 = 1 + 1^2 = 2$.
The correct statement is $(x + y)^2 = x^2 + 2xy + y^2$.

87. Let $x = y = 1$. $(7x)(7y) = (7 \cdot 1)(7 \cdot 1) = 49$ but $7xy = 7 \cdot 1 \cdot 1 = 7$.
The correct statement is $(7x)(7y) = 49xy$ or $(7x)y = 7xy$.

89. Let $y = 1$. $y + y + y = 1 + 1 + 1 = 3$ but $y^3 = 1^3 = 1$.
The correct statement is $y + y + y = 3y$ or $y \cdot y \cdot y = y^3$

91. Let $x = 0$. $(x-3)(x-2) = (0-3)(0-2) = 6$ but $x^2 - 5x - 6 = 0^2 - 5 \cdot 0 - 6 = -6$.
The correct statement is $(x-3)(x-2) = x^2 - 5x + 6$.

93. This is represented algebraically by $\dfrac{(x+1)^2 - (x-1)^2}{x}$. This simplifies to

$$\frac{(x^2 + 2x + 1) - (x^2 - 2x + 1)}{x} = \frac{4x}{x} = 4$$

Thus the result is always 4.

95. Take a number. Subtract 3 from it to get a first number. Now add three to the original number to get a second number. Multiply the first number and the second number. Add 9 to the result. Divide by the original number, and the result is the number that you started with.

1.C Factoring

1. $(x+2)(x-2)$

3. $9y^2 - 25 = (3y)^2 - 5^2 = (3y+5)(3y-5)$

5. $81x^2 + 36x + 4 = (9x)^2 + 2(9x)2 + 2^2 = (9x+2)^2$

7. $5 - x^2 = \left(\sqrt{5}\right)^2 - x^2 = \left(\sqrt{5} + x\right)\left(\sqrt{5} - x\right)$

9. $49 + 28z + 4z^2 = 7^2 + 2(7)(2z) + (2z)^2 = (7+2z)^2$

11. $x^4 - y^4 = \left(x^2\right)^2 - \left(y^2\right)^2 = \left(x^2 + y^2\right)\left(x^2 - y^2\right) = \left(x^2 + y^2\right)(x+y)(x-y)$

13. $(x+3)(x-2)$ **15.** $(z+1)(z+3)$ **17.** $(y+9)(y-4)$

19. $x^2 - 6x + 9 = x^2 - 2(3x) + 3^2 = (x-3)^2$ **21.** $(x+2)(x+5)$

23. $(x+2)(x+9)$ **25.** $(3x+1)(x+1)$ **27.** $(2z+3)(z+4)$

29. $9x(x-8)$ **31.** $10x^2 - 8x - 2 = 2\left(5x^2 - 4x - 1\right) = 2(5x+1)(x-1)$

33. $(4u-3)(2u+3)$ **35.** $4x^2 + 20xy + 25y^2 = (2x)^2 + 2(2x)(5y) + (5y)^2 = (2x+5y)^2$

37. $x^3 - 125 = x^3 - 5^3 = (x-5)\left(x^2 + x \cdot 5 + 5^2\right) = (x-5)\left(x^2 + 5x + 25\right)$

39. $x^3 + 6x^2 + 12x + 8 = x^3 + 3 \cdot x^2 \cdot 2 + 3 \cdot x \cdot 2^2 + 2^3 = (x+2)^3$

41. $8 + x^3 = 2^3 + x^3 = (2+x)(2^2 - 2x + x^2) = (2+x)(4 - 2x + x^2)$

43. $-x^3 + 15x^2 - 75x + 125 = (-x)^3 + 3 \cdot 5(-x)^2 + 3 \cdot 5^2(-x) + 5^3 = (-x+5)^3$ or $(5-x)^3$

45. $x^3 + 1 = x^3 + 1^3 = (x+1)(x^2 - x \cdot 1 + 1^2) = (x+1)(x^2 - x + 1)$

47. $8x^3 - y^3 = (2x)^3 - y^3 = (2x - y)[(2x)^2 + (2x)y + y^2] = (2x - y)(4x^2 + 2xy + y^2)$

49. $x^6 - 64 = (x^3)^2 - 8^2 = (x^3 + 8)(x^3 - 8) = (x+2)(x^2 - 2x + 4)(x-2)(x^2 + 2x + 4)$

51. $(y^2 + 2)(y^2 + 5)$ **53.** $81 - y^4 = 9^2 - (y^2)^2 = (9 + y^2)(9 - y^2) = (9 + y^2)(3 + y)(3 - y)$

55. $z^6 - 1 = (z^3)^2 - 1^2 = (z^3 + 1)(z^3 - 1) = (z+1)(z^2 - z + 1)(z-1)(z^2 + z + 1)$

57. $(x^2 + 3y)(x^2 - y)$

59. $x^2 - yz + xz - xy = x^2 + xz - yz - xy = x(x+z) - y(z+x) = (x+z)(x-y)$

61. $a^3 - 2b^2 + 2a^2b - ab = a^3 + 2a^2b - 2b^2 - ab$
$$= a^2(a + 2b) - b(2b + a) = (a^2 - b)(a + 2b)$$

63. $x^3 + 4x^2 - 8x - 32 = x^2(x + 4) - 8(x + 4) = (x + 4)(x^2 - 8)$
$$= (x + 4)(x + \sqrt{8})(x - \sqrt{8})$$

65. If there were real numbers c and d such that $x^2 + 1 = (x + c)(x + d)$ then
$x^2 + 1 = x^2 + (c + d)x + cd$. Then the two numbers would satisfy $cd = 1, c + d = 0$.
In other words, the two numbers would have a positive product, hence the same sign,
but would add to 0. This is impossible.

1.D Fractional Expressions

1. $\dfrac{63}{49} = \dfrac{7 \cdot 9}{7 \cdot 7} = \dfrac{9}{7}$ **3.** $\dfrac{13 \cdot 27 \cdot 22 \cdot 10}{6 \cdot 4 \cdot 11 \cdot 12} = \dfrac{13 \cdot 3 \cdot 3 \cdot 3 \cdot 2 \cdot 11 \cdot 2 \cdot 5}{3 \cdot 2 \cdot 2 \cdot 2 \cdot 11 \cdot 2 \cdot 2 \cdot 3} = \dfrac{13 \cdot 3 \cdot 5}{2 \cdot 2 \cdot 2} = \dfrac{195}{8}$

5. $\dfrac{x^2 - x - 2}{x^2 + 2x + 1} = \dfrac{(x+1)(x-2)}{(x+1)(x+1)} = \dfrac{x-2}{x+1}$

7. $\dfrac{a^2 - b^2}{a^3 - b^3} = \dfrac{(a-b)(a+b)}{(a-b)(a^2 + ab + b^2)} = \dfrac{a+b}{a^2 + ab + b^2}$

9. $\dfrac{(x+c)(x^2 - cx + c^2)}{x^4 + c^3 x} = \dfrac{x^3 + c^3}{x(x^3 + c^3)} = \dfrac{1}{x}$

11. $\dfrac{3}{7} + \dfrac{2}{5} = \dfrac{15}{35} + \dfrac{14}{35} = \dfrac{29}{35}$

13. $\left(\dfrac{19}{7} + \dfrac{1}{2}\right) - \dfrac{1}{3} = \dfrac{114}{42} + \dfrac{21}{42} - \dfrac{14}{42} = \dfrac{121}{42}$

15. $\dfrac{c}{d} + \dfrac{3c}{e} = \dfrac{ce}{de} + \dfrac{3cd}{de} = \dfrac{ce + 3cd}{de}$

17. $\dfrac{b}{c} - \dfrac{c}{b} = \dfrac{b^2}{bc} - \dfrac{c^2}{bc} = \dfrac{b^2 - c^2}{bc}$

19. $\dfrac{1}{x+1} - \dfrac{1}{x} = \dfrac{x}{x(x+1)} - \dfrac{1(x+1)}{x(x+1)} = \dfrac{x - (x+1)}{x(x+1)} = \dfrac{x - x - 1}{x(x+1)} = -\dfrac{1}{x(x+1)}$

21. $\dfrac{1}{x+4} + \dfrac{2}{(x+4)^2} - \dfrac{3}{x^2 + 8x + 16} = \dfrac{x+4}{(x+4)^2} + \dfrac{2}{(x+4)^2} - \dfrac{3}{(x+4)^2}$

$$= \dfrac{x + 4 + 2 - 3}{(x+4)^2} = \dfrac{x+3}{(x+4)^2}$$

23. $\dfrac{1}{x} - \dfrac{1}{3x - 4} = \dfrac{3x - 4}{x(3x-4)} - \dfrac{x}{x(3x-4)} = \dfrac{(3x-4) - x}{x(3x-4)} = \dfrac{2x - 4}{x(3x-4)}$

25. $\dfrac{1}{x+y} + \dfrac{x+y}{x^3 + y^3} = \dfrac{(x^2 - xy + y^2)}{(x+y)(x^2 - xy + y^2)} + \dfrac{x+y}{x^3 + y^3}$

$$= \dfrac{(x^2 - xy + y^2) + (x+y)}{x^3 + y^3} = \dfrac{x^2 - xy + y^2 + x + y}{x^3 + y^3}$$

27. $\dfrac{1}{4x(x+1)(x+2)^3} - \dfrac{6x+2}{4(x+1)^3} = \dfrac{(x+1)^2}{4x(x+1)^3(x+2)^3} - \dfrac{x(6x+2)(x+2)^3}{4x(x+1)^3(x+2)^3}$

$$= \dfrac{x^2 + 2x + 1 - (6x^2 + 2x)(x^3 + 6x^2 + 12x + 8)}{4x(x+1)^3(x+2)^3}$$

$$= \dfrac{x^2 + 2x + 1 - (6x^5 + 38x^4 + 84x^3 + 72x^2 + 16x)}{4x(x+1)^3(x+2)^3}$$

$$= \dfrac{-6x^5 - 38x^4 - 84x^3 - 71x^2 - 14x + 1}{4x(x+1)^3(x+2)^3}$$

29. $\dfrac{3}{4} \cdot \dfrac{12}{5} \cdot \dfrac{10}{9} = \dfrac{3 \cdot 2 \cdot 2 \cdot 3 \cdot 2 \cdot 5}{2 \cdot 2 \cdot 5 \cdot 3 \cdot 3} = 2$

31. $\dfrac{3a^2 c}{4ac} \cdot \dfrac{8ac^3}{9a^2 c^4} = \dfrac{24a^3 c^4}{36a^3 c^5} = \dfrac{2}{3c}$